P9-DYB-809

Heath Mathematics

Walter E. Rucker • Clyde A. Dilley • David W. Lowry

D. C. Heath and Company
Lexington, Massachusetts Toronto

Illustrations: Linda Bourke, Marc Tolon Brown, Maryann Cocca, Tom Cooke, Chris Czernota, Kathy Parkinson, Len Reno, Maxie Chambliss, George Ulrich

Photography: Jonathan Barkan: 132, 133, 345; Fredrik D. Bodin: 34, 35, 96, 305, 308; Andrew Brilliant/Carol Palmer: 3, 9, 13, 17, 26, 106, 118, 178, 213, 250, 315, 335; A. James Casner: 47, 135, 163, 167, 230, 323; Stuart Cohen/Stock, Boston: 39; John Coletti: 60, 103; Kevin Galvin: 21; David T. Hughes 1981/The Picture Cube: 23; Photo courtesy I.B.M.: 41; Paul Johnson: 10, 127, 225, 258, 293, 340 (bottom); Talbot D. Lovering: Cover, Title page, 1, 139, 215; Tom Magno: 2, 16, 105, 108, 109, 116, 120, 171, 208, 211, 226, 227, 236, 243, 295; Mike Malyszko: 152, 153, 154, 155, 164, 260, 266; Julie O'Neil: 319; James Rigney: 340 (top), 341; Slidemakers/ Dr. E.R. Degginer: 252, 253, 273; Deidra Delano Stead: 8, 37, 87, 122, 176, 177, 239, 286, 291; Spencer Swanger/Tom Stack & Associates: 151; Thomas Zimmerman/ Alpha: 266, 327

Published simultaneously in Canada.

Printed in the United States of America.

International Standard Book Number: 0-669-07416-0

Contents

Addition and Subtraction Facts

1

Basic addition facts

$$\begin{array}{r} 8 \\ +4 \\ \hline 12 \end{array}$$

8 and 4 ← addends
12 ← sum

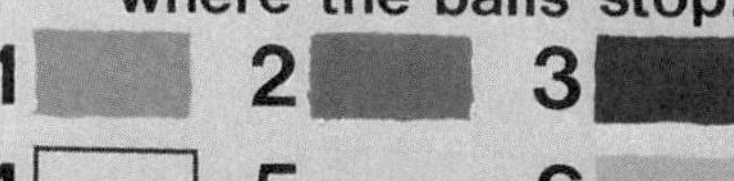

EXERCISES

Add.

1. $2 + 6$	2. $9 + 3$	3. $4 + 5$	4. $3 + 6$	5. $5 + 6$	6. $4 + 3$
7. $7 + 3$	8. $5 + 2$	9. $3 + 5$	10. $5 + 5$	11. $1 + 9$	12. $4 + 4$
13. $6 + 4$	14. $6 + 2$	15. $2 + 8$	16. $7 + 2$	17. $3 + 7$	18. $6 + 6$
19. $7 + 9$	20. $4 + 8$	21. $8 + 6$	22. $8 + 8$	23. $5 + 8$	24. $6 + 9$

25. $\begin{array}{r} 5 \\ +6 \\ \hline 11 \end{array}$	26. $\begin{array}{r} 3 \\ +8 \\ \hline 11 \end{array}$	27. $\begin{array}{r} 7 \\ +6 \\ \hline 13 \end{array}$	28. $\begin{array}{r} 7 \\ +8 \\ \hline 15 \end{array}$	29. $\begin{array}{r} 9 \\ +4 \\ \hline 13 \end{array}$	30. $\begin{array}{r} 9 \\ +8 \\ \hline 17 \end{array}$
31. $\begin{array}{r} 6 \\ +8 \\ \hline 14 \end{array}$	32. $\begin{array}{r} 4 \\ +6 \\ \hline 10 \end{array}$	33. $\begin{array}{r} 3 \\ +9 \\ \hline 12 \end{array}$	34. $\begin{array}{r} 6 \\ +5 \\ \hline 11 \end{array}$	35. $\begin{array}{r} 8 \\ +2 \\ \hline 10 \end{array}$	36. $\begin{array}{r} 5 \\ +9 \\ \hline 14 \end{array}$

37. 1 + 9 10	38. 9 + 5 14	39. 4 + 6 10	40. 7 + 5 12	41. 7 + 6 13
42. 4 + 4 8	43. 9 + 6 15	44. 3 + 4 7	45. 7 + 7 14	46. 9 + 9 18
47. 4 + 8 12	48. 4 + 5 9	49. 9 + 7 16	50. 7 + 2 9	51. 6 + 6 12
52. 5 + 8 13	53. 4 + 9 13	54. 3 + 3 6	55. 4 + 7 11	56. 8 + 8 16
57. 8 + 3 11	58. 9 + 8 17	59. 5 + 6 11	60. 8 + 7 15	61. 5 + 5 10
62. 6 + 9 15	63. 3 + 7 10	64. 5 + 9 14	65. 8 + 2 10	66. 8 + 9 17

Challenge!

Use the scoring rules from page 2 to find the scores for the following colors.

67.   **68.**   **69.**  

70. My score was 15. The first ball stopped on blue. What color did the second ball stop on?

71. Both the balls stopped on the same color. My score would be the same if I got orange and white. What color did the balls stop on?

Remembering basic facts

These ideas can help you remember the basic addition facts.

The Adding 0 Property
The sum of a number and 0 is the same number.

The Order Property
Changing the order of the addends does not change the sum.

EXERCISES

Find each sum.

1. $\begin{array}{r} 2 \\ +9 \\ \hline \end{array}$	**2.** $\begin{array}{r} 5 \\ +7 \\ \hline \end{array}$	**3.** $\begin{array}{r} 2 \\ +6 \\ \hline \end{array}$	**4.** $\begin{array}{r} 3 \\ +8 \\ \hline \end{array}$	**5.** $\begin{array}{r} 4 \\ +9 \\ \hline \end{array}$	**6.** $\begin{array}{r} 6 \\ +8 \\ \hline \end{array}$
7. $\begin{array}{r} 8 \\ +6 \\ \hline \end{array}$	**8.** $\begin{array}{r} 4 \\ +8 \\ \hline \end{array}$	**9.** $\begin{array}{r} 3 \\ +9 \\ \hline \end{array}$	**10.** $\begin{array}{r} 5 \\ +8 \\ \hline \end{array}$	**11.** $\begin{array}{r} 4 \\ +6 \\ \hline \end{array}$	**12.** $\begin{array}{r} 7 \\ +9 \\ \hline \end{array}$
13. $\begin{array}{r} 6 \\ +7 \\ \hline \end{array}$	**14.** $\begin{array}{r} 5 \\ +9 \\ \hline \end{array}$	**15.** $\begin{array}{r} 4 \\ +7 \\ \hline \end{array}$	**16.** $\begin{array}{r} 8 \\ +9 \\ \hline \end{array}$	**17.** $\begin{array}{r} 6 \\ +5 \\ \hline \end{array}$	**18.** $\begin{array}{r} 7 \\ +8 \\ \hline \end{array}$
19. $\begin{array}{r} 6 \\ +0 \\ \hline \end{array}$	**20.** $\begin{array}{r} 0 \\ +8 \\ \hline \end{array}$	**21.** $\begin{array}{r} 0 \\ +7 \\ \hline \end{array}$	**22.** $\begin{array}{r} 4 \\ +0 \\ \hline \end{array}$	**23.** $\begin{array}{r} 8 \\ +4 \\ \hline \end{array}$	**24.** $\begin{array}{r} 3 \\ +7 \\ \hline \end{array}$
25. $\begin{array}{r} 6 \\ +6 \\ \hline \end{array}$	**26.** $\begin{array}{r} 4 \\ +4 \\ \hline \end{array}$	**27.** $\begin{array}{r} 9 \\ +9 \\ \hline \end{array}$	**28.** $\begin{array}{r} 8 \\ +8 \\ \hline \end{array}$	**29.** $\begin{array}{r} 7 \\ +7 \\ \hline \end{array}$	**30.** $\begin{array}{r} 5 \\ +5 \\ \hline \end{array}$

31. **a.** $6 + 6$
b. $5 + 6$
c. $7 + 6$

32. **a.** $8 + 8$
b. $9 + 8$
c. $7 + 8$

33. **a.** $5 + 5$
b. $4 + 5$
c. $6 + 5$

34. $6 + (3 + 5)$

Note: The grouping symbols tell you to first add the 3 and 5.

35. $3 + (2 + 5)$

36. $5 + (2 + 6)$

37. $(2 + 3) + 5$

38. $(2 + 6) + 5$

39. **a.** $(4 + 3) + 6$
b. $4 + (3 + 6)$

40. **a.** $(5 + 2) + 6$
b. $5 + (2 + 6)$

41. **a.** $(7 + 2) + 6$
b. $7 + (2 + 6)$

42. **a.** $(3 + 5) + 4$
b. $3 + (5 + 4)$

43. **a.** $(6 + 3) + 5$
b. $6 + (3 + 5)$

44. **a.** $(4 + 3) + 5$
b. $4 + (3 + 5)$

More than two addends

$(6 + 3) + 2 = 6 + (3 + 2)$

The Grouping Property
Changing the grouping of the addends does not change the sum.

The order and grouping properties allow you to add numbers in any order.

EXERCISES

Add. ***Hint:*** **First look for a sum of ten.**

1. 3 5 +7	**2.** 5 4 +6	**3.** 2 3 +8	**4.** 3 7 +5	**5.** 4 2 +6	**6.** 2 8 +4
7. 1 4 +9	**8.** 7 3 +6	**9.** 3 6 +4	**10.** 2 7 +8	**11.** 5 3 +5	**12.** 3 7 +3
13. 8 2 6 +1	**14.** 3 7 2 +6	**15.** 8 7 2 +1	**16.** 7 4 2 +3	**17.** 8 4 3 +2	**18.** 1 2 9 +3
19. 3 2 9 +5	**20.** 4 5 3 +3	**21.** 3 4 4 +2	**22.** 2 6 2 +3	**23.** 1 5 7 +2	**24.** 3 2 4 +5

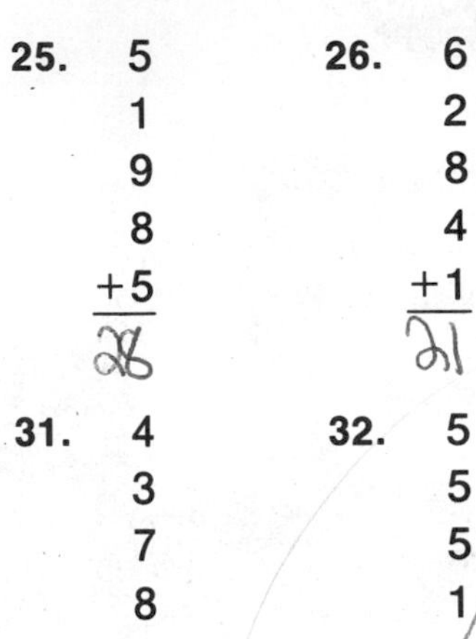

25.	26.	27.	28.	29.	30.
5	6	6	9	6	8
1	2	4	2	7	1
9	8	4	1	3	2
8	4	3	3	4	4
+5	+1	+7	+8	+5	+7
28	21	24	23	25	22

31.	32.	33.	34.	35.	36.
4	5	2	3	3	5
3	5	8	3	5	5
7	5	6	7	9	6
8	1	9	4	7	5
+2	+9	+4	+4	+1	+5
24	25	29	21	25	26

37.	38.	39.	40.	41.	42.
3	8	7	5	6	5
4	5	3	5	4	3
7	1	5	2	7	5
6	5	2	1	5	7
+3	+2	+8	+9	+3	+6
23	21	25	22	25	26

Challenge!

43.

44. Show the different ways you can order these addends.

45. One of these addend cards has been turned over. The sum of the four addends is 19. What addend is on the card that was turned over?

Missing addends

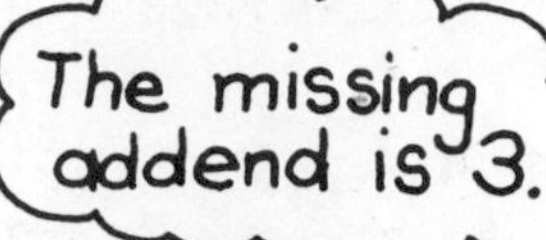

EXERCISES

Give each missing addend.

1. $3 + \underline{?} = 9$
2. $2 + \underline{?} = 4$
3. $3 + \underline{?} = 3$
4. $2 + \underline{?} = 10$
5. $3 + \underline{?} = 10$
6. $6 + \underline{?} = 9$
7. $0 + \underline{?} = 7$
8. $2 + \underline{?} = 9$
9. $2 + \underline{?} = 5$
10. $4 + \underline{?} = 10$
11. $5 + \underline{?} = 6$
12. $0 + \underline{?} = 6$
13. $3 + \underline{?} = 8$
14. $5 + \underline{?} = 9$
15. $1 + \underline{?} = 9$
16. $5 + \underline{?} = 7$
17. $2 + \underline{?} = 6$
18. $3 + \underline{?} = 10$
19. $2 + \underline{?} = 8$
20. $4 + \underline{?} = 8$
21. $1 + \underline{?} = 8$
22. $4 + \underline{?} = 6$
23. $7 + \underline{?} = 10$
24. $3 + \underline{?} = 6$

25. $5 + \underline{?} = 12$

26. $9 + \underline{?} = 12$

27. $4 + \underline{?} = 12$

28. $7 + \underline{?} = 14$

29. $6 + \underline{?} = 14$

30. $8 + \underline{?} = 15$

31. $7 + \underline{?} = 16$

32. $8 + \underline{?} = 13$

33. $5 + \underline{?} = 11$

34. $9 + \underline{?} = 13$

35. $7 + \underline{?} = 11$

36. $8 + \underline{?} = 17$

37. $9 + \underline{?} = 18$

38. $6 + \underline{?} = 15$

39. $8 + \underline{?} = 14$

40. $5 + \underline{?} = 13$

41. $7 + \underline{?} = 12$

42. $8 + \underline{?} = 16$

43. $\underline{?} + 9 = 16$

44. $\underline{?} + 9 = 14$

45. $\underline{?} + 6 = 13$

46. $\underline{?} + 7 = 15$

47. $\underline{?} + 9 = 15$

48. $\underline{?} + 9 = 17$

49. $\underline{?} + 6 = 12$

50. $\underline{?} + 3 = 11$

51. $\underline{?} + 8 = 12$

Challenge!

Guess my number.

52. If you add my number to 7, you get 16.

53. If you add mine to 8, you get 8.

54. If you add my number to itself, you get 16.

55. The sum of 6 and 9 is 2 more than my number.

56. The sum of 7 and 4 is 3 less than my number.

57. 4 less than the sum of 5 and 6 is 1 more than my number.

Subtraction

The answer to a subtraction is called the **difference.**

$$\begin{array}{r} 10 \\ -6 \\ \hline 4 \end{array} \leftarrow \text{difference}$$

EXERCISES

Subtract.

1. $\begin{array}{r} 6 \\ -1 \\ \hline \end{array}$
2. $\begin{array}{r} 5 \\ -3 \\ \hline \end{array}$
3. $\begin{array}{r} 9 \\ -4 \\ \hline \end{array}$
4. $\begin{array}{r} 8 \\ -7 \\ \hline \end{array}$
5. $\begin{array}{r} 6 \\ -2 \\ \hline \end{array}$
6. $\begin{array}{r} 9 \\ -3 \\ \hline \end{array}$
7. $\begin{array}{r} 10 \\ -7 \\ \hline \end{array}$
8. $\begin{array}{r} 10 \\ -8 \\ \hline \end{array}$
9. $\begin{array}{r} 5 \\ -2 \\ \hline \end{array}$
10. $\begin{array}{r} 7 \\ -4 \\ \hline \end{array}$
11. $\begin{array}{r} 9 \\ -5 \\ \hline \end{array}$
12. $\begin{array}{r} 8 \\ -6 \\ \hline \end{array}$
13. $\begin{array}{r} 12 \\ -6 \\ \hline \end{array}$
14. $\begin{array}{r} 12 \\ -7 \\ \hline \end{array}$
15. $\begin{array}{r} 14 \\ -7 \\ \hline \end{array}$
16. $\begin{array}{r} 13 \\ -7 \\ \hline \end{array}$
17. $\begin{array}{r} 15 \\ -8 \\ \hline \end{array}$
18. $\begin{array}{r} 10 \\ -3 \\ \hline \end{array}$
19. $\begin{array}{r} 9 \\ -5 \\ \hline \end{array}$
20. $\begin{array}{r} 15 \\ -6 \\ \hline \end{array}$
21. $\begin{array}{r} 18 \\ -9 \\ \hline \end{array}$
22. $\begin{array}{r} 17 \\ -8 \\ \hline \end{array}$
23. $\begin{array}{r} 14 \\ -8 \\ \hline \end{array}$
24. $\begin{array}{r} 11 \\ -3 \\ \hline \end{array}$
25. $\begin{array}{r} 12 \\ -4 \\ \hline \end{array}$
26. $\begin{array}{r} 11 \\ -7 \\ \hline \end{array}$
27. $\begin{array}{r} 13 \\ -9 \\ \hline \end{array}$
28. $\begin{array}{r} 14 \\ -6 \\ \hline \end{array}$
29. $\begin{array}{r} 12 \\ -9 \\ \hline \end{array}$
30. $\begin{array}{r} 15 \\ -7 \\ \hline \end{array}$

Give each difference.

31. $11 - 9$ 32. $12 - 6$ 33. $17 - 8$ 34. $13 - 4$ 35. $11 - 2$

36. $14 - 7$ 37. $15 - 6$ 38. $13 - 6$ 39. $11 - 7$ 40. $11 - 6$

41. $11 - 8$ 42. $12 - 7$ 43. $13 - 9$ 44. $11 - 5$ 45. $15 - 8$

46. $12 - 4$ 47. $14 - 5$ 48. $11 - 3$ 49. $15 - 7$ 50. $12 - 5$

51. $13 - 5$ 52. $12 - 9$ 53. $13 - 8$ 54. $16 - 8$ 55. $14 - 6$

56. $18 - 9$ 57. $11 - 4$ 58. $17 - 8$ 59. $14 - 9$ 60. $16 - 9$

61. $13 - 7$ 62. $16 - 7$ 63. $12 - 8$ 64. $15 - 9$ 65. $14 - 8$

66. $17 - 9$ 67. $11 - 7$ 68. $12 - 4$ 69. $12 - 7$ 70. $14 - 7$

Find the missing addend.

Hint: Subtract to find a missing addend.

$15 - 9 = 6$

71. $9 + \underline{?} = 15$ 72. $\underline{?} + 8 = 16$ 73. $7 + \underline{?} = 11$ 74. $8 + \underline{?} = 17$

75. $\underline{?} + 9 = 14$ 76. $7 + \underline{?} = 12$ 77. $\underline{?} + 9 = 16$ 78. $\underline{?} + 9 = 12$

79. $7 + \underline{?} = 15$ 80. $\underline{?} + 7 = 14$ 81. $3 + \underline{?} = 11$ 82. $6 + \underline{?} = 12$

Challenge!

Give each missing number.

83.

84.

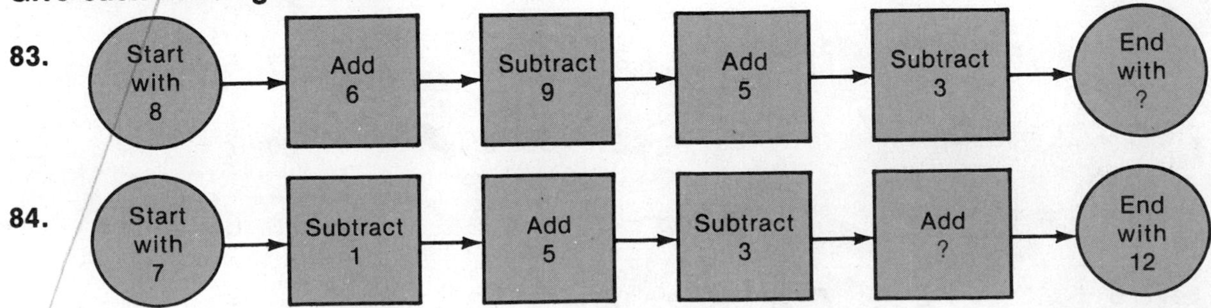

Addition and subtraction

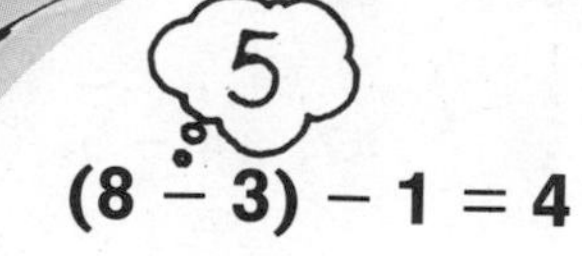

$(8 - 3) - 1 = 4$

$8 - (3 - 1) = 6$

EXERCISES

Compute.

1. $(5 + 3) + 2$
2. $(6 - 3) - 1$
3. $(8 + 2) - 5$
4. $(7 - 2) + 5$
5. $(6 + 3) - 4$
6. $(12 - 4) + 6$
7. $(15 - 6) - 4$
8. $15 - (6 - 4)$
9. $(14 - 8) + 6$
10. $(2 + 6) + 3$
11. $2 + (6 + 3)$
12. $(13 - 8) + 8$
13. $(12 - 8) - 4$
14. $12 - (8 - 4)$
15. $(11 - 6) + 6$
16. $(8 + 9) - 9$
17. $(18 - 9) - 5$
18. $(6 + 8) - 7$
19. $(7 + 6) - 5$
20. $7 + (6 - 5)$
21. $(9 - 4) + 2$
22. $9 - (4 + 2)$
23. $(16 - 7) + 8$
24. $(8 - 6) + 9$

Find the missing numbers.

25. $8 + 3 = 5 + \underline{?}$
26. $6 + 8 = 7 + \underline{?}$
27. $5 + 9 = 9 + \underline{?}$
28. $3 + \underline{?} = 6 + 5$
29. $4 + \underline{?} = 5 + 8$
30. $7 + \underline{?} = 8 + 8$
31. $5 + 3 = 14 - \underline{?}$
32. $13 - 6 = 3 + \underline{?}$
33. $6 + 3 = 12 - \underline{?}$
34. $15 - \underline{?} = 4 + 5$
35. $7 + \underline{?} = 12 - 3$
36. $13 - \underline{?} = 4 + 4$

Use only addition and subtraction. Can you use the three numbers to build the target number?

1. 8 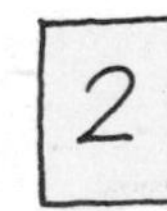 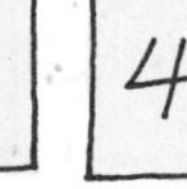 4 — target number 

8 2 4 — target number 6

2.  8 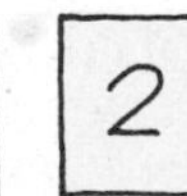 — target number 3

5 8 2 — target number 3

3. 7 4  — target number 5

7 4 6 — target number 5

4. 5 6 8 — target number 

5 6 8 — target number 4

Play the game.

1. Make number cards for 0 through 9.
2. Choose a leader and divide the class into two teams, team A and team B.
3. Without looking, the leader picks a target-number card and 3 other cards.
4. If team A can build the target number using the other 3 numbers, it scores 1 point.
5. The leader chooses new cards, and team B plays.
6. The first team to get 12 points wins the game.

Problem solving—planning what to do

Mrs. Allen opened 3 cans of food. Greg opened 2 cans of food. How many cans of food did they open?

Greg had 7 tent stakes. He broke 3 of them. How many good tent stakes did he have left?

Jill dug up 8 worms. Greg dug up 5 worms. How many more worms did Jill dig up?

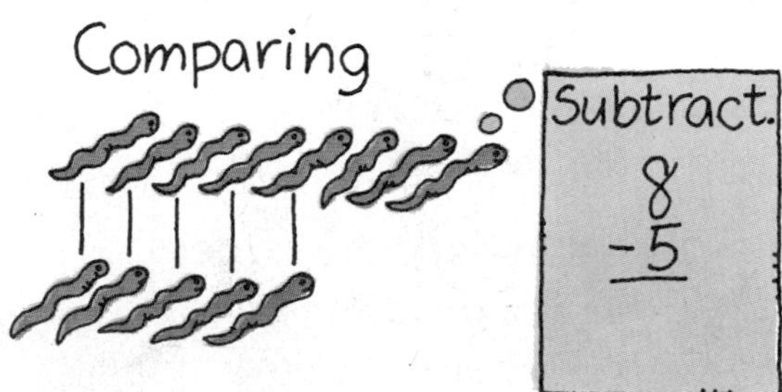

Jill caught 6 fish before lunch. She caught some more fish after lunch. She caught 10 fish in all. How many fish did she catch after lunch?

How will you solve the problem?
Will you add or subtract?

1. Jill caught 7 fish. Greg caught 5 fish. How many did they catch in all?

2. Mr. Allen baked 8 potatoes. His family ate 7 potatoes. How many were left?

3. Jill hiked 3 hours in the morning and 2 hours in the afternoon. How many more hours did she hike in the morning?

4. Mrs. Allen saw 4 red squirrels and 9 gray squirrels. How many squirrels did she see in all?

5. Mr. Allen took 6 pictures. Mrs. Allen took 8 pictures. How many pictures did they take in all?

6. Mr. Allen took 6 pictures. Mrs. Allen took 8 pictures. How many more pictures did Mrs. Allen take?

7. Greg had 6 arrowheads. He found some more. Then he had 9 arrowheads. How many did he find?

8. Greg gathered 5 pieces of firewood, and Jill gathered 3. How many more pieces did Greg gather?

9. Jill rowed 3 hours, and Greg rowed 4 hours. How many hours did they row in all?

10. Mr. Allen had 4 fishhooks. Greg gave him some more. Then Mr. Allen had 9 fishhooks. How many fishhooks did Greg give his father?

11. Mr. Allen caught 9 fish and threw 2 back. How many fish did he have then?

12. Mrs. Allen had 7 eggs and bought 6 more. How many eggs did she have then?

Challenge!

Make up a problem.

Problem solving

1. Study the problem until you understand it.

   15 students played softball. There were 8 on one team. How many were on the other team?

   Find the question.

   How many were on the other team?

   Find the information you need to answer the question.

   15 students, 8 on one team

2. Plan what to do and do it.

   Find a missing addend.

   $8 + ? = 15$

   Subtract.

   $$\begin{array}{r} 15 \\ -8 \\ \hline 7 \end{array}$$

3. Answer the question and check your answer.

   There were 7 on the other team.

   Check: $8 + 7 = 15$

Solve.

1. The first inning lasted 9 minutes. The second inning lasted 8 minutes. How many minutes longer was the first inning?

2. Allison struck out 9 batters. Dave struck out 5 batters. How many more batters did Allison strike out?

3. One team had 6 softball gloves. The other team had 7 softball gloves. How many gloves did they have in all?

4. The winning team scored 9 runs. The losing team scored 3 fewer runs than the winning team. How many runs did the losing team score?

5. Debra caught 8 fly balls. Terry caught 7 more fly balls than Debra. How many fly balls did Terry catch?

6. Alex hit the ball 5 times during warm-up. He hit the ball 6 times during the game. How many times did he hit the ball?

7. In the last inning, 8 hits were made. One team made 3 hits. How many hits did the other team make?

8. The team members had 9 bats at the game. They lost 2 bats. How many bats did they have left?

9. Bill scored 8 points during the first half of a basketball game. He scored 4 points during the second half. How many points did he score?

10. Craig took 9 long shots and 7 short shots at the basket. How many shots did he take?

11. Linda took 8 shots at the basket. She made 6 shots. How many shots did she miss?

★ 12. The team members had 7 basketballs. They lost 2 of them. Then they bought 6 new ones. How many basketballs did they have then?

CHAPTER CHECKUP

Give each sum. [pages 2–9,12–13]

1. $\begin{array}{r}5\\+8\\\hline\end{array}$	2. $\begin{array}{r}9\\+6\\\hline\end{array}$	3. $\begin{array}{r}7\\+7\\\hline\end{array}$	4. $\begin{array}{r}3\\+9\\\hline\end{array}$	5. $\begin{array}{r}9\\+5\\\hline\end{array}$	6. $\begin{array}{r}5\\+5\\\hline\end{array}$
7. $\begin{array}{r}4\\+9\\\hline\end{array}$	8. $\begin{array}{r}8\\+6\\\hline\end{array}$	9. $\begin{array}{r}8\\+8\\\hline\end{array}$	10. $\begin{array}{r}4\\+8\\\hline\end{array}$	11. $\begin{array}{r}5\\+6\\\hline\end{array}$	12. $\begin{array}{r}9\\+7\\\hline\end{array}$
13. $\begin{array}{r}9\\+9\\\hline\end{array}$	14. $\begin{array}{r}4\\+7\\\hline\end{array}$	15. $\begin{array}{r}6\\+6\\\hline\end{array}$	16. $\begin{array}{r}8\\+7\\\hline\end{array}$	17. $\begin{array}{r}6\\+7\\\hline\end{array}$	18. $\begin{array}{r}6\\+8\\\hline\end{array}$

19. $5 + 9$	20. $8 + 9$	21. $6 + 9$	22. $6 + 4$
23. $9 + 8$	24. $7 + 8$	25. $7 + 9$	26. $7 + 5$

Give each difference. [pages 10–13]

27. $\begin{array}{r}15\\-8\\\hline\end{array}$	28. $\begin{array}{r}11\\-8\\\hline\end{array}$	29. $\begin{array}{r}14\\-7\\\hline\end{array}$	30. $\begin{array}{r}13\\-8\\\hline\end{array}$	31. $\begin{array}{r}14\\-5\\\hline\end{array}$	32. $\begin{array}{r}16\\-8\\\hline\end{array}$
33. $\begin{array}{r}12\\-9\\\hline\end{array}$	34. $\begin{array}{r}16\\-7\\\hline\end{array}$	35. $\begin{array}{r}12\\-6\\\hline\end{array}$	36. $\begin{array}{r}17\\-9\\\hline\end{array}$	37. $\begin{array}{r}11\\-7\\\hline\end{array}$	38. $\begin{array}{r}15\\-6\\\hline\end{array}$
39. $\begin{array}{r}17\\-8\\\hline\end{array}$	40. $\begin{array}{r}12\\-8\\\hline\end{array}$	41. $\begin{array}{r}14\\-8\\\hline\end{array}$	42. $\begin{array}{r}13\\-6\\\hline\end{array}$	43. $\begin{array}{r}11\\-9\\\hline\end{array}$	44. $\begin{array}{r}13\\-9\\\hline\end{array}$

45. $14 - 6$	46. $18 - 9$	47. $14 - 9$
48. $15 - 9$	49. $15 - 7$	50. $16 - 9$

Solve.

51. Ames School has 9 computers. Banneker School has 6 computers. How many computers do they have altogether?

52. Washburn School has 16 computers. McNair School has 9 computers. How many more computers does Washburn School have?

CHAPTER PROJECT

1. Get two cubes.
2. Write 3, 4, 5, 6, 7, and 8 on each cube.
3. Toss the cubes and graph the sum.

Write the title of your graph at the top.

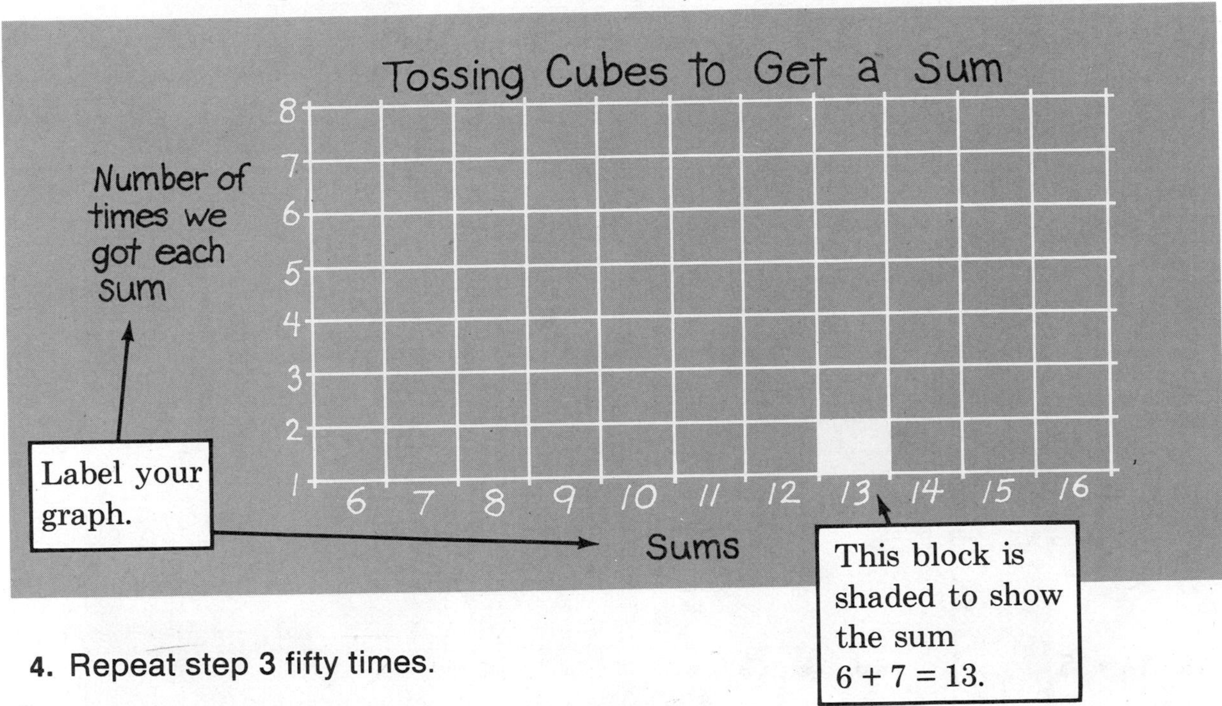

Label your graph.

This block is shaded to show the sum $6 + 7 = 13$.

4. Repeat step 3 fifty times.
5. Answer these questions about your graph.
   a. Which sum appears most often?
   b. Which sum appears least often?
   c. Did any sums appear the same number of times? Which ones were they?
   d. If you were asked to guess a sum before you tossed the cubes, what sum would you guess? Why?

★ 6. This table shows all the ways you can toss a sum of 11. Make similar tables for the other sums.

First cube	3	4	5	6	7	8
Second cube	8	7	6	5	4	3

CHAPTER REVIEW

1. Copy and complete this addition table.

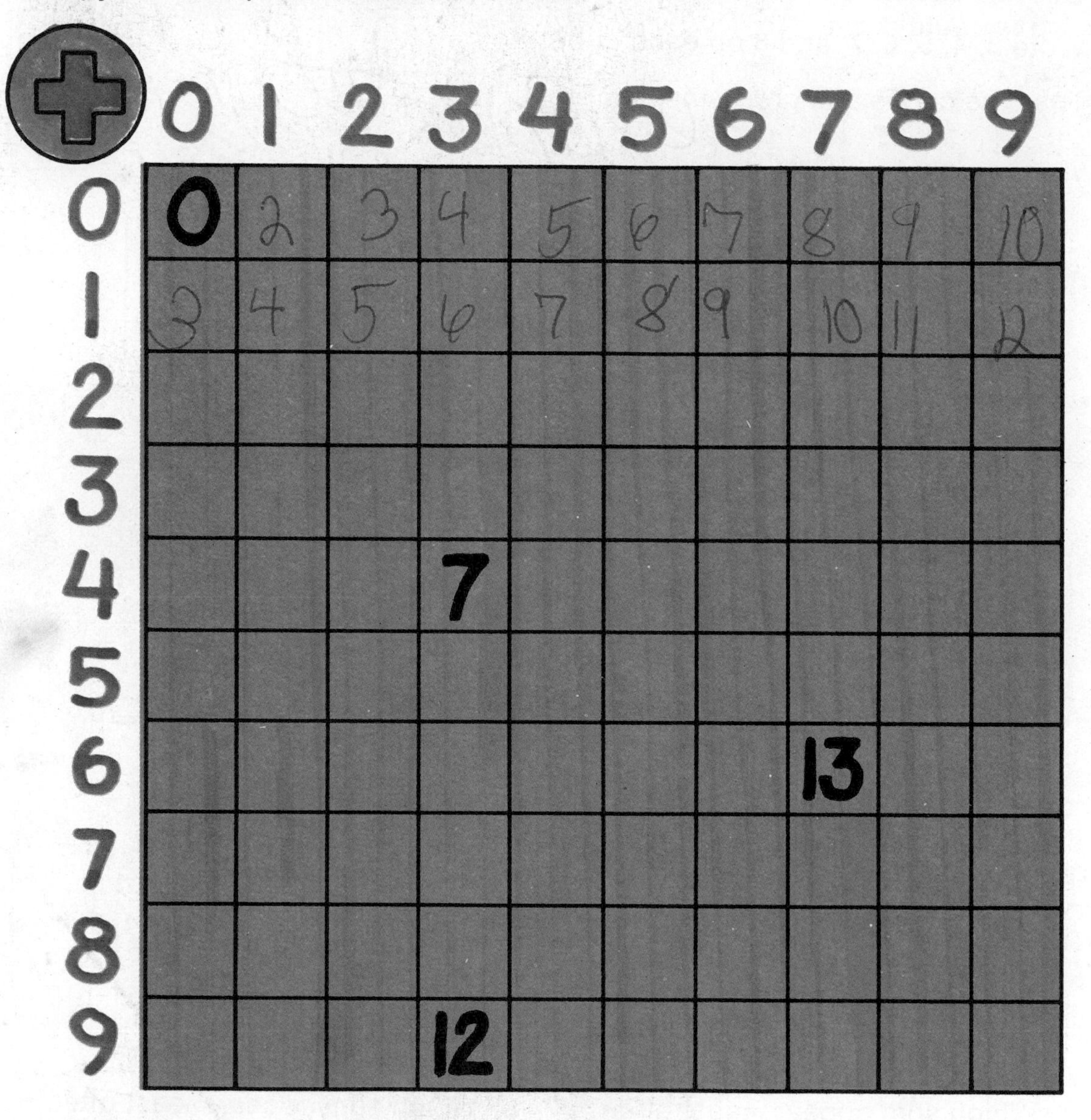

2. Keep your addition table to review your facts.

CHAPTER CHALLENGE

Copy and complete.

Add across. Add down.

1. (+ across, + down)

5	2	7
3	4	7
8	6	14

2. (+ across, + down)

3	?	7
?	2	8
?	?	?

Subtract across. Subtract down.

3. (− across, − down)

15	6	?
8	2	?
?	?	?

4. (− across, − down)

17	?	9
?	3	?
8	?	?

Add across. Subtract down.

5. (+ across, − down)

9	6	?
3	4	?
?	?	?

6. (+ across, − down)

9	8	?
3	5	?
?	?	?

7. (+ across, − down)

?	5	13
2	?	6
?	?	?

8. (+ across, − down)

7	?	9
?	?	?
4	?	5

Choose the correct letter.

1. Add.

$$\begin{array}{r} 7 \\ +3 \\ \hline \end{array}$$

a. 4
b. 10
c. 9
d. none of these

2. Add.

$$\begin{array}{r} 6 \\ +8 \\ \hline \end{array}$$

a. 15
b. 17
c. 14
d. none of these

3. Add.
$9 + 7$

a. 16
b. 15
c. 2
d. none of these

4. Add.

$$\begin{array}{r} 3 \\ 4 \\ +6 \\ \hline \end{array}$$

a. 10
b. 12
c. 14
d. none of these

5. Add.

$$\begin{array}{r} 2 \\ 7 \\ +8 \\ \hline \end{array}$$

a. 16
b. 17
c. 18
d. none of these

6. Complete.
$6 + \underline{?} = 9$

a. 3
b. 15
c. 16
d. none of these

7. Complete.
$5 + \underline{?} = 6$

a. 11
b. 1
c. 12
d. none of these

8. Subtract.

$$\begin{array}{r} 9 \\ -5 \\ \hline \end{array}$$

a. 5
b. 3
c. 4
d. none of these

9. Subtract.

$$\begin{array}{r} 16 \\ -7 \\ \hline \end{array}$$

a. 9
b. 7
c. 8
d. none of these

10. Subtract.
$12 - 7$

a. 6
b. 5
c. 9
d. none of these

11. There were 3 boys and 8 girls at the party. How many more girls than boys were at the party?

a. 5 **b.** 12
c. 11 **d.** none of these

12. Charles had 6 marbles. Sarah had 7 more marbles than Charles. How many did Sarah have?

a. 13 **b.** 14
c. 1 **d.** none of these

Place Value

Hundreds

We use the digits 0, 1, 2, 3, 4, 5, 6, 7, 8, 9 in our place-value system.

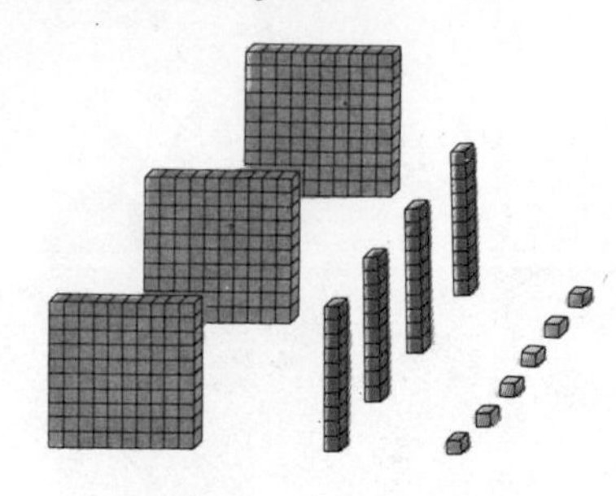

Hundreds	Tens	Ones
3	**4**	**6**

The 3 stands for 3 hundreds, or 300.

The 4 stands for 4 tens, or 40.

The 6 stands for 6 ones, or 6.

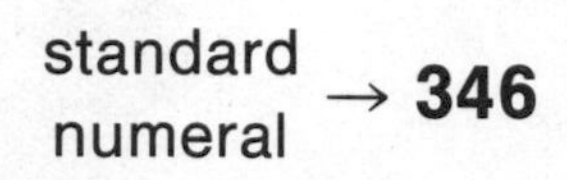

standard numeral → **346**

three hundred forty-six

EXERCISES

How many blocks?

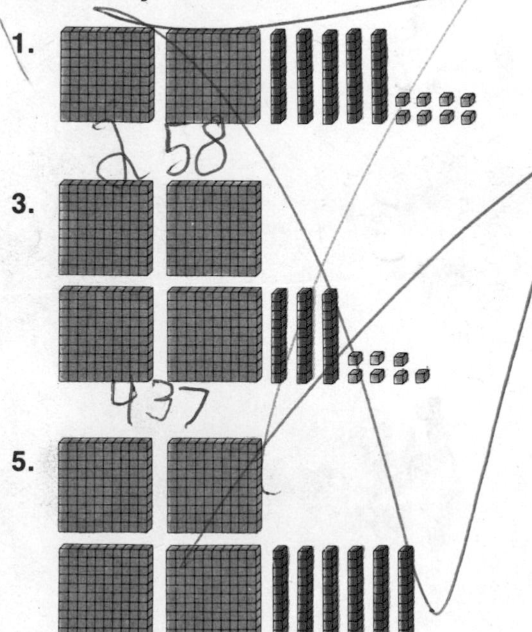

1. 258

3. 437

5. 460

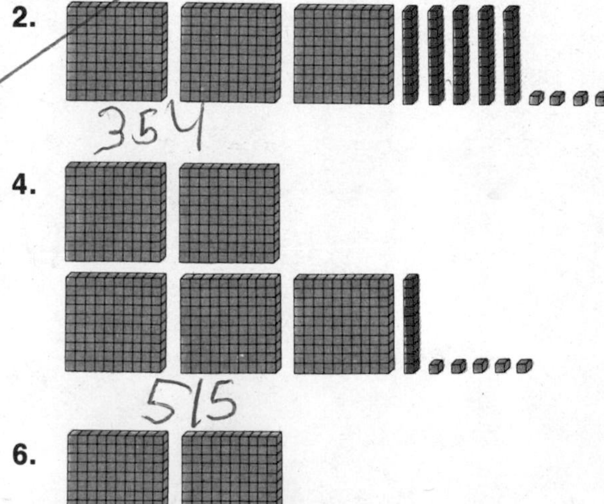

2. 354

4. 515

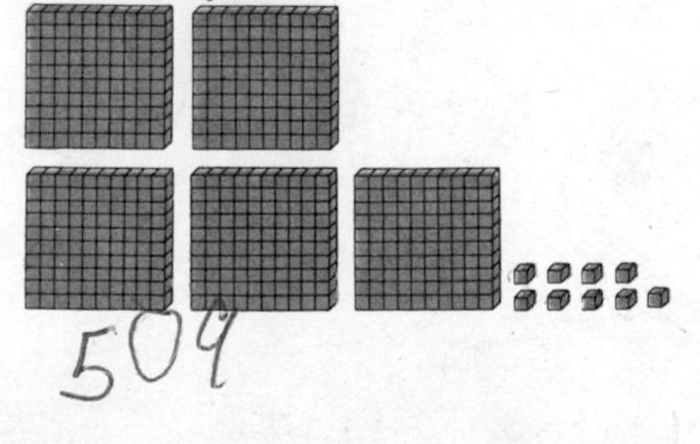

6. 509

What does the red digit stand for?
Give two answers.

7. 374	**8.** 596	**9.** 708	**10.** 312
11. 640	**12.** 666	**13.** 666	**14.** 666

Give the standard numeral.

15. forty-three

16. one hundred twenty-four

17. three hundred thirty-eight

18. three hundred eight

19. six hundred twelve

20. eight hundred fifty-six

21. nine hundred seventy

22. four hundred twenty-two

23. five hundred fifty-five

24. two hundred thirteen

Write in words.

25. 400	**26.** 430	**27.** 436	**28.** 375	**29.** 621
30. 511	**31.** 819	**32.** 370	**33.** 506	**34.** 617

Challenge!

Give each score.

Rounding to the nearest ten

Here is how to round numbers to the nearest ten.

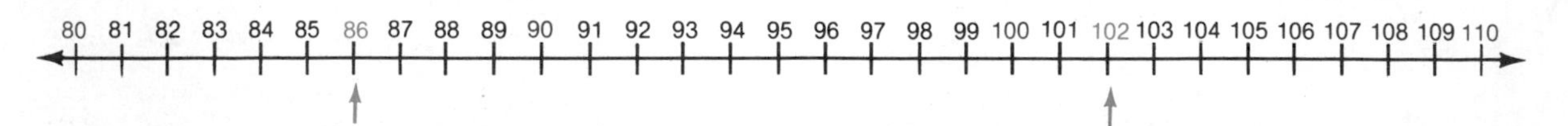

Round 86 to the nearest ten.
86 is between 80 and 90.
It is nearer 90.
So, round to 90.

Round 102 to the nearest ten.
102 is between 100 and 110.
It is nearer 100.
So, round to 100.

When the number is halfway between, round up.

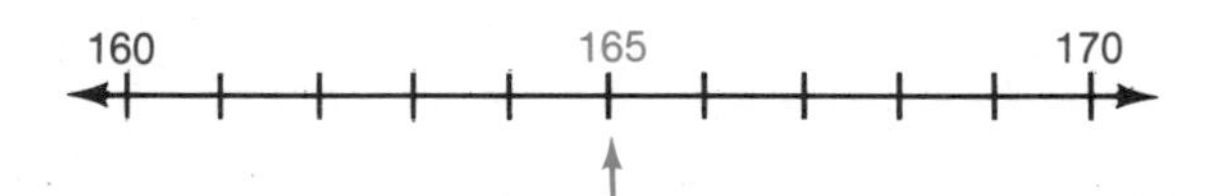

Round 165 to the nearest ten.
165 is halfway between 160 and 170.
So, round to 170.

EXERCISES

Round to the nearest ten.

1. 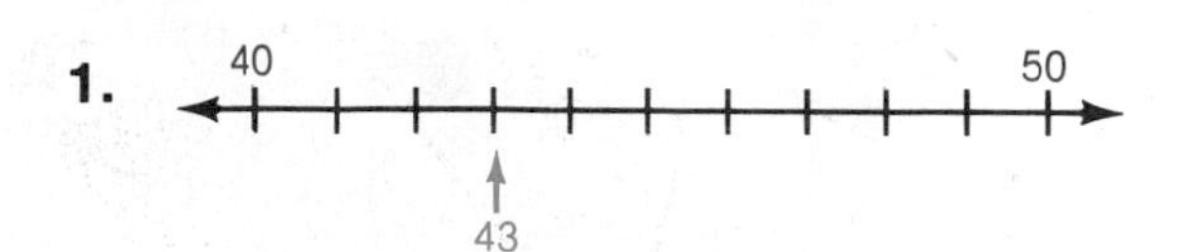

2. 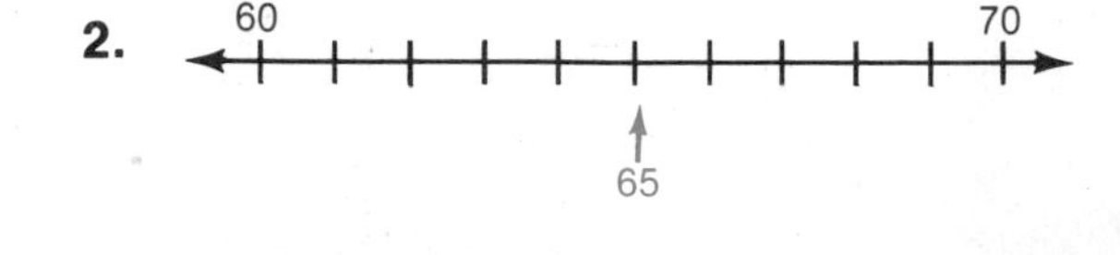

3. 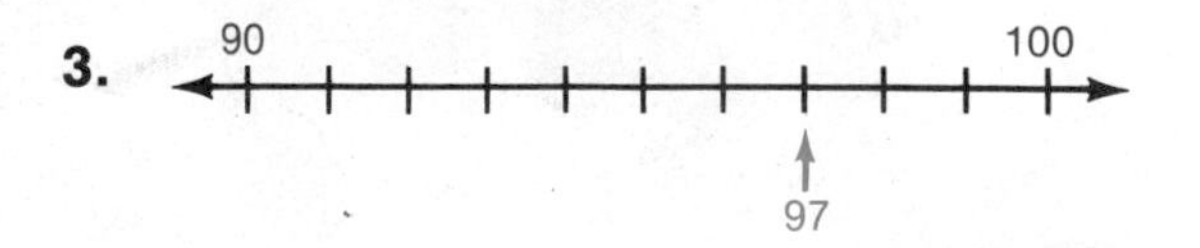

4. 

Round to the nearest ten.

5. 8	6. 14	7. 15	8. 19	9. 26	10. 29
11. 30	12. 35	13. 38	14. 41	15. 53	16. 55
17. 57	18. 68	19. 76	20. 82	21. 88	22. 93
23. 132	24. 167	25. 245	26. 118	27. 434	28. 150
29. 474	30. 183	31. 495	32. 699	33. 207	34. 512
35. 529	36. 458	37. 675	38. 326	39. 875	40. 980

Round each top running speed to the nearest ten.

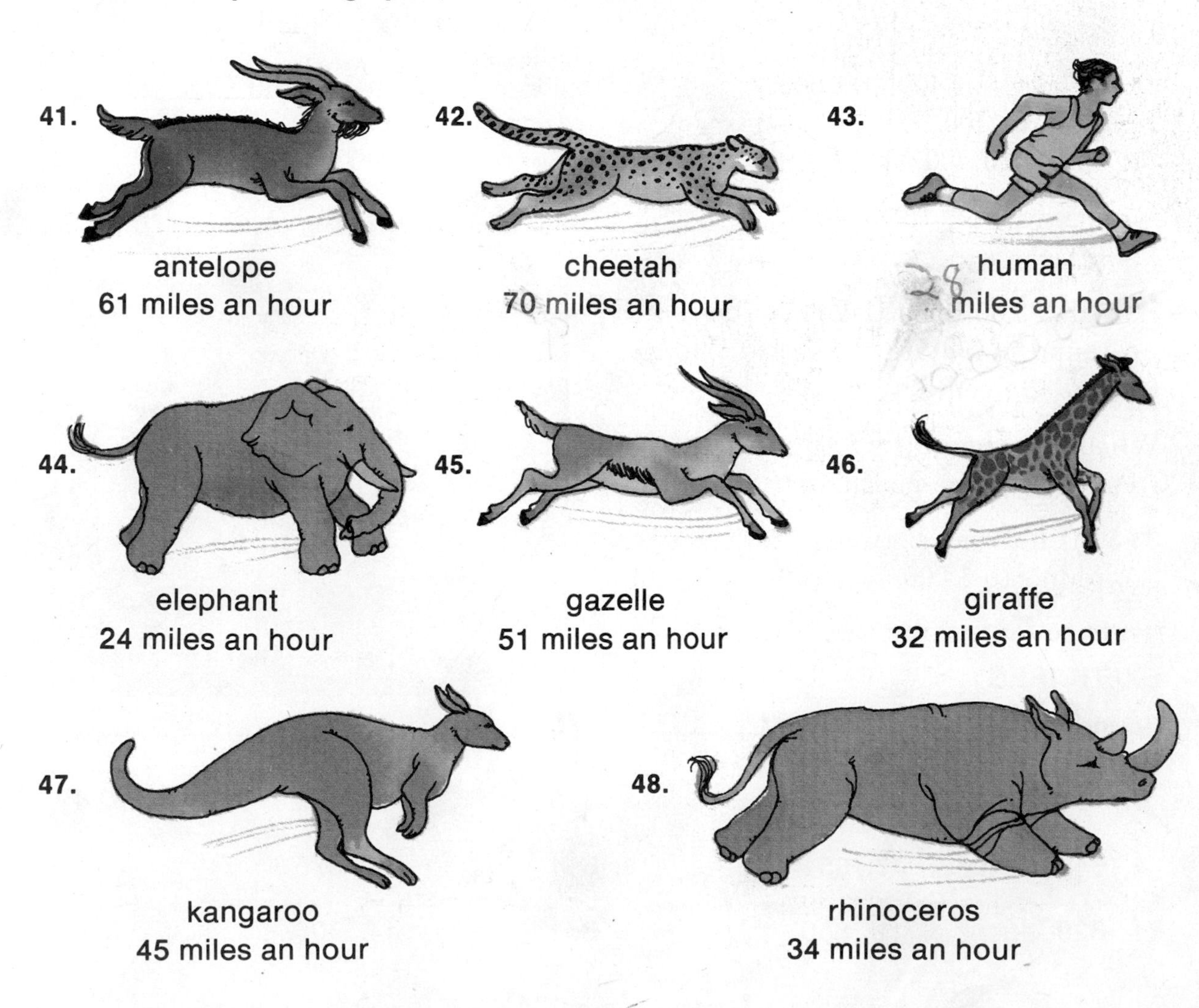

Rounding to the nearest hundred

The zebra weighs 447 pounds.

The baby elephant weighs 583 pounds.

Here is how to round to the nearest hundred.

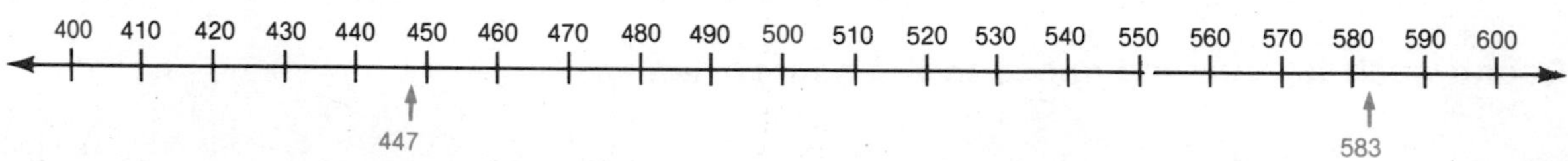

447 is between 400 and 500.
It is nearer 400.
So, round to 400.

583 is between 500 and 600.
It is nearer 600.
So, round to 600.

It costs $3.50 for a ticket to the zoo.
To round money to the nearest dollar, look at the cents.

When the amount is $.50 or more, round up to the next dollar.

$3.50 is halfway between $3.00 and $4.00.
So, round to $4.00.

EXERCISES

Round to the nearest hundred.

1. 160	**2.** 250	**3.** 329	**4.** 465	**5.** 498
6. 540	**7.** 551	**8.** 611	**9.** 678	**10.** 719
11. 829	**12.** 850	**13.** 851	**14.** 906	**15.** 950

Round each weight to the nearest hundred.

16. 

750 pounds

17. 

426 pounds

18. 

291 pounds

19. 

835 pounds

Round to the nearest dollar.

20. \$2.38	**21.** \$2.79	**22.** \$3.11
23. \$3.50	**24.** \$5.23	**25.** \$8.09
26. \$9.47	**27.** \$5.80	**28.** \$2.83
29. \$1.64	**30.** \$8.89	**31.** \$2.35
32. \$7.02	**33.** \$9.38	**34.** \$6.50
35. \$3.80	**36.** \$3.08	**37.** \$8.30
38. \$2.91	**39.** \$2.19	**40.** \$9.21

Challenge!

Guess my number.

41. If you round my number to the nearest ten, you get 400. It is the smallest such number.

42. If you round my number to the nearest hundred, you get 600. It is the greatest such amount.

KEEPING SKILLS SHARP

1. 9 +8

2. 7 +9

3. 5 +8

4. 6 +9

5. 8 +6

6. 9 +4

7. 5 5 +4

8. 6 7 +4

9. 7 8 +2

10. 5 9 +5

11. 6 1 +9

12. 8 5 +2

13. 9 6 4 +1

14. 5 7 3 +2

15. 8 5 2 7 +3

16. 4 6 5 4 +5

Thousands

When you put 10 hundreds together, you get 1 thousand.

10 hundreds = 1 thousand = 1000

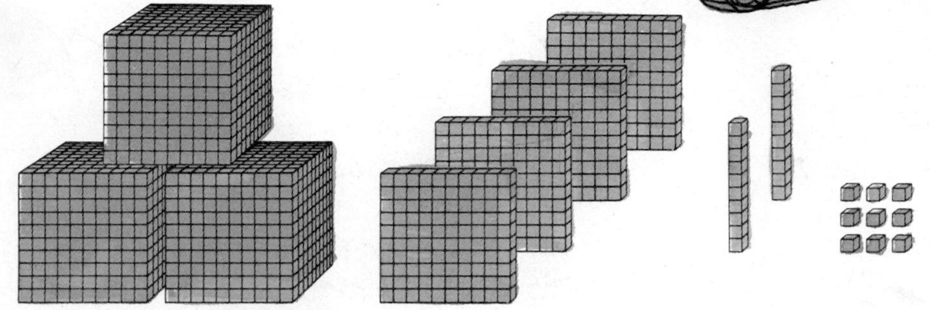

Thousands	Hundreds	Tens	Ones	
3	**4**	**2**	**9**	**3429**

EXERCISES

How many blocks?

1. **2.** **3.** **4.**

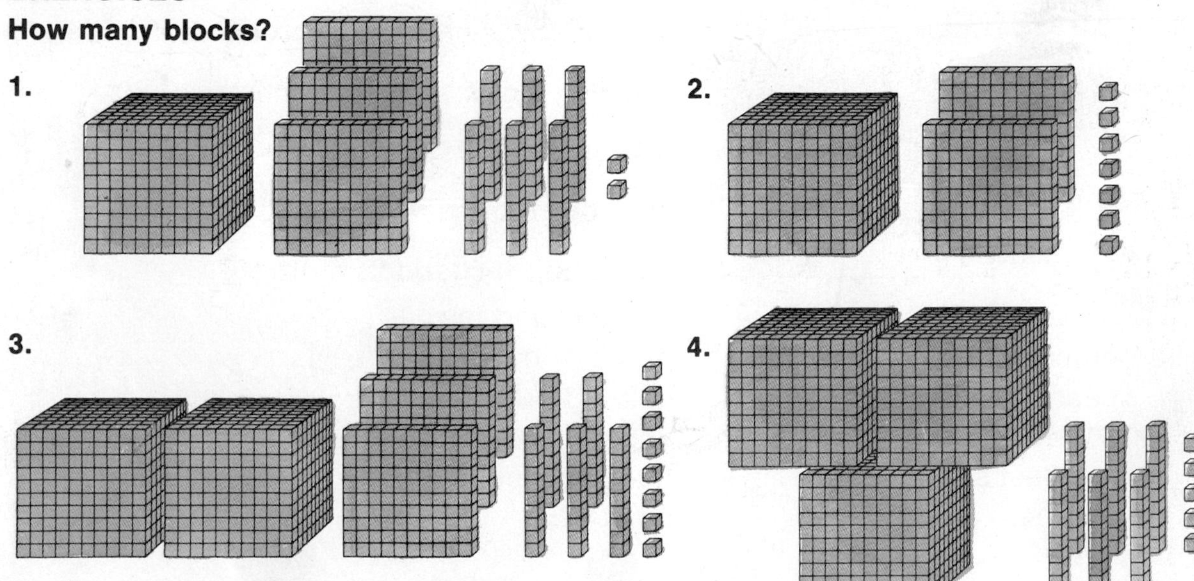

What does the red digit stand for?
Give two answers.

5. 3821	**6.** 5604	**7.** 9715	**8.** 7392
9. 8615	**10.** 2911	**11.** 8362	**12.** 9541
13. 2222	**14.** 2222	**15.** 2222	**16.** 2222

Build a numeral.

17. **4** in the **tens** place
6 in the **ones** place
7 in the **thousands** place
3 in the **hundreds** place

18. **2** in the **hundreds** place
5 in the **tens** place
9 in the **ones** place
4 in the **thousands** place

19. **6** in the **ones** place
1 in the **thousands** place
7 in the **tens** place
4 in the **hundreds** place

20. **2** in the **ones** place
5 in the **hundreds** place
3 in the **thousands** place
8 in the **tens** place

Give the standard numeral.

21. eight thousand five hundred fifty-four

22. six thousand two hundred ninety-three

23. five thousand four hundred eleven

24. two thousand six hundred twenty

25. four thousand five hundred seven

26. nine thousand thirty-five

Write in words.

27. 5000	**28.** 5300	**29.** 5380	**30.** 5386
31. 7802	**32.** 9051	**33.** 3012	**34.** 6009

Thousands

The number below tells about how many times Alan's heart beats in a week.

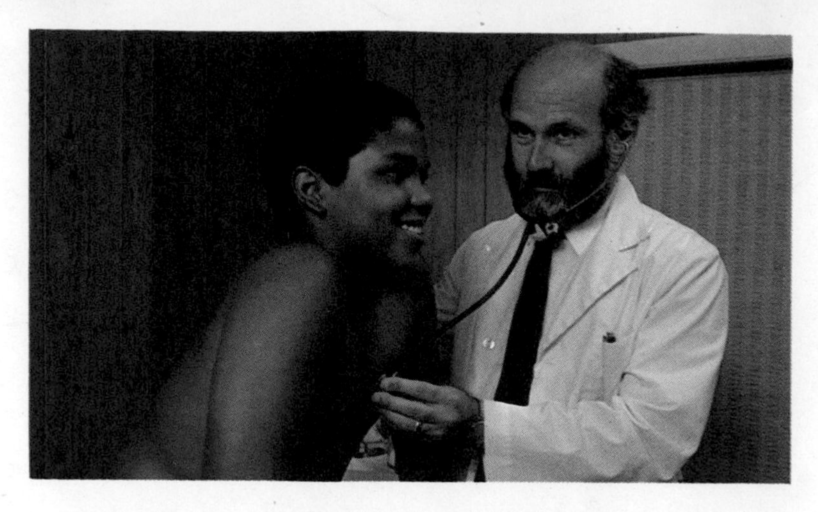

Thousands					
Hundreds	Tens	Ones	Hundreds	Tens	Ones
7	4	5	9	2	0

The 7 stands for 7 hundred thousands, or 700,000.

The 4 stands for 4 ten thousands, or 40,000.

The 5 stands for 5 thousands, or 5000.

standard numeral → **745,920**

seven hundred forty-five thousand, nine hundred twenty

A comma may be used to mark off the thousands.

EXERCISES

Read each numeral.

1. 3217
2. 74,286
3. 539,160
4. 26,358
5. 473,519
6. 682,307
7. 85,620
8. 34,316
9. 52,086
10. 129,340
11. 610,725
12. 736,130
13. 405,274
14. 342,070
15. 250,300
16. 510,000
17. 946,003
18. 804,012
19. 606,600
20. 660,060
21. 600,606

First copy and put in the comma.
Then read the numeral.

22. 93812 | **23.** 56074 | **24.** 90316

25. 72655 | **26.** 57376 | **27.** 44781

28. 581402 | **29.** 290390 | **30.** 836596

31. 602265 | **32.** 764431 | **33.** 189207

34. 495147 | **35.** 315280 | **36.** 351008

Study this 6-digit numeral.

Give the digit that is in the

37. hundreds place.

38. tens place.

39. thousands place.

40. hundred thousands place.

41. ones place.

42. ten thousands place.

What does the red digit stand for?
Give two answers.

43. 16,753 | **44.** 52,814 | **45.** 96,360 | **46.** 59,840

47. 444,444 | **48.** 444,444 | **49.** 444,444 | **50.** 444,444

Give the standard numeral.

51. twenty-six thousand, four hundred sixty-two

52. five hundred seventy-two thousand, two hundred ten

53. six hundred forty thousand, one hundred twelve

54. nineteen thousand, sixty-one

Write in words.

55. 72,346 | **56.** 59,307 | **57.** 82,150 | **58.** 65,040

59. 378,162 | **60.** 829,460 | **61.** 703,401 | **62.** 800,075

Using numbers

Richard and Susan visited the Statue of Liberty. They learned many number facts about the Statue of Liberty.

Read each fact. Then write the standard numeral.

1. France gave the statue to the United States in eighteen hundred eighty-four.
2. It arrived in New York City in two hundred fourteen packing crates.
3. Workmen put the statue together in about three thousand seven hundred sixty days.

4. The statue weighs four hundred fifty thousand pounds.
5. The total cost was five hundred thousand dollars.
6. Richard and Susan climbed one hundred sixty-eight steps to get to the head of the statue.
7. They counted thirty-six people in the head.
8. They had never seen such a large statue. The length of the index finger is ninety-six inches.
9. The length of the hand is one hundred ninety-seven inches.
10. The nose is as long as Susan is tall, fifty-four inches.

KEEPING SKILLS SHARP

1. $\begin{array}{r} 12 \\ -8 \\ \hline \end{array}$
2. $\begin{array}{r} 13 \\ -4 \\ \hline \end{array}$
3. $\begin{array}{r} 14 \\ -7 \\ \hline \end{array}$
4. $\begin{array}{r} 13 \\ -5 \\ \hline \end{array}$
5. $\begin{array}{r} 13 \\ -9 \\ \hline \end{array}$
6. $\begin{array}{r} 15 \\ -6 \\ \hline \end{array}$
7. $\begin{array}{r} 15 \\ -9 \\ \hline \end{array}$
8. $\begin{array}{r} 13 \\ -8 \\ \hline \end{array}$
9. $\begin{array}{r} 11 \\ -2 \\ \hline \end{array}$
10. $\begin{array}{r} 15 \\ -8 \\ \hline \end{array}$
11. $\begin{array}{r} 17 \\ -8 \\ \hline \end{array}$
12. $\begin{array}{r} 11 \\ -5 \\ \hline \end{array}$
13. $\begin{array}{r} 11 \\ -9 \\ \hline \end{array}$
14. $\begin{array}{r} 16 \\ -7 \\ \hline \end{array}$
15. $\begin{array}{r} 14 \\ -8 \\ \hline \end{array}$
16. $\begin{array}{r} 13 \\ -6 \\ \hline \end{array}$
17. $\begin{array}{r} 14 \\ -5 \\ \hline \end{array}$
18. $\begin{array}{r} 16 \\ -8 \\ \hline \end{array}$
19. $\begin{array}{r} 18 \\ -9 \\ \hline \end{array}$
20. $\begin{array}{r} 12 \\ -7 \\ \hline \end{array}$

Comparing and ordering numbers

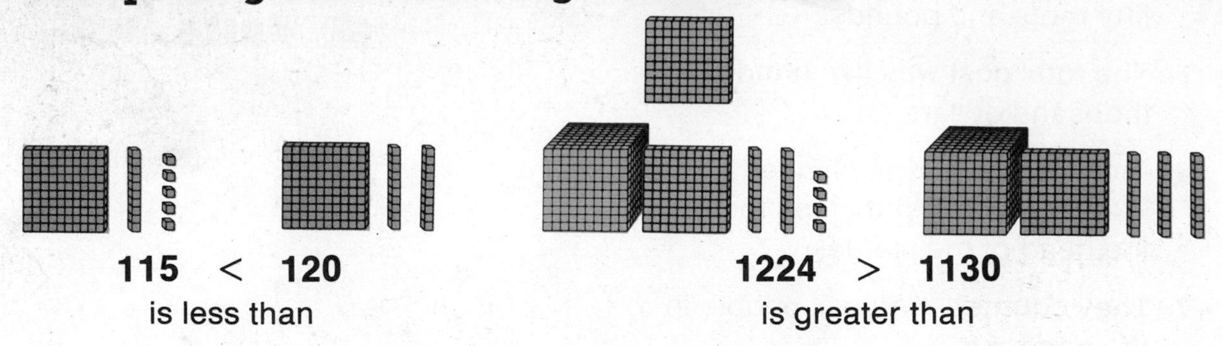

Notice that each symbol points to the smaller number.

Here is how to compare larger numbers.

EXAMPLE. Compare 35,286 and 36,410.

Step 1. Compare the digits in the ten-thousands place. Are they the same?

35,286 **36,410**
↑ ↑

Step 2. Next compare the digits in the thousands place. Are they the same?

35,286 **36,410**
↑—— **5 < 6** —↑

Step 3. Since 5 is less than 6, 35,286 is less than 36,410.

35,286 < 36,410

Less than (<) or greater than (>)?

1. 59 ○ 58
2. 829 ○ 836
3. 743 ○ 745
4. 7346 ○ 7410
5. 5382 ○ 5374
6. 9346 ○ 8123
7. 53,742 ○ 53,821
8. 69,385 ○ 60,381
9. 42,478 ○ 42,500
10. 274,369 ○ 235,782
11. 763,510 ○ 763,780
12. 593,261 ○ 593,099

Order from least to greatest.

13. 36, 29, 45, 50
14. 59, 37, 84, 66
15. 153, 207, 187, 168
16. 594, 672, 629, 579
17. 2174, 2096, 2147
18. 8620, 8753, 8044
19. 32,163; 23,988; 27,429
20. 59,999; 61,235; 60,000

Which number is greatest?

1.

2.

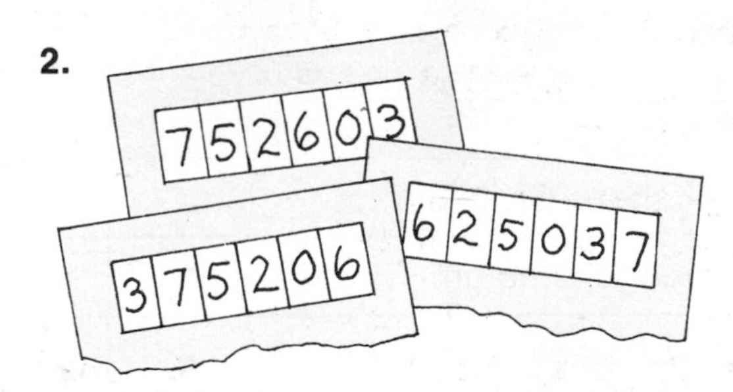

Play the game.

1. Make ten cards, each with a different digit on it.
2. Each player should make a table like this on a piece of paper:

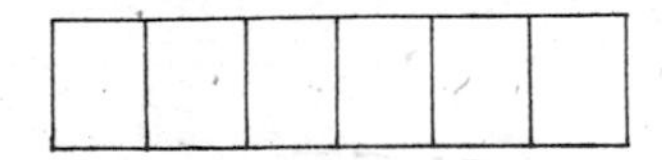

3. The game leader picks a card. Each player writes the digit in a section of the table.
4. Repeat step 3 until six digits have been picked.
5. The player who has built the greatest number is the winner.

Rounding

The Boston Garden can seat 15,314 people for a basketball game.

Here is how to round numbers to the nearest thousand.

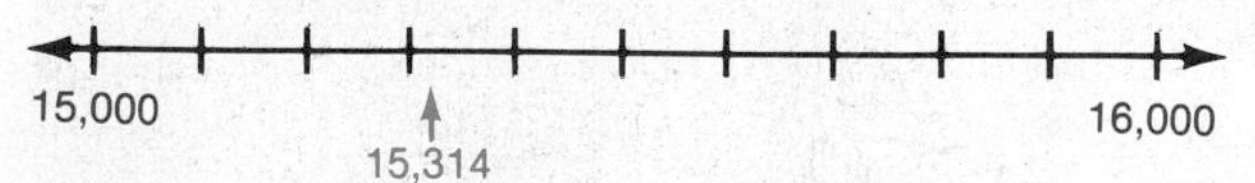

15,314 is between 15,000 and 16,000.
It is nearer 15,000.
So, round to 15,000.

Remember, when the number is halfway between, round up.

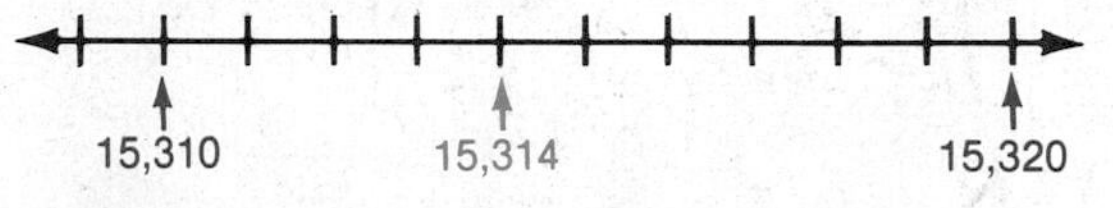

15,314 rounded to the nearest ten is 15,310.

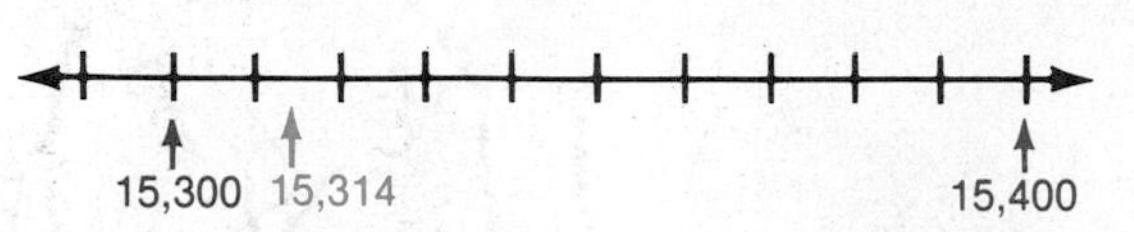

15,314 rounded to the nearest hundred is 15,300.

EXERCISES

Complete.

Basketball Arena	Seating Capacity	Rounded to nearest ten	Rounded to nearest hundred	Rounded to nearest thousand
Baltimore Civic Center	12,289	**1.**	**2.**	**3.**
Chicago Stadium	17,374	**4.**	**5.**	**6.**
Cow Palace, San Francisco	14,500	**7.**	**8.**	**9.**
Dallas Memorial Auditorium	8,088	**10.**	**11.**	**12.**
Kiel Auditorium, St. Louis	10,574	**13.**	**14.**	**15.**
Los Angeles Sports Arena	15,333	**16.**	**17.**	**18.**
Milwaukee Arena	10,756	**19.**	**20.**	**21.**
Nassau Coliseum, Hempstead, New York	15,367	**22.**	**23.**	**24.**
Sam Houston Coliseum, Houston	8,925	**25.**	**26.**	**27.**
Veterans Memorial Coliseum, Phoenix	12,535	**28.**	**29.**	**30.**

The exact number was rounded to the nearest hundred.
Pick the rounded number.

31. The seating capacity of the basketball arena was 17,183.

a. 17,000 **b.** 17,190 **c.** 17,200 **d.** 17,900

32. There were 14,219 people at the basketball game.

a. 14,000 **b.** 14,200 **c.** 14,300 **d.** 15,000

33. There were 2964 empty seats in the basketball arena.

a. 2700 **b.** 2900 **c.** 2970 **d.** 3000

Millions

The number **283,996,800** tells how many seconds you will have lived by your ninth birthday.

The **2** stands for 2 hundred millions, or 200,000,000.

The **8** stands for 8 ten millions, or 80,000,000.

The **3** stands for 3 millions, or 3,000,000.

Millions			Thousands					
Hundreds	Tens	Ones	Hundreds	Tens	Ones	Hundreds	Tens	Ones
2	8	3	9	9	6	8	0	0

standard numeral → **283,996,800**

two hundred eighty-three million, nine hundred ninety-six thousand, eight hundred

Commas may be used to mark off thousands and millions.

EXERCISES

Read aloud.

1. 8,265,431
2. 9,058,162
3. 26,371,420
4. 58,296,095
5. 321,716,595
6. 874,263,815
7. 734,216,300
8. 426,385,196
9. 582,306,039
10. 750,000,465
11. 853,071,000
12. 639,211,007
13. 206,500,317
14. 500,023,700
15. 375,005,041
16. 458,072,005

These numerals were printed by a computer.
First copy the numeral and put in the commas.
Then read the numeral.

17. 5834219
18. 5186324
19. 5951672
20. 1628957
21. 56583215
22. 27491651
23. 30659481
24. 67215943
25. 125537691
26. 840635910
27. 700325462
28. 917834607

Computers and dates

Computers used in business must keep a record of the date. This is done by making the date in the form of a six-digit number like this:

11 23 56

The first two digits are the **month.** January is 01, February is 02, so November is 11.

The middle two digits are the **day** of the month.

The last two digits are the **year.**

So, 112356 is November 23, 1956.

EXERCISES

1. Give the month, day, and year of the following computer dates:

   a. 020981 b. 111736 c. 083093 d. 041976 e. 050268 f. 101074

2. Give your birthday as a computer date.
3. Give Thanksgiving Day as a computer date.
4. Give New Year's Day as a computer date.
5. Give today's date as a computer date.

CHAPTER CHECKUP

Round to the nearest ten. [pages 24–27]

1. 74
2. 55
3. 167
4. 203
5. 345

Round to the nearest hundred or nearest dollar. [pages 28–29]

6. 75
7. 126
8. 348
9. $5.50
10. $6.19

What does the red digit stand for? [pages 30–31]
Give two answers.

11. 9682
12. 3574
13. 6172
14. 4839

[pages 32–35]
Which digit is in the

15. ten thousands place?
16. thousands place?
17. hundred thousands place?

< **or** >**?** [pages 36–37]

18. 838  829
19. 2385  2386
20. 9748 ⬤ 9874
21. 35,261  35,172
22. 24,078  24,178
23. 369,152 ⬤ 369,143

Round to the nearest thousand. [pages 38–39]

24. 7953
25. 128,581
26. 396,483

CHAPTER PROJECT

You can make a code by assigning a number pattern to the alphabet.

1. Complete the number pattern for all letters of the alphabet.

A	B	C	D	E	F	G	H
2	4	6	8	10	12	14	16

2. Using the number pattern above, decode this message:

SECRET MESSAGE

50|30|42☆28|30|46☆22|28|30|46☆

16|30|46☆40|30☆8|10|6|30|8|10☆

2☆26|10|38|38|2|14|10

3. a. Here is a more difficult number pattern. Complete the code. *Hint:* Add 1, add 2, add 1, add 2, etc.

A	B	C	D	E	F	G	H	I	J	K
1	2	4	5	7	8	10	11	13	14	16

   b. Write the message with this code.

4. a. With a partner, make up a number pattern and code.
   b. Send a message to each other in code.
   c. See if anyone else can decode the message.

CHAPTER REVIEW

Copy and complete.

1. There are ? hundreds.
2. The 6 stands for 6 tens, or ? .
3. 368 rounded to the nearest hundred is ? .
4. 368 rounded to the nearest ten is ? .

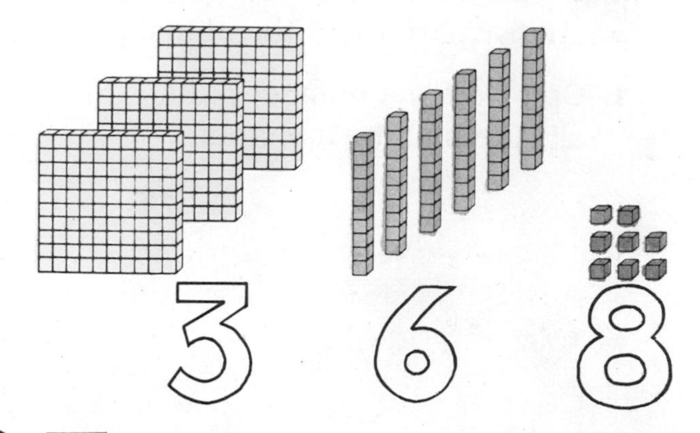

5. The 1 stands for ? .
6. The 8 stands for ? .

8125

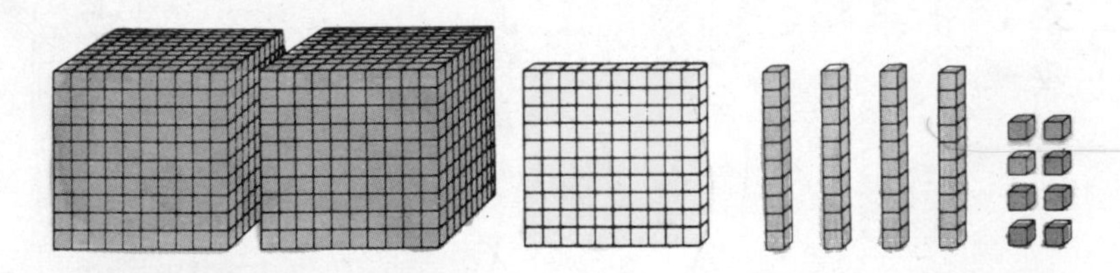

7. Pictured above are two thousand one hundred ? blocks.
8. 8146 rounded to the nearest thousand is ? .

Which digit is in the

9. ten thousands place?
10. hundred thousands place?

Less than (<) or greater than (>)?

11. 5267 ● 5248
12. 537 ● 548
13. 906 ● 839
14. 2178 ● 2164
15. 3752 ● 3706
16. 29,135 ● 29,456
17. 674,291 ● 674,189

CHAPTER CHALLENGE

Roman numerals have been used for hundreds of years. Roman numerals are written with letters. The first twelve Roman numerals are shown on the clock.

Look at the Roman numerals for 4 and 9. Since the I comes before the V and the X, its value is subtracted. IV means 5 − 1. IX means 10 − 1.

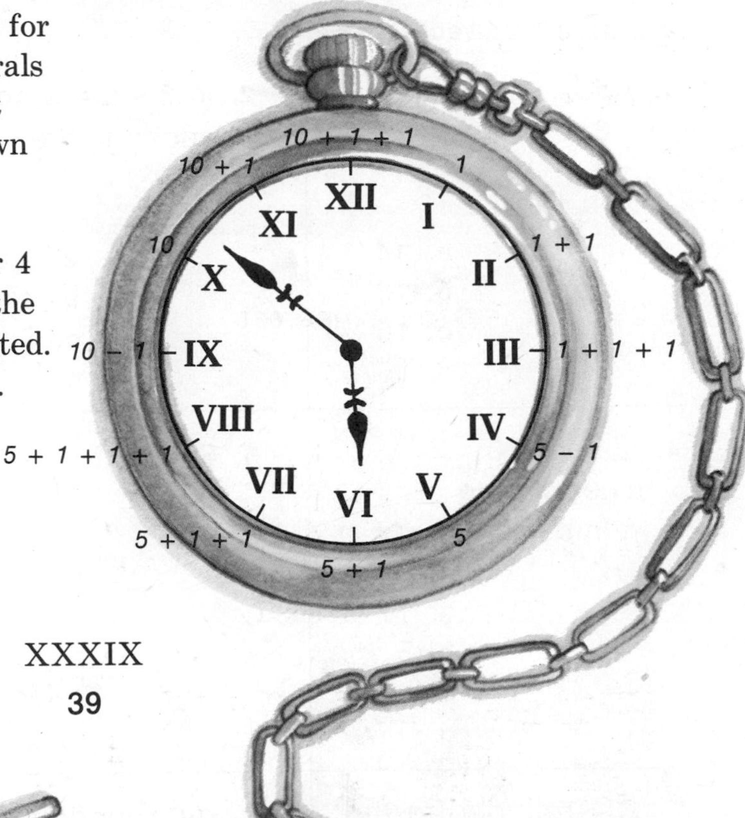

Study these Roman numerals.

XI	XIX	XX	XXIV	XXXIX
11	19	20	24	39

Write Roman numerals.

1. 4 **2.** 5 **3.** 10 **4.** 9 **5.** 8 **6.** 13

7. 14 **8.** 17 **9.** 25 **10.** 26 **11.** 38 **12.** 29

Write standard numerals.

13. XII **14.** XVI **15.** XVIII **16.** XXI **17.** XXIV **18.** XXIX

19. XXX **20.** XXXII **21.** XXXV **22.** XXXVI **23.** XXXIV **24.** XXXIX

Answer with Roman numerals.

25. How old are you?

26. How many boys are in your class?

27. How many girls are in your class?

28. What day of the month is today?

2\. a b c d
5\. a b c d
6\. a b c
8\. 9. 10. 11. 12.

MAJOR CHECKUP

STANDARDIZED FORMAT

Choose the correct letter.

1\. Add.

$$\begin{array}{r} 6 \\ +9 \\ \hline \end{array}$$

a. 14
b. 16
c. 15
d. none of these

2\. In 3 + 5 = 8, the addends are

a. 3 and 5
b. 3 and 8
c. 5 and 8
d. none of these

3\. Add.

$$\begin{array}{r} 6 \\ 3 \\ 4 \\ 7 \\ +2 \\ \hline \end{array}$$

a. 24
b. 23
c. 21
d. 22

4\. _?_ + 3 = 8
The missing number is

a. 11
b. 3
c. 5
d. none of these

5\. Subtract.

$$\begin{array}{r} 16 \\ -9 \\ \hline \end{array}$$

a. 7
b. 6
c. 8
d. 9

6\. 13 − (6 − 2) = _?_
The missing number is

a. 5
b. 9
c. 4
d. none of these

7\. 234 rounded to the nearest ten is

a. 230
b. 240
c. 200
d. none of these

8\. 750 rounded to the nearest hundred is

a. 760
b. 700
c. 800
d. none of these

9\. The standard numeral for four thousand two hundred two is

a. 4202
b. 4022
c. 4220
d. none of these

10\. In 328,174 the 2 stands for

a. 20
b. 2000
c. 200
d. none of these

11\. 13 students played baseball. There were 7 on one team. How many were on the other team?

a. 6 b. 20
c. 5 d. none of these

12\. There were 9 boys and 4 girls at a party. How many children were there?

a. 5 b. 6
c. 12 d. none of these

Addition and Subtraction

3

READY OR NOT?

Add.

1.	$\begin{array}{r} 6 \\ +6 \\ \hline \end{array}$	**2.**	$\begin{array}{r} 9 \\ +7 \\ \hline \end{array}$
3.	$\begin{array}{r} 7 \\ +7 \\ \hline \end{array}$	**4.**	$\begin{array}{r} 3 \\ +8 \\ \hline \end{array}$
5.	$\begin{array}{r} 2 \\ +9 \\ \hline \end{array}$	**6.**	$\begin{array}{r} 3 \\ +7 \\ \hline \end{array}$
7.	$\begin{array}{r} 7 \\ +8 \\ \hline \end{array}$	**8.**	$\begin{array}{r} 8 \\ +9 \\ \hline \end{array}$
9.	$\begin{array}{r} 9 \\ +6 \\ \hline \end{array}$	**10.**	$\begin{array}{r} 8 \\ +8 \\ \hline \end{array}$

Subtract.

11.	$\begin{array}{r} 15 \\ -8 \\ \hline \end{array}$	**12.**	$\begin{array}{r} 11 \\ -5 \\ \hline \end{array}$
13.	$\begin{array}{r} 13 \\ -4 \\ \hline \end{array}$	**14.**	$\begin{array}{r} 16 \\ -7 \\ \hline \end{array}$
15.	$\begin{array}{r} 18 \\ -9 \\ \hline \end{array}$	**16.**	$\begin{array}{r} 12 \\ -4 \\ \hline \end{array}$
17.	$\begin{array}{r} 10 \\ -4 \\ \hline \end{array}$	**18.**	$\begin{array}{r} 14 \\ -6 \\ \hline \end{array}$
19.	$\begin{array}{r} 15 \\ -6 \\ \hline \end{array}$	**20.**	$\begin{array}{r} 14 \\ -5 \\ \hline \end{array}$

Adding without regrouping

You can find the total number of marbles by adding in columns.

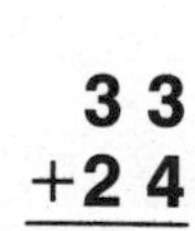

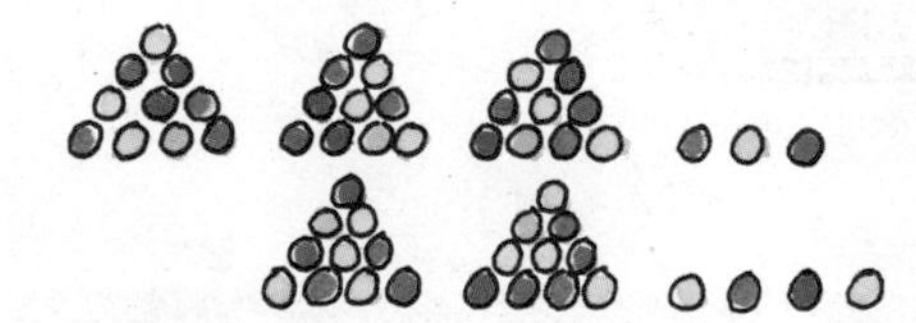

Step 1. Add ones.

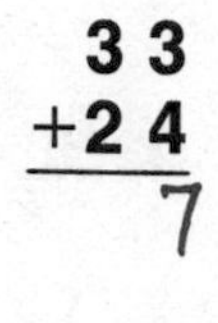

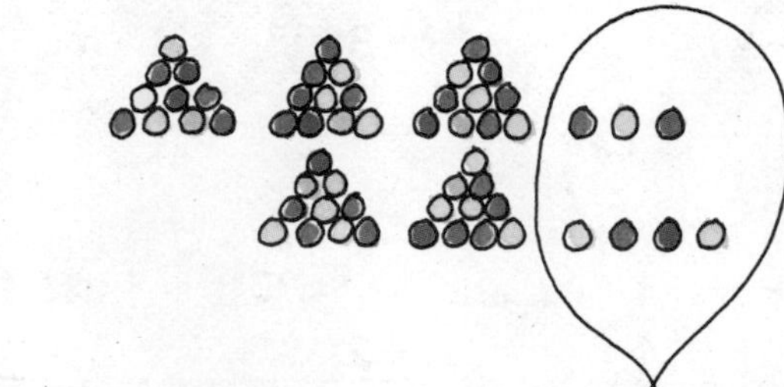

Step 2. Add tens.

Study these examples.

$$\begin{array}{r} 213 \\ +145 \\ \hline 358 \end{array} \qquad \begin{array}{r} 38,291 \\ +10,304 \\ \hline 48,595 \end{array}$$

EXERCISES

Add.

1. 42 +21	**2.** 38 +30	**3.** 56 +41	**4.** 65 +24	**5.** 76 +13
6. 356 +221	**7.** 403 +174	**8.** 206 +21	**9.** 405 +390	**10.** 226 +430
11. 205 +483	**12.** 274 +425	**13.** 674 +215	**14.** 511 +46	**15.** 46 +511
16. 5214 +653	**17.** 3608 +2101	**18.** 1284 +8510	**19.** 7653 +304	**20.** 5326 +2401
21. 32,864 +5,031	**22.** 87,400 +2,405	**23.** 43,215 +453	**24.** 64,102 +23,243	**25.** 54,103 +10,782
26. 583,243 +415,214	**27.** 402,032 +453,647	**28.** 727,134 +52,601	**29.** 43,856 +914,022	**30.** 642,354 +150,410

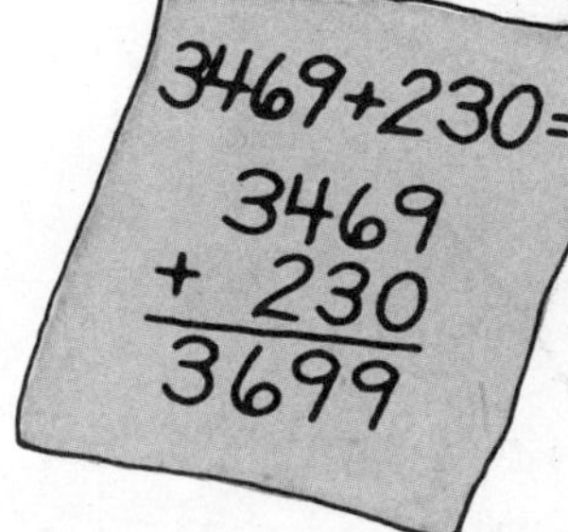

Give each sum.

31. 54 + 21
32. 36 + 42
33. 251 + 306
34. 740 + 38
35. 3912 + 5003
36. 4235 + 421
37. 53,614 + 21,052
38. 376,042 + 11,823

Adding with regrouping

Adams Elementary School was having a school fair. Marissa sold 27 adult tickets and 16 children's tickets. You can add to find the total number of tickets she sold.

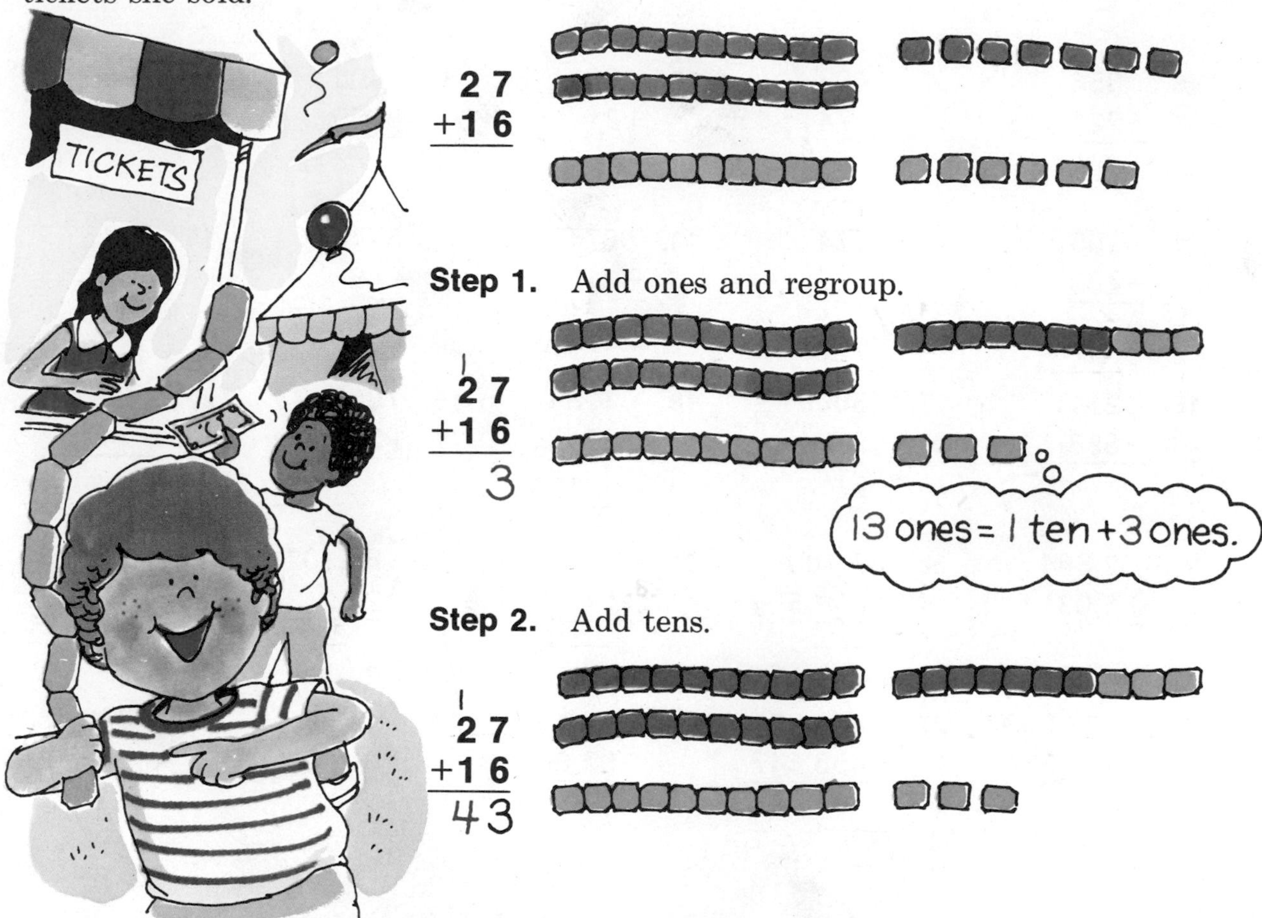

When adding 152 and 174, 10 tens are regrouped for 1 hundred.

Step 1. Add ones.

$$\begin{array}{r} 152 \\ +174 \\ \hline 6 \end{array}$$

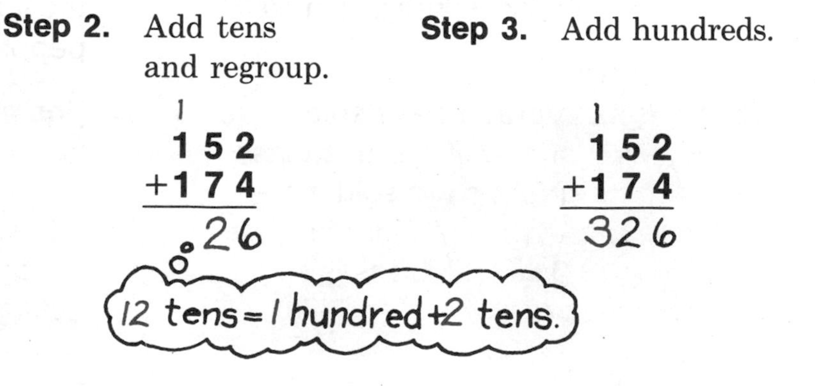

EXERCISES

Add.

1. $38 + 25$
2. $56 + 39$
3. $21 + 49$
4. $28 + 69$
5. $55 + 7$
6. $48 + 21$
7. $56 + 29$
8. $56 + 39$
9. $37 + 45$
10. $45 + 37$
11. $68 + 91$
12. $53 + 72$
13. $52 + 80$
14. $75 + 43$
15. $86 + 72$
16. $253 + 481$
17. $162 + 284$
18. $363 + 576$
19. $492 + 184$
20. $273 + 581$
21. $329 + 408$
22. $133 + 557$
23. $205 + 366$
24. $429 + 534$
25. $437 + 419$
26. $842 + 93$
27. $68 + 51$
28. $74 + 18$
29. $39 + 448$
30. $536 + 147$

Solve.

31. The fourth-grade class sold 128 tickets for the school fair. The fifth-grade class sold 119. How many tickets did they sell in all?

32. Three hundred fifty-two people came to the fair the first night. Two hundred ninety-one came the second night. How many people came on the two nights?

33. The fourth-grade class sold more tickets than the fifth-grade class. The fifth-grade class sold more tickets than the sixth-grade class. Which of the three classes sold the most tickets?

★ 34. The first-grade class sold 54 adult tickets and 37 children's tickets. The second-grade class sold 49 adult tickets and 40 children's tickets. Which class sold more tickets?

Regrouping more than once

Sometimes you have to regroup more than once when adding. Study these examples.

EXAMPLE 1.

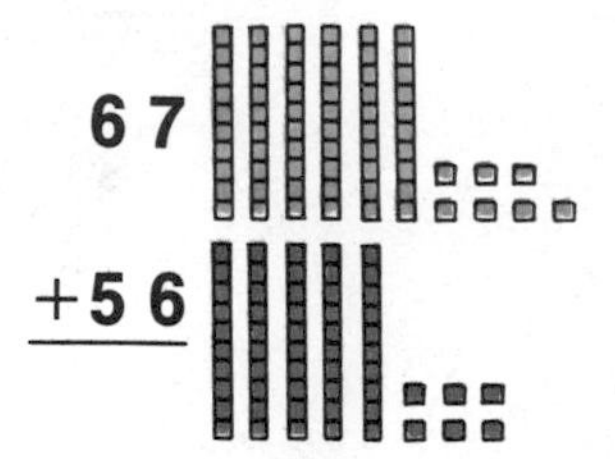

Step 1. Add ones and regroup.

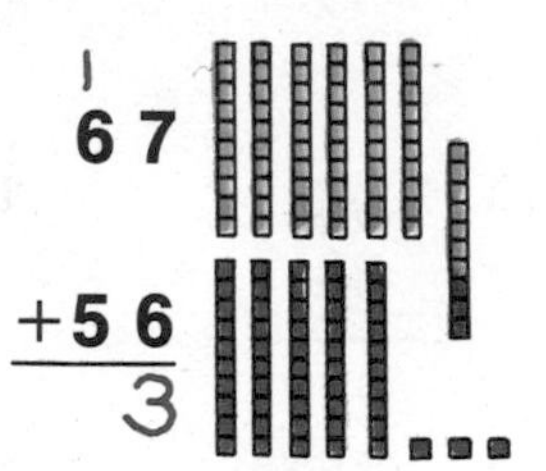

Step 2. Add tens and regroup.

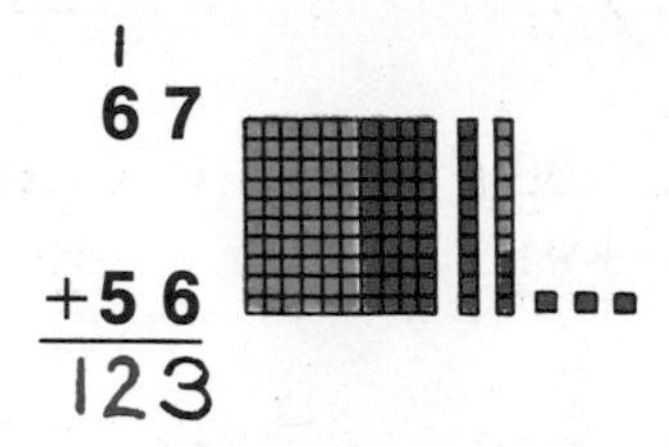

EXAMPLE 2.

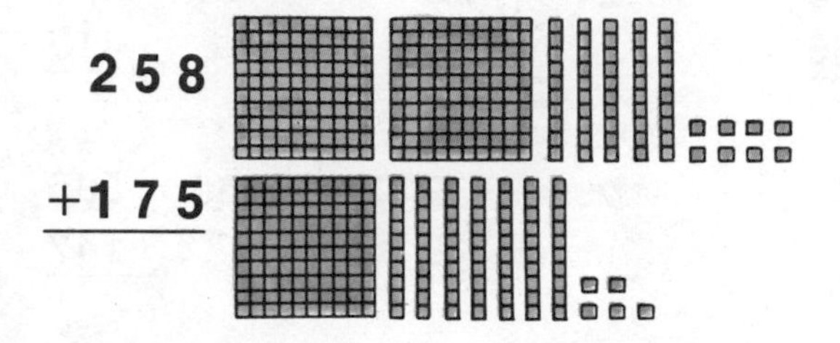

Step 1. Add ones and regroup.

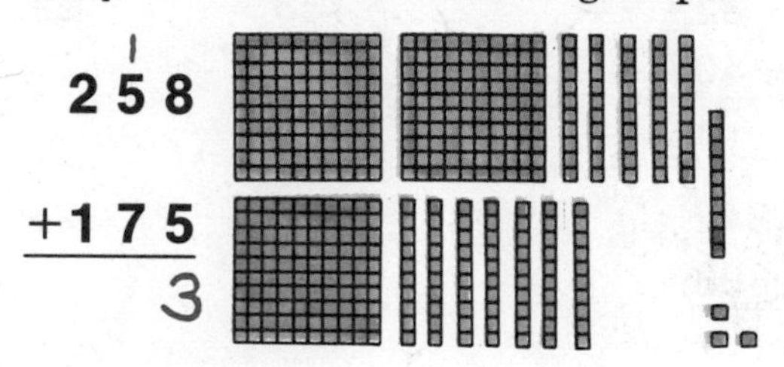

Step 2. Add tens and regroup.

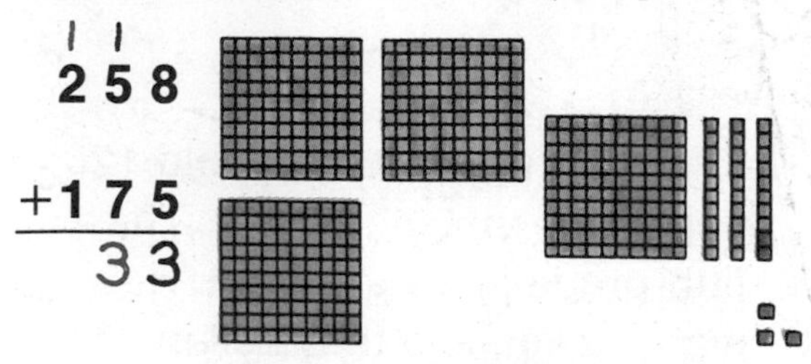

Step 3. Add hundreds.

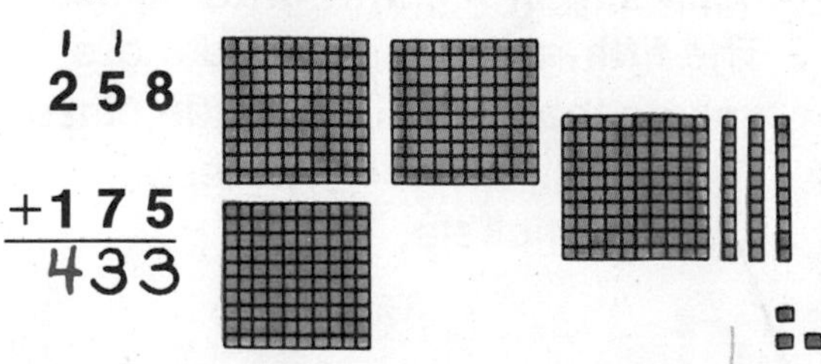

EXERCISES

Add.

1. 75 + 68	**2.** 39 + 83	**3.** 76 + 79	**4.** 97 + 44	**5.** 69 + 58
6. 486 + 95	**7.** 567 + 93	**8.** 48 + 387	**9.** 59 + 469	**10.** 258 + 68
11. 275 + 438	**12.** 579 + 285	**13.** 368 + 245	**14.** 693 + 208	**15.** 572 + 198
	16. $3.75 + 1.21	**17.** $5.98 + 2.62	**18.** $5.37 + 1.95	**19.** $2.99 + 4.08
	20. $4.29 + 1.82	**21.** $6.20 + 2.95	**22.** $4.08 + 3.69	**23.** $7.49 + 1.87

Remember that money is added in the same way.

Solve.

24. John's class collected cans. The first week they collected 389, and the second week they collected 496. How many cans did they collect in these two weeks?

25. The class set a goal of 950 cans during the third and fourth weeks. They collected 419 cans the third week and 537 the fourth week. Did they reach their goal?

Challenge!

Find the missing digits.

26. 3 ■ 6 + 2 4 9 = 6 0 5	**27.** 4 8 6 + 1 7 ■ = 6 ■ 5	**28.** 3 7 ■ + 5 ■ 4 = 9 0 2	**29.** 4 ■ 9 + 4 3 6 = 8 4 ■	**30.** 1 4 ■ + 2 ■ 3 = 4 0 8

Estimating sums

When you estimate an answer, you think about what number the answer is near. Sometimes an estimate is called an "educated guess." An estimate can help you decide whether or not you have made a mistake.

Estimate these sums.

58	Round to nearest ten	60
+29	Round to nearest ten	+30
		90

526	Round to nearest hundred	500
+385	Round to nearest hundred	+400
		900

\$4.85	Round to nearest dollar	\$5.00
+3.27	Round to nearest dollar	+3.00
		\$8.00

EXERCISES

Round each addend to the nearest ten. Estimate the sum.

1. 38 + 53 **2.** 47 + 19 **3.** 56 + 20 **4.** 25 + 54 **5.** 62 + 19 **6.** 63 + 81

7. 65 + 53 **8.** 31 + 83 **9.** 76 + 64 **10.** 52 + 82 **11.** 62 + 78 **12.** 93 + 89

Round each addend to the nearest hundred.
Estimate the sum.

13. 574 + 238	14. 621 + 308	15. 482 + 278	16. 609 + 82	17. 59 + 342	18. 259 + 340
19. 150 + 438	20. 346 + 590	21. 439 + 551	22. 623 + 106	23. 387 + 605	24. 317 + 481
25. 195 + 376	26. 204 + 385	27. 350 + 125	28. 736 + 148	29. 295 + 376	30. 403 + 231

Round each addend to the nearest dollar.
Estimate the sum.

31. $2.42 + 1.73	32. $4.39 + 2.70	33. $2.72 + 6.81	34. $5.21 + 3.90	35. $1.43 + 6.81
36. $4.36 + 3.50	37. $3.94 + 2.29	38. $5.98 + 2.99	39. $6.15 + .89	40. $5.29 + .99

KEEPING SKILLS SHARP

Less than (<) or greater than (>)?

1. 456 ● 465
2. 312 ● 409
3. 856 ● 850
4. 3617 ● 3528
5. 5263 ● 5438
6. 2674 ● 3200
7. 26,174 ● 26,274
8. 38,061 ● 38,056
9. 59,274 ● 58,374
10. 618,321 ● 617,493
11. 583,274 ● 590,003
12. 99,999 ● 100,000

Adding larger numbers

> Remember to regroup when a sum is 10 or greater.

Let's add 573 and 785.

Step 1.
Add ones.

$$\begin{array}{r} 573 \\ +785 \\ \hline 8 \end{array}$$

Step 2.
Add tens and regroup.

$$\begin{array}{r} \scriptstyle 1 \\ 573 \\ +785 \\ \hline 58 \end{array}$$

Step 3.
Add hundreds and regroup.

$$\begin{array}{r} \scriptstyle 1 \\ 573 \\ +785 \\ \hline 1358 \end{array}$$

When we add 5287 and 4963, we have to regroup in each step.

Step 1.

$$\begin{array}{r} \scriptstyle 1 \\ 5287 \\ +4963 \\ \hline 0 \end{array}$$

Step 2.

$$\begin{array}{r} \scriptstyle 1\,1 \\ 5287 \\ +4963 \\ \hline 50 \end{array}$$

Step 3.

$$\begin{array}{r} \scriptstyle 1\,1\,1 \\ 5287 \\ +4963 \\ \hline 250 \end{array}$$

Step 4.

$$\begin{array}{r} \scriptstyle 1\,1\,1 \\ 5287 \\ +4963 \\ \hline 10{,}250 \end{array}$$

EXERCISES

Add.

1. $\begin{array}{r} 638 \\ +257 \\ \hline \end{array}$
2. $\begin{array}{r} 519 \\ +324 \\ \hline \end{array}$
3. $\begin{array}{r} 743 \\ +896 \\ \hline \end{array}$
4. $\begin{array}{r} 547 \\ +968 \\ \hline \end{array}$
5. $\begin{array}{r} 658 \\ +729 \\ \hline \end{array}$
6. $\begin{array}{r} 472 \\ +298 \\ \hline \end{array}$
7. $\begin{array}{r} 799 \\ +948 \\ \hline \end{array}$
8. $\begin{array}{r} 973 \\ +918 \\ \hline \end{array}$

9. 4618
+3812

10. 3957
+384

11. 3219
+2537

12. 5674
+6384

13. 7038
+8299

14. 8078
+9951

15. $63.52
+74.98

16. $78.56
+59.74

17. $95.26
+6.97

18. 32,781
+14,625

19. 83,175
+9,654

20. 75,263
+7,491

21. 381,724
+97,382

22. 658,725
+234,983

23. 493,785
+176,594

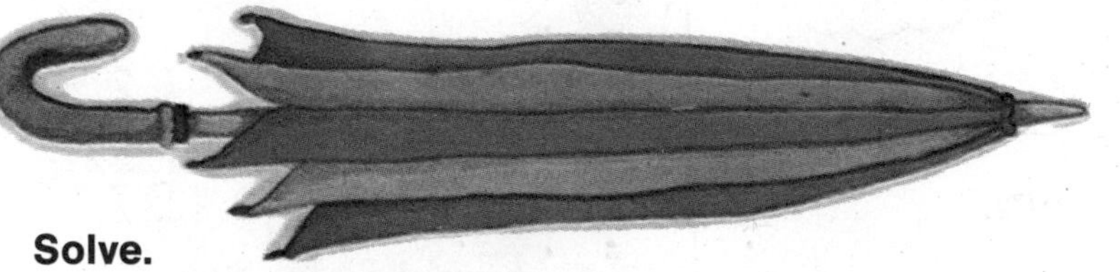

Solve.

24. The computer weather report says that it rained 16 days in October and 7 days in November. How many more days did it rain in October?

25. Through October, an umbrella factory had made 25,603 umbrellas. In November, the factory made 2670 umbrellas. How many umbrellas had the factory made in all?

26. The factory also makes plastic rain hats. They had one order for 9525 rain hats and another order for 10,156 rain hats. How many rain hats are needed to fill both orders?

KEEPING SKILLS SHARP

1. 14
−8

2. 11
−7

3. 15
−8

4. 13
−6

5. 12
−7

6. 16
−7

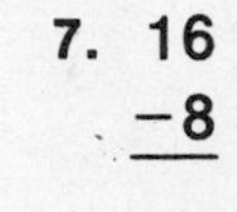

7. 16
−8

8. 10
−6

9. 14
−7

10. 17
−9

11. 13
−5

12. 12
−6

13. 15
−9

14. 11
−3

15. 12
−8

16. 11
−5

17. 14
−9

18. 18
−9

19. 11
−2

20. 14
−6

Adding three or more numbers

EXAMPLE.

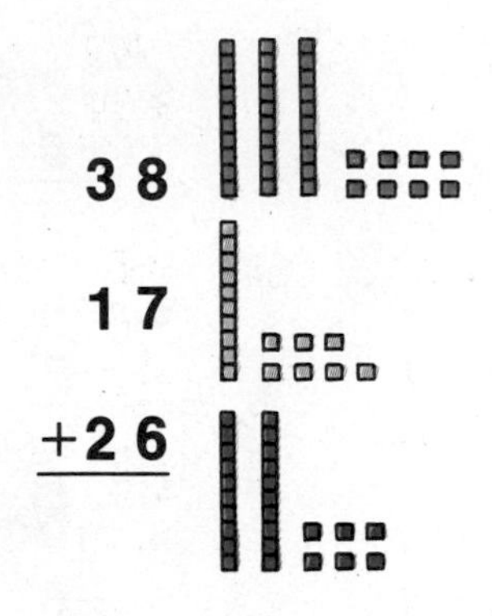

Step 1. Add ones and regroup.

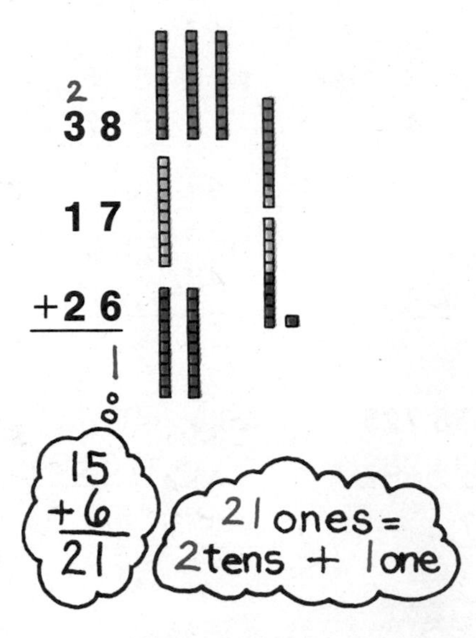

Step 2. Add tens.

$$\begin{array}{r} {}^{2} \\ 38 \\ 17 \\ +26 \\ \hline 81 \end{array}$$

EXERCISES

Add.

1.	2.	3.	4.	5.	6.
4	24	5	35	8	48
3	13	6	46	9	19
+8	+38	+7	+17	+6	+26

7.	8.	9.	10.	11.	12.
7	37	8	38	7	87
4	24	9	59	6	96
+9	+39	+9	+79	+6	+56

13.	14.	15.	16.	17.	18.
5	25	8	48	5	85
6	36	7	57	8	68
9	79	8	78	9	49
+4	+54	+9	+59	+6	+36

Find the sum.

19. 43 56 +19	20. 43 78 +62	21. 53 79 +18	22. 59 48 +7	23. 38 74 +16
24. 62 35 +17	25. 85 74 +91	26. 46 27 +58	27. 38 19 +8	28. 49 35 +27
29. 13 38 44 +22	30. 85 6 71 +53	31. 97 34 48 +71	32. 19 22 55 +63	33. 46 36 84 +8
34. 128 426 +234	35. 373 519 +267	36. 286 452 +635	37. 578 341 +824	38. 769 33 +247
39. 231 742 591 +280	40. 162 303 244 +15	41. 425 103 241 +153	42. \$2.43 2.56 1.45 +1.58	43. \$3.59 2.16 3.05 +.56

Solve.

44. 

How much for both?

45. Maria had 478 stamps. She bought 36 stamps, and a friend gave her 19. How many stamps did she have then?

46. Alex had 279 stamps. His father gave him 24 stamps each week for 3 weeks. How many stamps did he have then?

47. Maria's new stamp album will hold 600 stamps. She has 283 U.S. stamps and 309 Canadian stamps. Can she put all her stamps in the album?

Problem solving

Many people work for airlines. Some fly the planes. Others work with passengers. Still others work on the planes.

Modern jets can fly long distances. The chart below shows the air distance in kilometers between certain cities of the world.

Air Distance in Kilometers between Cities

City	Chicago	Hong Kong	London	Montreal	Moscow	New York	Peking	San Francisco
Chicago		12,475	6,333	1,192	7,979	1,142	10,566	2,974
Hong Kong	12,475		9,584	12,378	7,099	12,896	1,947	11,048
London	6,333	9,584		5,206	2,502	5,550	8,118	8,587
Montreal	1,192	12,378	5,206		7,042	530	10,422	4,069
Moscow	7,979	7,099	2,502	7,042		7,493	5,771	9,416
New York	1,142	12,896	5,550	530	7,493		10,950	4,115
Peking	10,566	1,947	8,118	10,422	5,771	10,950		9,469
San Francisco	2,974	11,048	8,587	4,069	9,416	4,115	9,469	

Solve.

How far is it from

1. Montreal to Chicago?
2. Peking to Moscow?
3. London to San Francisco?
4. New York to Hong Kong?

Which city is

5. 1192 kilometers from Chicago?
6. 5550 kilometers from London?
7. 9469 kilometers from Peking?
8. over 12,000 kilometers from New York?
9. almost 8000 kilometers from Moscow?

Which city is farther from San Francisco,

10. Montreal or New York?
11. Moscow or Peking?

Give the total kilometers for the following flights.

12. New York–London–Hong Kong
13. San Francisco–Chicago–New York
14. Moscow–London–Chicago
15. Montreal–Moscow–Peking
16. New York–San Francisco–Peking–Hong Kong
17. London–Montreal–Chicago–San Francisco

KEEPING SKILLS SHARP

Round to the nearest ten.

1. 18	2. 32
3. 41	4. 25
5. 63	6. 81
7. 79	8. 86
9. 92	10. 95
11. 103	12. 165
13. 183	14. 274
15. 318	16. 357
17. 495	18. 623
19. 2816	20. 4203

Round to the nearest hundred.

21. 321	22. 418
23. 463	24. 509
25. 546	26. 585
27. 632	28. 651
29. 795	30. 703
31. 819	32. 875
33. 906	34. 945
35. 991	36. 998
37. 2567	38. 3574
39. 4067	40. 5294

Subtracting without regrouping

You can find differences by subtracting in columns.

EXAMPLE 1.

$$\begin{array}{r} 68 \\ -23 \\ \hline \end{array}$$

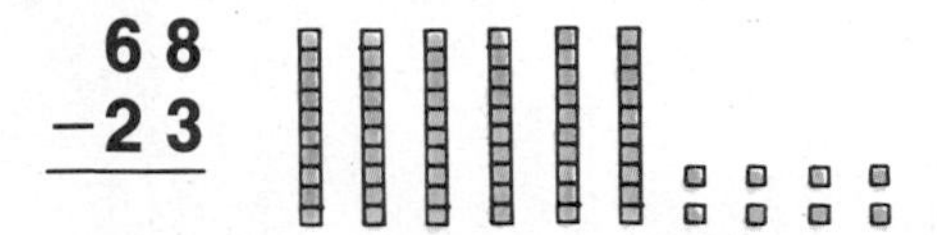

Step 1. Subtract in ones column.

$$\begin{array}{r} 68 \\ -23 \\ \hline 5 \end{array}$$

Step 2. Subtract in tens column.

$$\begin{array}{r} 68 \\ -23 \\ \hline 45 \end{array}$$

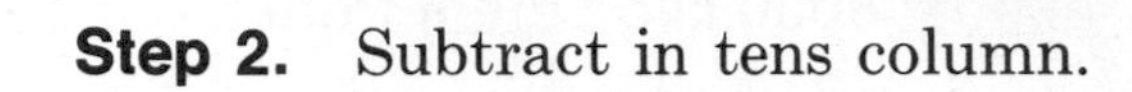

EXAMPLE 2.

$$\begin{array}{r} 457 \\ -243 \\ \hline \end{array}$$

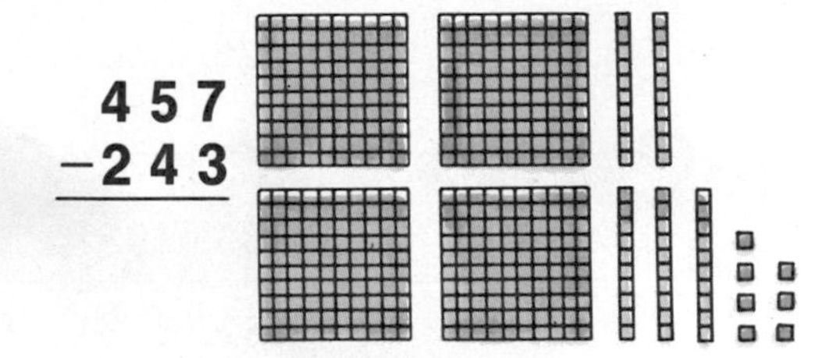

Step 1. Subtract ones.

$$\begin{array}{r} 457 \\ -243 \\ \hline 4 \end{array}$$

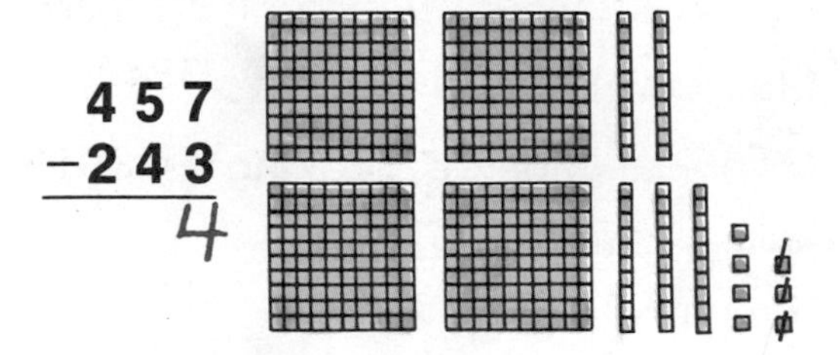

Step 2. Subtract tens.

$$\begin{array}{r} 457 \\ -243 \\ \hline 14 \end{array}$$

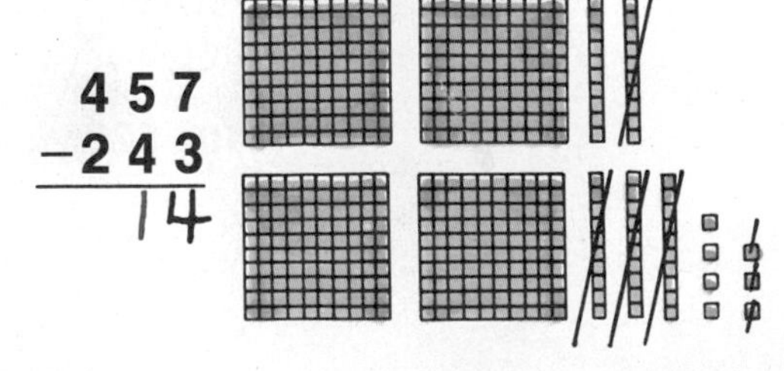

Step 3. Subtract hundreds.

$$\begin{array}{r} 457 \\ -243 \\ \hline 214 \end{array}$$

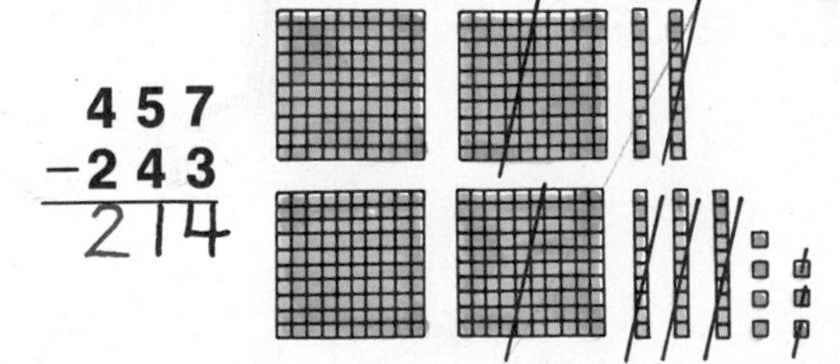

EXERCISES

Subtract.

1. 85 − 32
2. 81 − 41
3. 69 − 33
4. 77 − 12
5. 98 − 56
6. 359 − 246
7. 478 − 123
8. 527 − 205
9. 694 − 423
10. 975 − 125
11. 753 − 42
12. 826 − 301
13. 588 − 26
14. 590 − 150
15. 967 − 245
16. 5634 − 3500
17. 7827 − 4305
18. 9740 − 6240
19. 6558 − 2123
20. 4369 − 1023
21. 78,293 − 2,030
22. 57,608 − 3,604
23. 92,735 − 30,402
24. 46,918 − 12,604
25. 37,925 − 12,404

Give each difference.

26. 78 − 23
27. 59 − 14
28. 594 − 211
29. 465 − 15
30. 3842 − 1510
31. 5864 − 324
32. 74,381 − 22,130
33. 894,637 − 162,514

Challenge!

Find the missing input or output.

	Input	Output
	1747	1780
34.	1850	?
35.	945	?
36.	?	486

Subtracting with regrouping

In this example, 1 ten is regrouped for 10 ones.

EXAMPLE 1.

```
 54
-29
```

Step 1. Not enough ones. Regroup 1 ten for 10 ones.

```
 4 14
 5  4
-2  9
```

Step 2. Subtract ones. Subtract tens.

```
 4 14
 5  4
-2  9
 2  5
```

In this example, 1 hundred is regrouped for 10 tens.

EXAMPLE 2.

Step 1. Subtract ones.

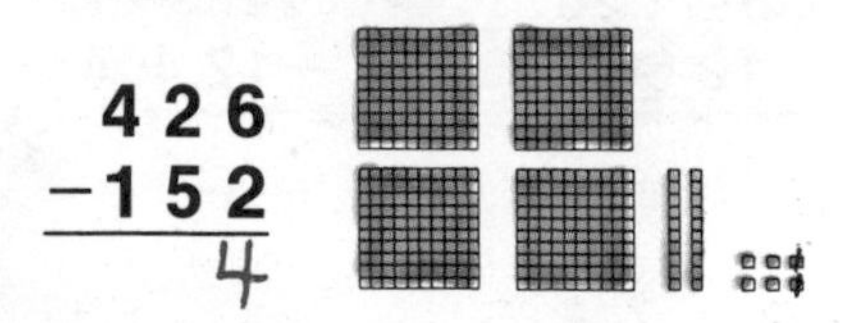

Step 2. Not enough tens. Regroup 1 hundred for 10 tens.

```
  3 12
  4  2 6
- 1  5 2
       4
```

Step 3. Subtract tens.

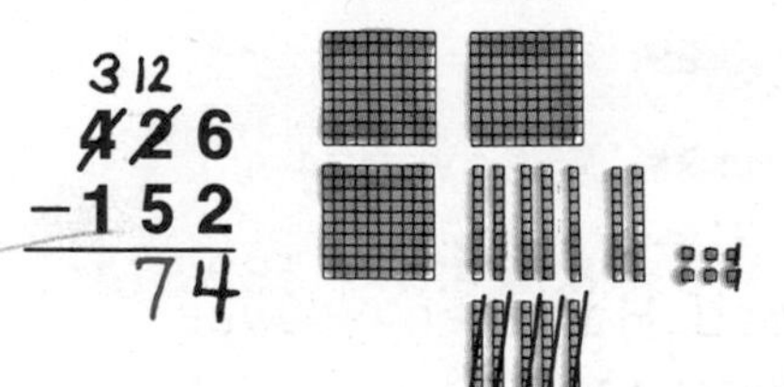

Step 4. Subtract hundreds.

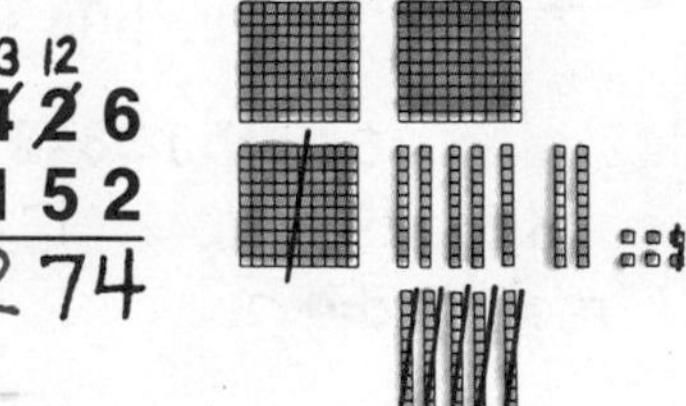

EXERCISES

Subtract.

1. $\begin{array}{r} 24 \\ -18 \\ \hline \end{array}$ 2. $\begin{array}{r} 73 \\ -44 \\ \hline \end{array}$ 3. $\begin{array}{r} 92 \\ -46 \\ \hline \end{array}$ 4. $\begin{array}{r} 70 \\ -25 \\ \hline \end{array}$ 5. $\begin{array}{r} 84 \\ -39 \\ \hline \end{array}$

6. $\begin{array}{r} 78 \\ -29 \\ \hline \end{array}$ 7. $\begin{array}{r} 87 \\ -26 \\ \hline \end{array}$ 8. $\begin{array}{r} 65 \\ -19 \\ \hline \end{array}$ 9. $\begin{array}{r} 93 \\ -58 \\ \hline \end{array}$ 10. $\begin{array}{r} 81 \\ -74 \\ \hline \end{array}$

11. $\begin{array}{r} 742 \\ -216 \\ \hline \end{array}$ 12. $\begin{array}{r} 586 \\ -48 \\ \hline \end{array}$ 13. $\begin{array}{r} 753 \\ -628 \\ \hline \end{array}$ 14. $\begin{array}{r} 978 \\ -49 \\ \hline \end{array}$ 15. $\begin{array}{r} 684 \\ -212 \\ \hline \end{array}$

16. $\begin{array}{r} 528 \\ -152 \\ \hline \end{array}$ 17. $\begin{array}{r} 406 \\ -231 \\ \hline \end{array}$ 18. $\begin{array}{r} 753 \\ -61 \\ \hline \end{array}$ 19. $\begin{array}{r} 517 \\ -303 \\ \hline \end{array}$ 20. $\begin{array}{r} 856 \\ -574 \\ \hline \end{array}$

21. $\begin{array}{r} 753 \\ -219 \\ \hline \end{array}$ 22. $\begin{array}{r} 647 \\ -481 \\ \hline \end{array}$ 23. $\begin{array}{r} 958 \\ -223 \\ \hline \end{array}$ 24. $\begin{array}{r} 526 \\ -373 \\ \hline \end{array}$ 25. $\begin{array}{r} 838 \\ -529 \\ \hline \end{array}$

26. $\begin{array}{r} 678 \\ -595 \\ \hline \end{array}$ 27. $\begin{array}{r} 895 \\ -462 \\ \hline \end{array}$ 28. $\begin{array}{r} 463 \\ -229 \\ \hline \end{array}$ 29. $\begin{array}{r} 742 \\ -715 \\ \hline \end{array}$ 30. $\begin{array}{r} 527 \\ -275 \\ \hline \end{array}$

Solve.

31. Louis brought $1.50 to the bake sale. He spent $1.29 for bread. How much money did he have left?

32. There were 24 loaves of white bread. There were 36 loaves of rye bread. How many loaves were there in all?

33. There were 24 loaves of white bread. There were 36 loaves of rye bread. How many more loaves of rye bread were there?

34. 127 cupcakes were made. 53 of them were sold. How many were left?

★ 35. 156 rolls were made. 28 of them were sold. Then 35 more of them were sold. How many were left?

★ 36. There were 56 oatmeal cookies. There were 128 peanut butter cookies. 39 of the cookies were sold. How many cookies were left?

Regrouping more than once

Sometimes you have to regroup more than once when subtracting.

EXAMPLE 1.

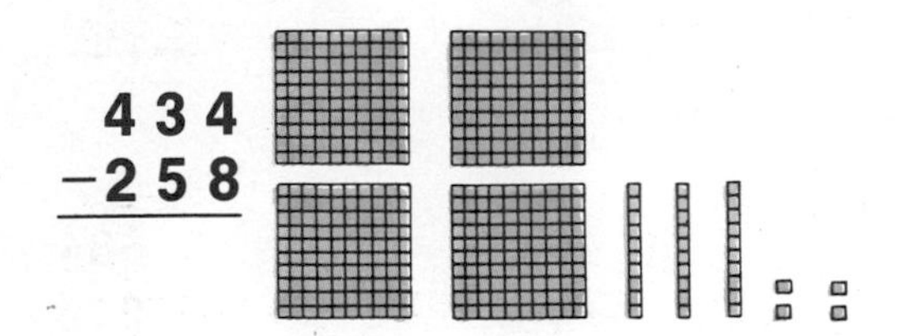

Step 1. Not enough ones, so regroup 1 ten for 10 ones.

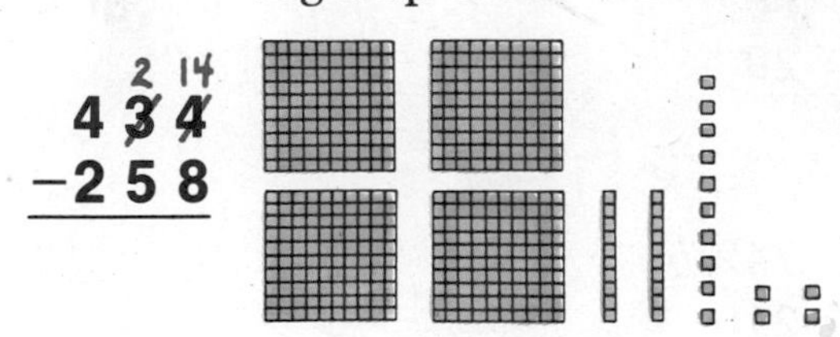

Step 2. Subtract ones.

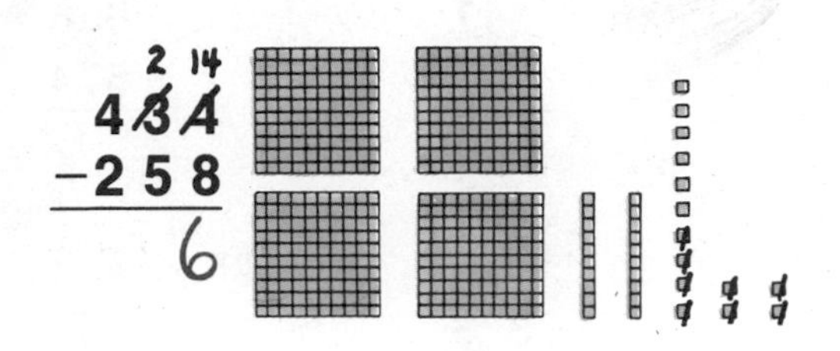

Step 3. Not enough tens, so regroup 1 hundred for 10 tens.

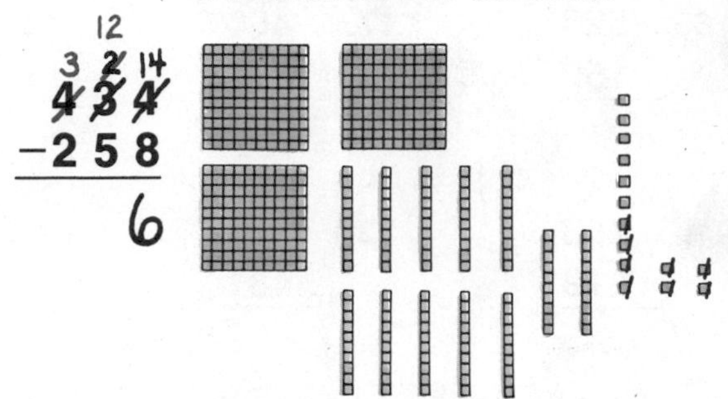

Step 4. Subtract tens.

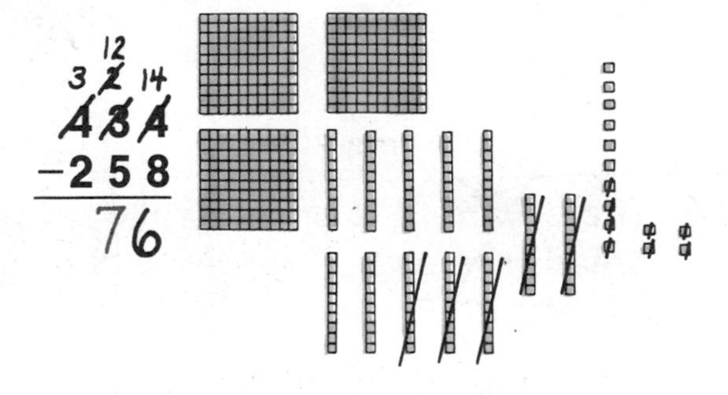

Step 5. Subtract hundreds.

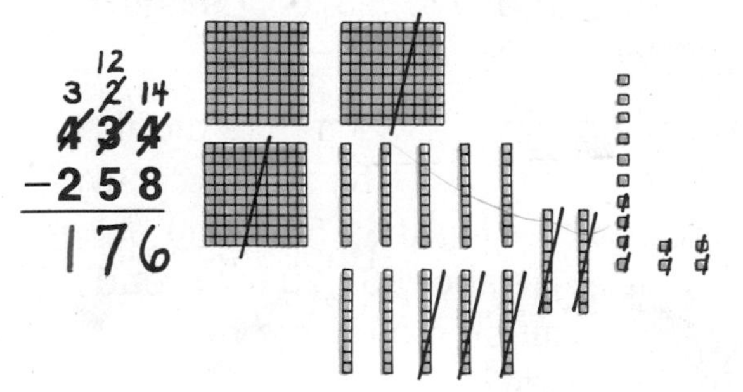

EXAMPLE 2.

Step 1.	**Step 2.**	**Step 3.**	**Step 4.**	**Step 5.**
620	620	620	620	620
−438	−438	−438	−438	−438

EXERCISES

Subtract.

1. 641 − 259
2. 381 − 193
3. 823 − 485
4. 426 − 358
5. 750 − 467
6. 712 − 359
7. 512 − 458
8. 633 − 275
9. 510 − 394
10. 746 − 268
11. 635 − 367
12. 816 − 608
13. 930 − 718
14. 924 − 501
15. 842 − 375

You subtract money in the same way that you subtract whole numbers.

16. \$8.23 − 4.58
17. \$7.52 − 3.69
18. \$6.31 − 1.74
19. \$9.75 − 6.29
20. \$7.30 − 4.26
21. \$5.52 − 2.75
22. \$9.25 − 3.58
23. \$8.14 − 6.29

Solve.

24. The Allens drove 452 kilometers the first day and 379 kilometers the second day.
    - **a.** How many kilometers did they drive during the first two days?
    - **b.** How much farther did they drive the first day?

25. On their trip Nancy saw this road sign:

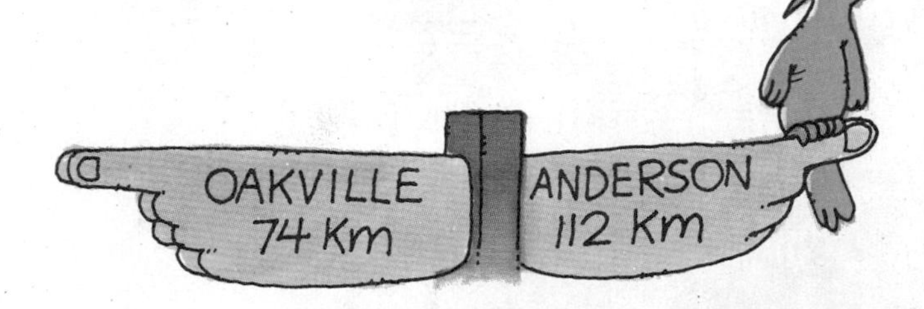

How far apart are the two cities?

Challenge!

Give the missing digits.

26. 5 2 1 − 3 7 ▢ = ▢ 4 7

27. 6 ▢ 1 − ▢ 5 8 = 1 6 3

28. ▢ 2 2 − 4 ▢ 9 = 2 0 ▢

Zeros in subtraction

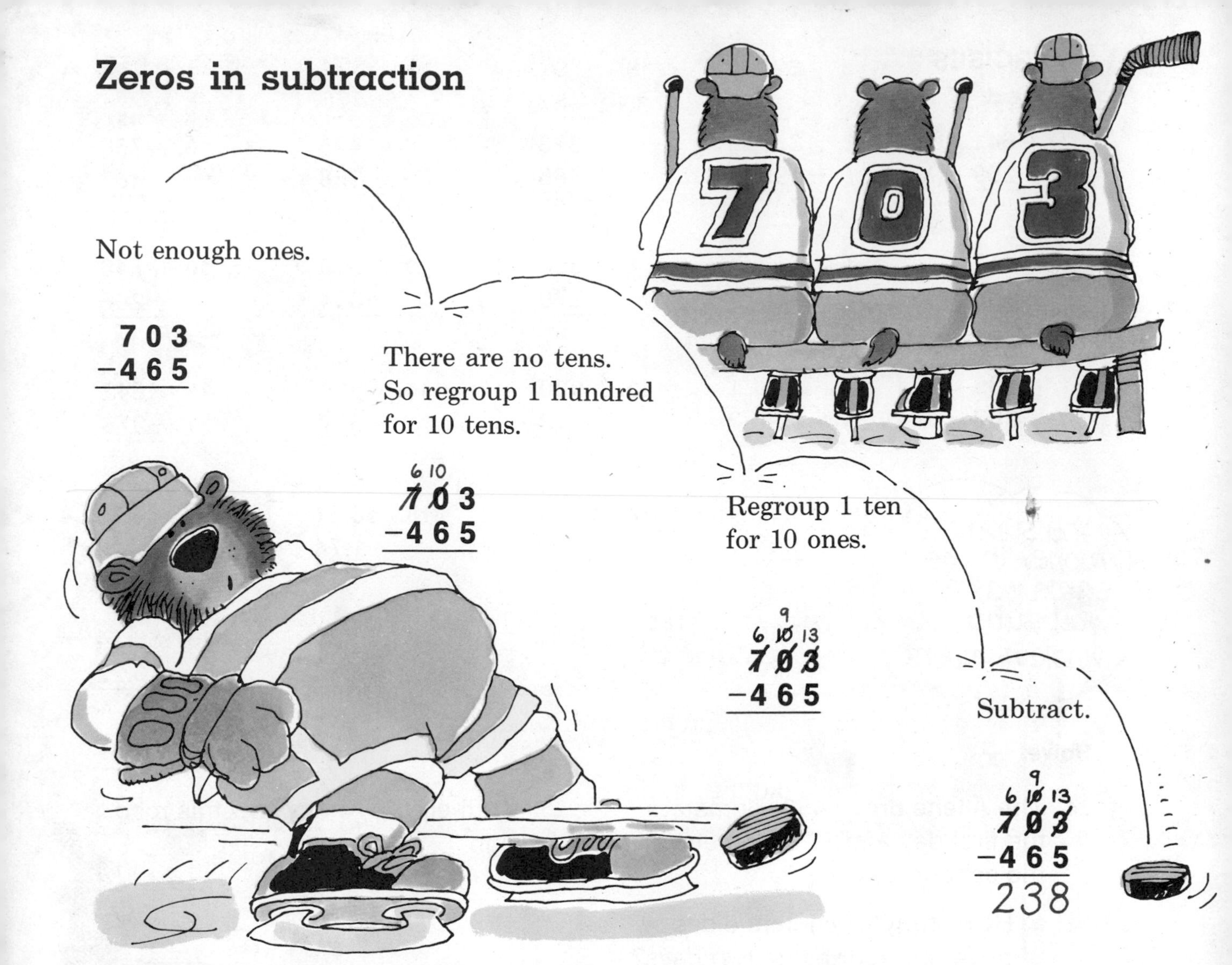

Not enough ones.

$$\begin{array}{r} 703 \\ -465 \\ \hline \end{array}$$

There are no tens.
So regroup 1 hundred
for 10 tens.

$$\begin{array}{r} {\scriptstyle 6\,10} \\ \not{7}\not{0}3 \\ -465 \\ \hline \end{array}$$

Regroup 1 ten
for 10 ones.

$$\begin{array}{r} {\scriptstyle 9} \\ {\scriptstyle 6\,\not{10}\,13} \\ \not{7}\not{0}\not{3} \\ -465 \\ \hline \end{array}$$

Subtract.

$$\begin{array}{r} {\scriptstyle 9} \\ {\scriptstyle 6\,\not{10}\,13} \\ \not{7}\not{0}\not{3} \\ -465 \\ \hline 238 \end{array}$$

EXERCISES

Subtract.

1. 802 − 536	**2.** 604 − 478	**3.** 701 − 263	**4.** 506 − 477	**5.** 903 − 286
6. 705 − 167	**7.** 801 − 496	**8.** 600 − 348	**9.** 402 − 317	**10.** 300 − 189
11. 824 − 617	**12.** 732 − 138	**13.** 613 − 219	**14.** 700 − 308	**15.** 407 − 199

16. 701 −358	**17.** 628 −319	**18.** 507 −438	**19.** 594 −276	**20.** 958 −659
21. 873 −496	**22.** 900 −527	**23.** 746 −185	**24.** 609 −349	**25.** 653 −209
26. $8.96 −4.23	**27.** $6.04 −3.69	**28.** $7.01 −1.75	**29.** $9.05 −6.29	**30.** $8.00 −6.37
31. $7.21 −4.10	**32.** $5.02 −2.75	**33.** $6.00 −4.75	**34.** $8.40 −3.00	**35.** $4.06 −3.89

Solve.

36. The Bruins hockey team scored 308 goals this year. Last year they scored 279 goals.
- **a.** How many more goals did they score this year?
- **b.** How many goals did they score during the two years?

37. The Bruins led the league in scoring with 308 goals. They scored 35 more goals than the Sabres.
- **a.** How many goals did the Sabres score?
- **b.** Did the two teams score more than 600 goals together?

38. A ticket to a Bruins game costs $5.75 for a rinkside seat and $3.50 for a general admission seat.
- **a.** How much more does a rinkside seat cost than a general admission seat?
- **b.** Could you buy 3 general admission tickets with a $10 bill?

★ **39.**

Estimating differences

You learned to round to estimate sums. Rounding can help you estimate differences too. An estimate can help you decide whether you made a mistake subtracting.

Each number was rounded to the nearest ten. Then the difference was estimated.

78	Round to	80
−19	Round to	−20
		60

This difference is about 30.

83	Round to	80
−47	Round to	−50
		30

This difference is about 50.

92	Round to	90
−35	Round to	−40
		50

Each of these numbers was rounded to the nearest hundred.

This difference is about 300.

742	Round to	700
−388	Round to	−400
		300

Is the difference less than or greater than 300?

Each of these amounts was rounded to the nearest dollar.

This difference is about $3.00.

$8.03	Round to	$8.00
−4.75	Round to	−5.00
		$3.00

Is the difference less than or greater than $3.00?

EXERCISES

Round each number to the nearest ten. Then estimate the difference.

1. $58 - 17$
2. $63 - 39$
3. $53 - 21$
4. $81 - 47$
5. $75 - 57$
6. $65 - 18$
7. $75 - 47$
8. $80 - 53$
9. $78 - 32$
10. $60 - 26$

Round each number to the nearest hundred. Then estimate the difference.

11. $529 - 242$
12. $782 - 365$
13. $604 - 427$
14. $813 - 385$
15. $922 - 409$
16. $723 - 575$
17. $682 - 429$
18. $900 - 675$
19. $519 - 372$
20. $857 - 689$

Round each amount to the nearest dollar. Then estimate the difference.

21. \$5.51 − 2.56
22. \$6.25 − 3.78
23. \$8.03 − 4.96
24. \$6.58 − 1.59
25. \$7.25 − 2.78
26. \$6.93 − 1.27
27. \$8.41 − 2.85
28. \$6.03 − .89
29. \$9.13 − 7.67
30. \$8.35 − .89

Solve.

31. Estimate the total price.
32. Is your estimate greater than or less than the exact total? How much greater or less?
33. You gave the clerk \$8.00. Estimate your change.
34. Is your estimate greater than or less than the exact amount of change? How much greater or less?

Problem solving—estimating

1. Study and understand.
2. Plan and do.
3. Answer and check.

Mark had 149 baseball cards.
Jeanette gave him 32 baseball cards.
How many baseball cards did
Mark have then?

$$\begin{array}{r} 149 \\ +\ 32 \\ \hline 181 \end{array}$$

Mark had 181 baseball cards.

Estimate.

The answer seems right.

Kathy had 82 comic books.
She sold 38 of them. How
many did she have then?

$$\begin{array}{r} 82 \\ -\ 38 \\ \hline 56 \end{array}$$ 44

kathy had 56 comic books.

Estimate.

Something is wrong!

Estimate.

Does the answer seem right?

1.  Marie jogs **21** kilometers each week. How far does she jog in **3** weeks?

Answer: 83 kilometers

2. 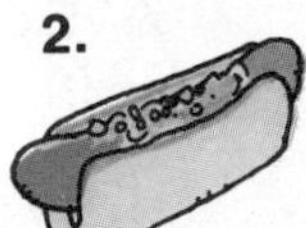 A hot dog costs **$.65**. Mike bought **2** hot dogs. How much did they cost?

Answer: $1.30

3.  Cindy read **203** pages one week and **194** pages the next week. How many pages did she read in all?

Answer: 297

4. 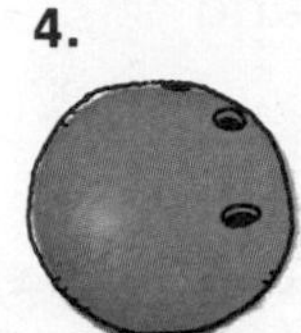 Alice's bowling score was **109**. Denise's bowling score was **82**. How much higher was Alice's score?

Answer: 47

5.  One week Julia watched TV for **164** minutes. The next week she watched TV for **116** minutes. How long did she watch in all?

Answer: 280 minutes

6.  Julia watched TV for **164** minutes. Jeff watched TV for **119** minutes. How many more minutes did Julia watch TV?

Answer: 65

7.  Mariko had **$10**. She bought a shirt for **$6.95**. How much money did she have left?

Answer: $4.95

8. John bought some tennis shoes for **$9.08** and some socks for **$1.89**. What was the total cost?

Answer: $10.97

9. 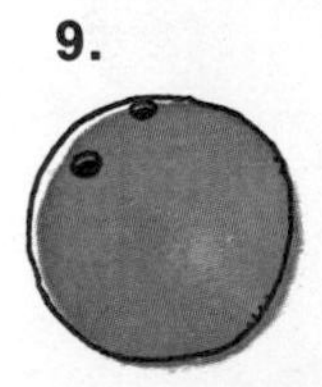 Brian's bowling scores for **3** games were **83**, **95**, and **100**. What was his total score for the **3** games?

Answer: 278

10. Andrew had **89¢**. He bought a hamburger for **68¢**. How much money did he have left?

Answer: 31¢

Problem solving—two-step problems

Sarah had $9. She bought one package of stamps for $3.45 and another for $2.78. How much money did she have then?

One way to solve.

Step 1.

Add to find the total amount spent.

$$\begin{array}{r} \$3.45 \\ +2.78 \\ \hline \$6.23 \end{array}$$

Step 2.

Subtract to find the amount left.

$$\begin{array}{r} \$9.00 \\ -6.23 \\ \hline \$2.77 \end{array}$$

She had $2.77 left.

Another way to solve.

Step 1.

Subtract to find the amount left after buying the first package.

$$\begin{array}{r} \$9.00 \\ -3.45 \\ \hline \$5.55 \end{array}$$

Step 2.

Subtract to find the amount left after buying the second package.

$$\begin{array}{r} \$5.55 \\ -2.78 \\ \hline \$2.77 \end{array}$$

She had $2.77 left.

EXERCISES

Solve.

1. Peter had $8. He bought a package of stamps for $2.75. Later he bought another package of stamps for $2.25. How much money did he have left?

2. Sarah had 587 stamps. Then she bought 28 new stamps and sold 37 stamps. How many stamps did she have then?

3. Ken had 385 stamps in one book, 463 stamps in another book, and 256 stamps in a third book. How many stamps did he have in all?

4. Jill had 463 stamps. She traded 83 of her stamps for 57 new stamps. How many stamps did she have then?

5. Mary had $5. She bought a set of stamps for $3.45. She sold 2 of the stamps for $.75. How much money did she have then?

6. Larry had 873 stamps. He gave 56 stamps to Karen and 35 stamps to John. How many stamps did he have then?

7. Earl had 576 stamps. He bought 245 more. Then he traded 47 of them for 58 new ones. How many stamps did he have then?

8. Tamara had 865 stamps. She sold 27 of them. Then she traded 45 stamps for 36 new ones. How many stamps did she have then?

★ 9. Sue's stamp collection was worth $250 five years ago. Now it is worth $375 more. Today she sold some stamps for $28. What is her stamp collection worth now?

★ 10. Jack had 831 stamps. He bought 48 and sold 34. Then he traded 21 stamps for 32 new ones. How many stamps did he have then?

Subtracting larger numbers

Larger numbers are subtracted in the same way as smaller numbers. Study these examples.

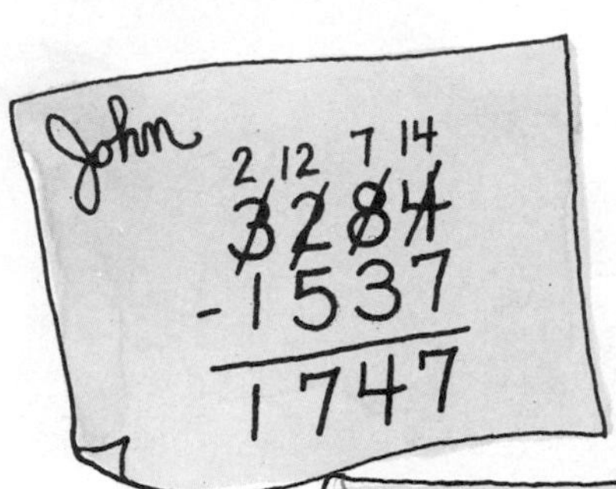

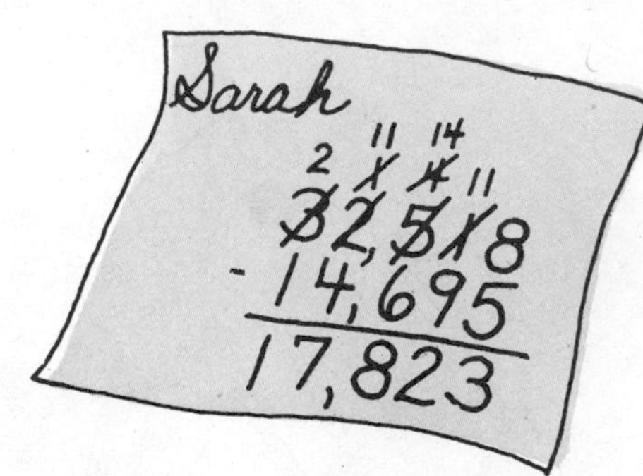

Patricia remembered that addition and subtraction are related. Subtraction is finding a missing addend. So, to check her subtraction, she did this addition:

Patricia

785,062	750,465
− 34,597	+ 34,597
750,465	785,062

It checks!

What numbers would you add to check John's subtraction? To check Sarah's?

EXERCISES

Subtract.

1. 5735 −2168	**2.** 3529 −1674	**3.** 8375 −2694	**4.** 9782 −3597
5. 6581 −4259	**6.** 5029 −1382	**7.** 3406 −1658	**8.** 5003 −4475

First subtract.
Then check your work by addition.

9. 86,215 − 34,802

10. 73,052 − 58,629

11. 94,160 − 41,387

12. 75,384 − 38,965

13. 62,415 − 5,658

14. 90,346 − 2,918

15. 78,002 − 35,146

16. 50,000 − 16,742

Subtract.

17. 378,219 − 3,948

18. 564,381 − 64,592

19. 901,276 − 340,561

20. 828,391 − 352,675

21. 756,550 − 437,891

22. 544,602 − 257,284

23. 637,082 − 305,674

24. 710,028 − 529,674

25. 999,999 − 538,215

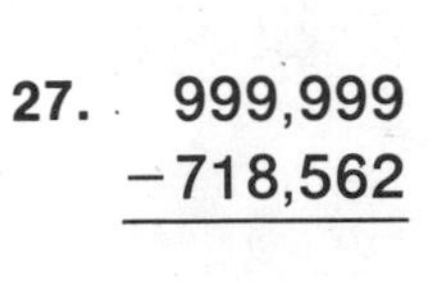

★ **26.** 1,000,000 − 538,216

27. 999,999 − 718,562

★ **28.** 1,000,000 − 718,563

Challenge!

What number was entered first?

29. ? + 3872 =

30. ? + 17,586

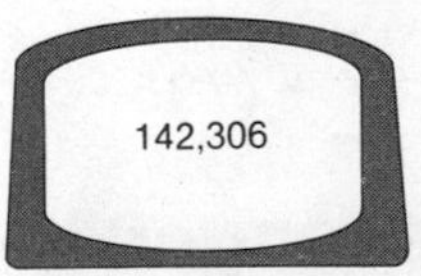

31. ? − 14,974 =

3865

32. ? − 2045 =

Addition and subtraction

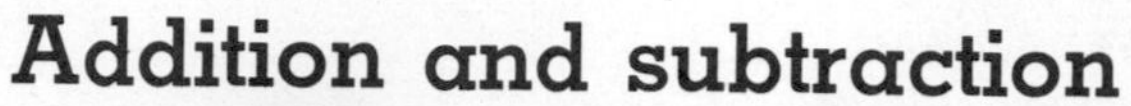

EXAMPLE. **(278 + 456) − 532 = 202**

Remember: Do the work inside the grouping symbols first.

$$\begin{array}{r} 278 \\ +456 \\ \hline 734 \\ -532 \\ \hline 202 \end{array}$$

EXERCISES

Compute.

1. (375 + 296) + 492
2. 375 + (296 + 492)
3. (693 − 281) − 105
4. 693 − (281 − 105)
5. (856 + 395) − 399
6. (408 − 275) + 856
7. (927 + 384) − 384
8. (759 − 257) + 578
9. 867 + (795 − 426)
10. 876 − (253 + 125)

Challenge!

Make up a problem.

NUMBER NEWS 253 − (135 + 26) = ☐

NAILS

Build the greatest difference.

SKILL GAME

Tell which difference is greater.

1. $\begin{array}{r} 8075 \\ -\,3921 \\ \hline \end{array}$ $\quad$ $\begin{array}{r} 9781 \\ -\,2305 \\ \hline \end{array}$

2. $\begin{array}{r} 6385 \\ -\,1294 \\ \hline \end{array}$ $\quad$ $\begin{array}{r} 9831 \\ -\,5246 \\ \hline \end{array}$

3. $\begin{array}{r} 6407 \\ -\,3124 \\ \hline \end{array}$ $\quad$ $\begin{array}{r} 7044 \\ -\,2613 \\ \hline \end{array}$

4. $\begin{array}{r} 4685 \\ -\,1563 \\ \hline \end{array}$ $\quad$ $\begin{array}{r} 8316 \\ -\,5456 \\ \hline \end{array}$

Play the game.

1. Make two cards for each of the digits.
2. Choose a leader.
3. 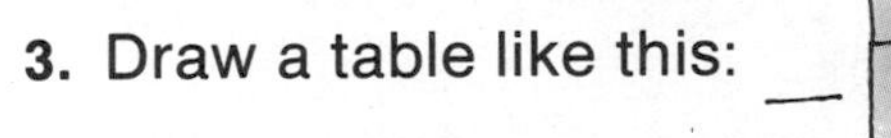 Draw a table like this: 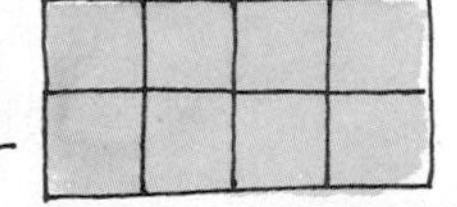
4. As the leader picks the digits, fill in your table.
   Caution: Try to have a greater number on top. If the greater number is on the bottom, you are out of the game.
5. Repeat step 4 until your table is filled in.
6. The player who has the greatest difference wins the game.

Problem solving

How many years ago were these things invented?

1. adding machine
2. ENIAC computer
3. airplane
4. microprocessor
5. slide rule
6. steam car
7. How many years after the steam car was invented was the gasoline car invented?
8. How many years before the airplane was invented was the gasoline-powered car invented?
9. How many years before the personal computer was invented was the slide rule invented?

airplane

Inventors: Orville and Wilbur Wright
Year: 1903

analytical engine

Inventor: Babbage
Year: 1830

automobile (gasoline-powered)

Inventor: Daimler
Year: 1886

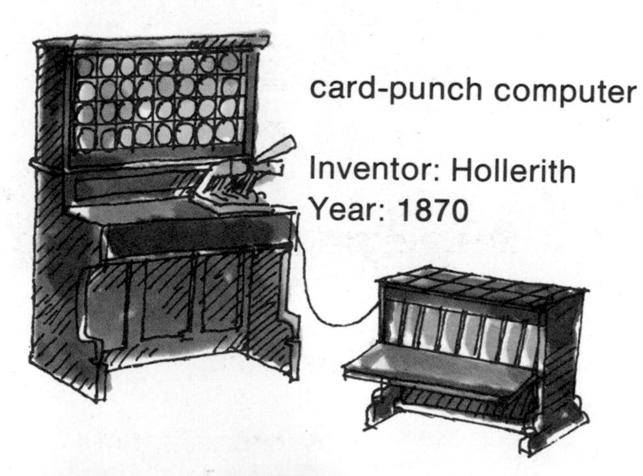

card-punch computer

Inventor: Hollerith
Year: 1870

bicycle

Inventor: Starley
Year: 1871

abacus

Inventor: Unknown
Year: Before recorded history

personal computer

Year: 1976

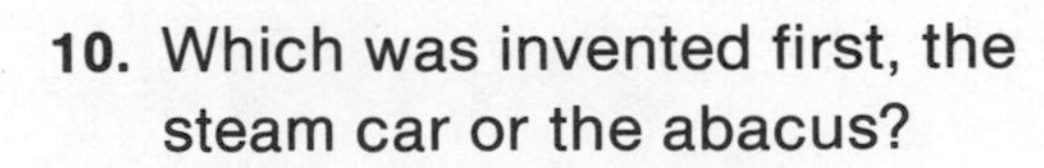

10. Which was invented first, the steam car or the abacus?

11. The gasoline-powered car was invented 181 years after the steam-piston engine. In what year was the steam-piston engine invented?

Pascal's counting machine

Inventor: Pascal
Year: 1642

ENIAC computer

Inventors: Eckert and Mauchly
Year: 1946

slide rule

Inventor: Oughtred
Year: 1620

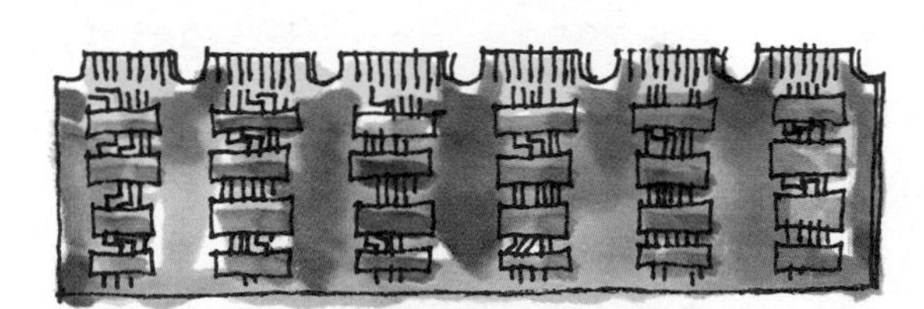

transistor

Inventors: Bell Laboratories
Year: 1947

steam car

Inventor: Cugnot
Year: 1770

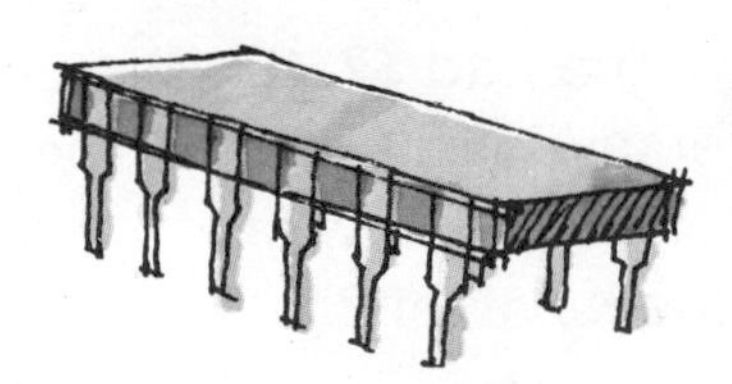

microprocessor

Inventors: Kilby and Noyce
Year: 1951

CHAPTER CHECKUP

Add. [pages 48–59, 78]

1. $58 + 21$
2. $65 + 29$
3. $58 + 96$
4. $243 + 164$
5. $593 + 369$
6. $758 + 695$
7. $3921 + 7865$
8. $5934 + 8476$
9. $35,891 + 26,748$
10. $532,168 + 293,857$
11. $694 + 359 + 786$
12. $428 + 139 + 357 + 275$

Subtract. [pages 62–71, 76–79]

13. $95 - 42$
14. $73 - 45$
15. $456 - 283$
16. $572 - 156$
17. $603 - 429$
18. $800 - 614$
19. $5916 - 2358$
20. $7146 - 3958$
21. $63,941 - 28,056$
22. $90,131 - 28,464$
23. $529,108 - 256,795$
24. $650,081 - 281,374$

Solve. [pages 60–61, 72–75, 80–81]

25. 

What is the total price?

26. Jerry had $12.80. He bought a basketball for $9.75. How much money did he have left?

27. Maura had $7.40. She bought a jersey for $3.98 and some socks for $1.75. How much money did she have left?

CHAPTER PROJECT

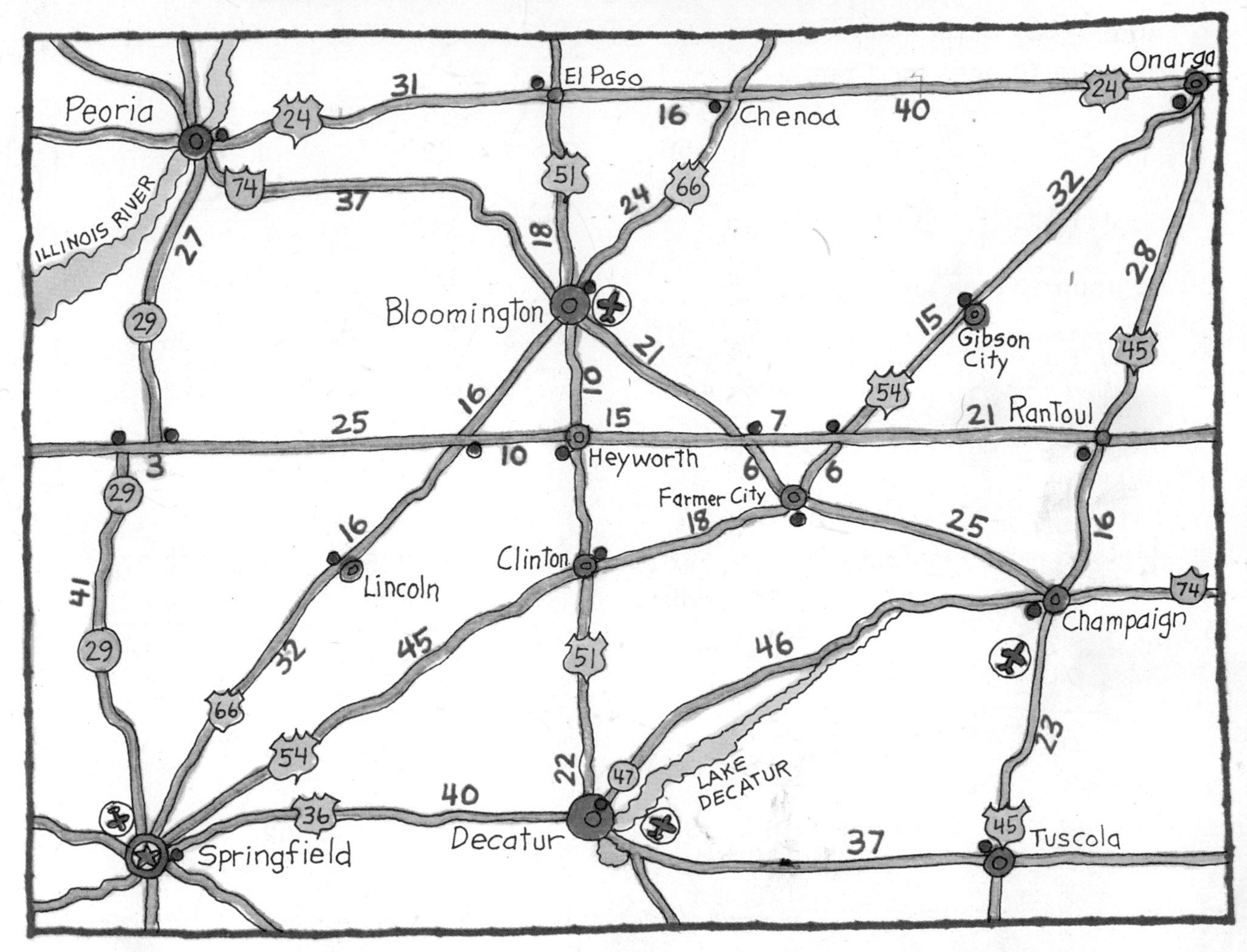

1. How far is it from Bloomington to Peoria?
2. What is the distance from Champaign to Bloomington?
3. Which cities have airline stops?
4. How far is it from the state capital to El Paso?
5. List the state highway numbers that are shown on the map.
6. List four cities in your own state that you would like to visit. Use a map to see how far you would have to travel to visit all four cities and return home.

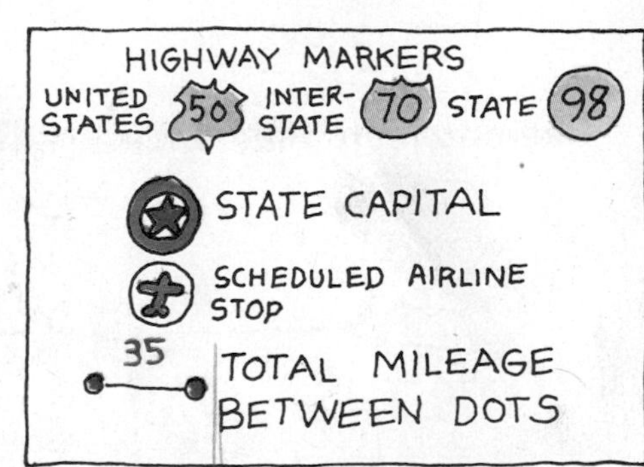

7. Plan some other trips. See how far you would travel on each trip.

CHAPTER REVIEW

Add.

Regroup 10 ones for 1 ten.

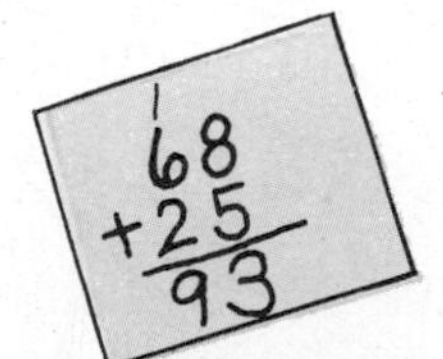

1. 56 + 19
2. 48 + 48
3. 259 + 128

Regroup 10 tens for 1 hundred.

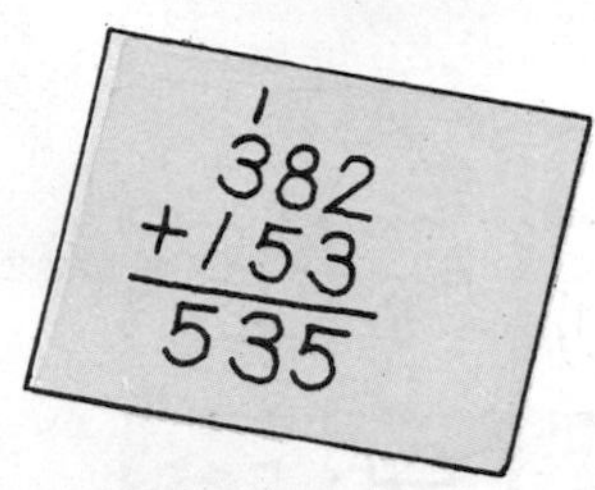

4. 572 + 253
5. 621 + 195
6. 463 + 384

Regroup more than once.

1 1 1
3596
+2458
6054

7. 4678 + 3917
8. 35,925 + 65,426
9. 389,275 + 654,897

Subtract.

Regroup 1 ten for 10 ones.

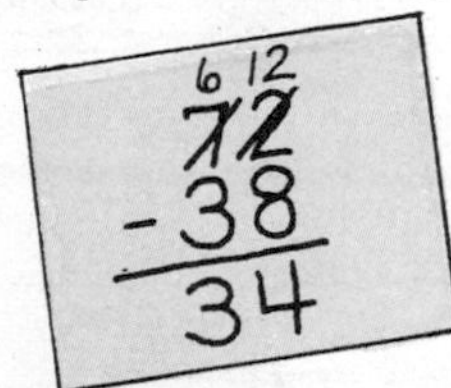

10. 92 − 56
11. 80 − 36
12. 262 − 129

Regroup more than once.

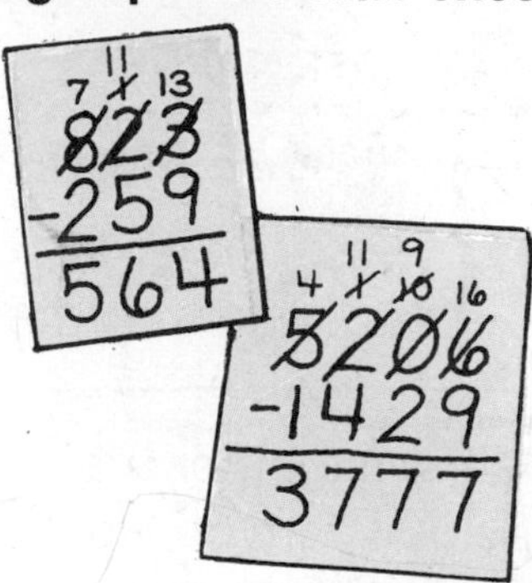

13. 752 − 386
14. 2914 − 1859
15. 6243 − 4858
16. 56,234 − 38,175
17. 76,048 − 41,293
18. 320,413 − 154,289

CHAPTER CHALLENGE

1. a. Add the numbers in each row.
   b. Add the numbers in each column.
   c. Add the numbers along each diagonal.

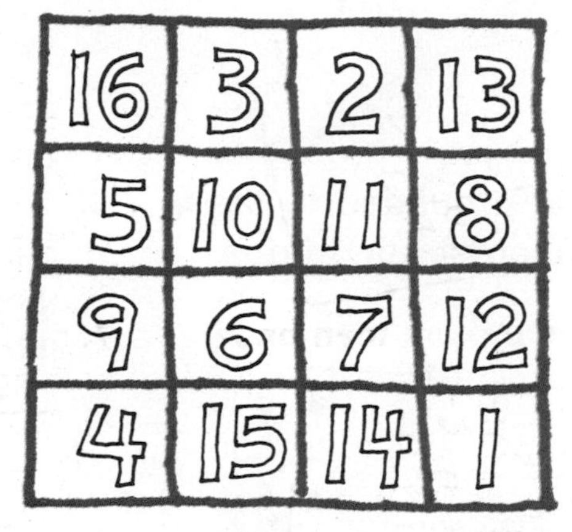

16	3	2	13
5	10	11	8
9	6	7	12
4	15	14	1

Since all the sums are the same, it is a magic square.

2. Copy and complete this magic square.
   Hint: First find the magic sum.

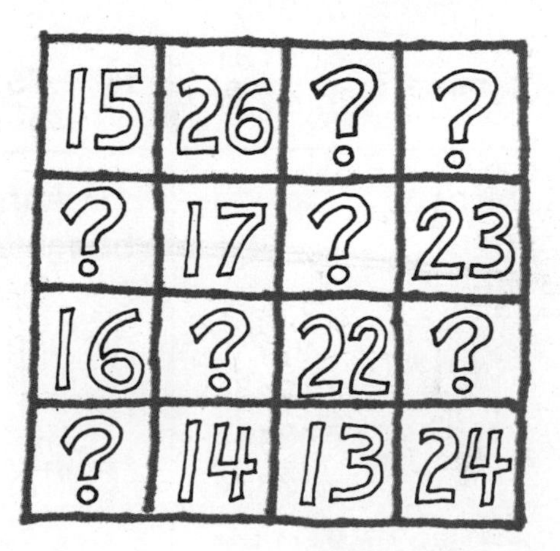

15	26	?	?
?	17	?	23
16	?	22	?
?	14	13	24

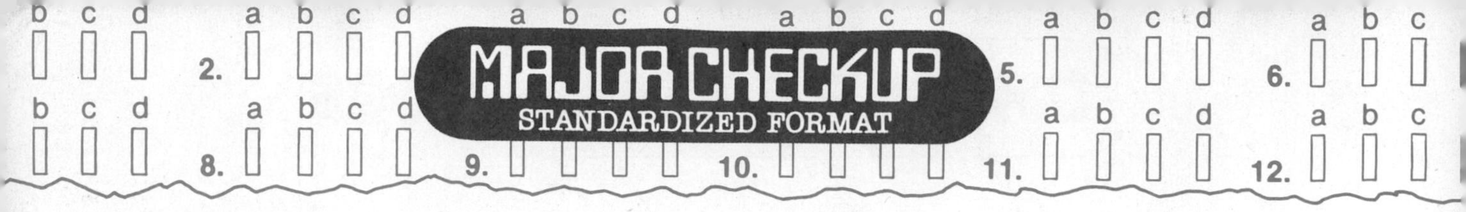

Choose the correct letter.

1. In 7 + 2 = 9, which number is the sum?

a. 7
b. 2
c. 9
d. none of these

2. 4 + ? = 9
The missing number is

a. 13
b. 5
c. 14
d. none of these

3. 426 rounded to the nearest ten is

a. 420
b. 430
c. 400
d. none of these

4. 559 rounded to the nearest hundred is

a. 500
b. 550
c. 560
d. 600

5. The standard numeral for three thousand thirty-three is

a. 3033
b. 3330
c. 3303
d. none of these

6. Which number is greatest?

a. 537,824
b. 521,979
c. 540,793
d. 541,988

7. Add.

$$\begin{array}{r} 5962 \\ +3879 \\ \hline \end{array}$$

a. 9841
b. 9831
c. 8731
d. 9731

8. Add.

$$\begin{array}{r} 526 \\ 348 \\ +789 \\ \hline \end{array}$$

a. 1643
b. 1663
c. 1543
d. none of these

9. Subtract.

$$\begin{array}{r} 8216 \\ -2549 \\ \hline \end{array}$$

a. 5667
b. 6333
c. 5777
d. none of these

10. Subtract.

$$\begin{array}{r} 4028 \\ -1694 \\ \hline \end{array}$$

a. 2434
b. 3434
c. 3334
d. none of these

11. Anne had 92¢. She bought a sandwich for 65¢. How much money did she have left?

a. 37¢ **b.** 27¢
c. 33¢ **d.** none of these

12. Mike had $9. He bought a game for $4 and a book for $3. How much money did he have then?

a. $16 **b.** $10
c. $2 **d.** none of these

Time and Money

4

Hours and minutes

There are 60 minutes in 1 hour. The examples show different ways to tell time.

We can say two thirty
or thirty minutes after two
or half past two.

We write 2:30.

We can say four forty-two
or eighteen minutes to five.

We write 4:42.

We can say ten fifteen
or fifteen minutes after ten
or quarter past ten.

We write 10:15.

We can say eleven forty-five
or fifteen minutes to twelve
or quarter to twelve.

We write 11:45.

EXERCISES

Write each time.

1.

2.

3.

Copy and complete.

4.

a. nine _?_
b. twenty-six minutes _?_ _?_

5.

a. four _?_
b. thirty minutes _?_ _?_
c. half _?_ _?_

6.

a. seven _?_
b. fifteen minutes _?_ _?_
c. quarter _?_ _?_

7.

a. one _?_
b. fifteen minutes _?_ _?_
c. quarter _?_ _?_

8.

a. eleven _?_
b. _?_ minutes to twelve

9.

a. two _?_
b. _?_ minutes to three

Match.

10.	4:12	**a.**	ten minutes to five
11.	4:50	**b.**	quarter to three
12.	2:45	**c.**	eight minutes after four
13.	3:48	**d.**	twelve minutes to four
14.	3:15	**e.**	quarter past three
15.	4:08	**f.**	twelve minutes after four

A.M. and P.M.

The **second hand** goes around once in one minute.

There are 60 seconds in one minute.

The **minute hand** goes around once in one hour.

There are 60 minutes in one hour.

The **hour hand** goes around twice in one day.

There are 24 hours in 1 day.

A.M. is used for times after 12:00 midnight and before 12:00 noon.

P.M. is used for times after 12:00 noon and before 12:00 midnight.

EXERCISES

Daylight or dark?

1. 12:00 midnight **2.** 12:00 noon **3.** 3:00 A.M. **4.** 3:00 P.M.

5. 11:45 A.M. **6.** 11:45 P.M. **7.** 2:36 A.M. **8.** 2:36 P.M.

9. 10:30 A.M. **10.** 10:30 P.M.

How many

11. hours in a day?

12. minutes in an hour?

13. seconds in a minute?

14. seconds in two minutes?

15. minutes in a quarter hour?

16. minutes in a half hour?

Use A.M. or P.M. in your answers. What time did you (or will you)

17. get up today?

18. eat breakfast?

19. leave for school?

20. arrive at school?

21. have morning recess?

22. eat lunch?

23. have afternoon recess?

24. get out of school?

25. get home?

26. eat dinner?

Challenge!

Who is right?

27.

36 minutes is more than 2150 seconds.

Tom

Wrong! 2150 seconds is more than 36 minutes.

Ron

28.

8400 seconds is more than 3 hours.

Jane

Maria

KEEPING SKILLS SHARP

1. $26 + 13$

2. $37 + 58$

3. $96 + 57$

4. $243 + 159$

5. $575 + 868$

6. $351 + 999$

7. $786 + 954$

8. $3821 + 4675$

9. $3982 + 7465$

10. $8974 + 5628$

More about time

Maria woke up at

She had to be in school at

How much time did she have?

Here are two ways to get the answer.

Wake-up time — School time

7:15 → [15 minutes] → **7:30** → [1 hour] → **8:30**

1 hour and 15 minutes

Wake-up time — School time

7:15 → [45 minutes] → **8:00** → [30 minutes] → **8:30**

Add the minutes.

$$\begin{array}{rl} 45 & \text{minutes} \\ +30 & \text{minutes} \\ \hline 75 & \text{minutes} \end{array}$$

Regroup 60 minutes for 1 hour.

1 hour and 15 minutes

EXERCISES

Give the time that is

1. 20 minutes later than 3:10.
2. 15 minutes later than 2:50.
3. 23 minutes later than 11:45.
4. 10 minutes earlier than 8:25.
5. 15 minutes earlier than 6:47.
6. 20 minutes earlier than 11:12.

How many minutes from

7. 2:15 to 3:00?
8. 7:45 to 8:30?
9. 11:15 to 12:15?
10. 4:25 to 5:00?
11. 6:03 to 7:00?
12. 1:05 to 1:57?

Solve.

13. School begins at 8:20. Susan got to school at 8:05. How many minutes early was she?
14. The television show started at 7:30. Mark turned it on at 7:45. How many minutes did he miss?
15. Jon left for the ball game at 7:45. He arrived at 8:10. How long did it take him to get there?
16. Julie started raking leaves at 4:15. She finished at 5:35. How long did she work?

KEEPING SKILLS SHARP

1. $58 - 23$
2. $62 - 47$
3. $458 - 165$
4. $842 - 356$
5. $503 - 248$
6. $600 - 475$
7. $3916 - 2489$
8. $6203 - 1457$
9. $8300 - 3479$
10. $5211 - 1378$

Money

twenty-dollar bill

$20 or **$20.00**

ten-dollar bill

$10 or **$10.00**

half-dollar

50¢ or **$.50**

quarter

25¢ or **$.25**

five-dollar bill

$5 or **$5.00**

one-dollar bill

$1 or **$1.00**

dime

10¢ or **$.10**

nickel

5¢ or **$.05**

penny

1¢ or **$.01**

EXERCISES

How much money?

Give each total.

										TOTAL
7.	0	0	1	1	1	0	1	0	3	?
8.	0	1	0	3	0	2	0	1	0	?
9.	1	0	1	0	1	0	2	0	4	?
10.	1	1	0	0	0	1	3	2	3	?
11.	1	0	1	2	0	2	4	1	1	?
12.	1	1	2	3	1	0	2	3	5	?

Challenge!

Tell what is in each bank.

13.  35¢ 3 coins

14.  $1.54 1 bill 6 coins

15.  $2.45 1 bill 6 coins

Making change

Sales clerks must know how to use a cash register. They must also be able to give customers the correct change. The example shows the method that most clerks use to count change.

Terry bought this shirt for $3.79. She gave the clerk a five-dollar bill.

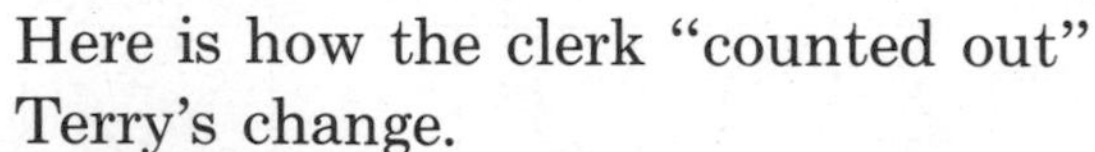

Here is how the clerk "counted out" Terry's change.

Terry's change was $1.21.

ORAL EXERCISES

Touch the money needed as you "count out" the change.

1. You have  You spend 

2. You have   You spend 

3. You have  You spend 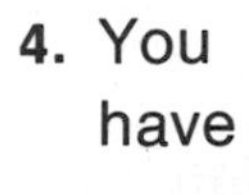 

4. You have  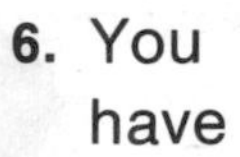 You spend 

5. You have  You spend 

6. You have  You spend 

7. You have You spend 

8. You have  You spend 

Problem solving—too much information

1. Study and understand.
2. Plan and do.
3. Answer and check.

In some problems there are more facts than you need. You have to pick the facts you need to answer the question.

On November 17, Carl delivered 126 morning papers and 93 evening papers. Janet delivered 108 morning papers and 100 evening papers. How many morning papers did they deliver in all?

Question: How many morning papers did they deliver in all?

Facts needed: Carl delivered 126 morning papers.
Janet delivered 108 morning papers.

They delivered 234 morning papers in all.

EXERCISES

Solve.

1. Alan collected $145.80 in one week and $126.40 in the next week. He paid $204.15 to the newspaper company. How much did he collect in the 2 weeks?

2. To win a free trip to an amusement park, Alan had to get 24 new customers. He spoke to 36 people and got 11 new customers on the first day. How many more did he need in order to win the trip?

3. On October 15, Joyce delivered 145 morning papers and 102 evening papers. On November 15, she delivered 137 morning papers and 97 evening papers.

   a. How many more morning papers than evening papers did she deliver on November 15?

   b. How many papers did Joyce deliver on October 15?

   c. How many more morning papers did Joyce deliver on October 15 than on November 15?

4. One week Carl collected $95.70 and paid the newspaper company $59.64. The next week he collected $108.60 and paid the newspaper company $67.93.

   a. How much did he collect in the two weeks?

   b. How much more did Carl collect in the second week than in the first week?

   c. How much did Carl earn in the second week?

5. One week Mr. Thomas gave Joyce a $5 bill to pay $1.35 for his morning papers, $1.35 for his evening papers, and $.25 for a tip.

   a. How much did he pay in all?

   b. How much change did Joyce give Mr. Thomas?

★ 6. Mrs. Jacobson owed $4.45 for her papers. She gave Carl a $5 bill. He gave her 4 coins in change. What were the coins?

CHAPTER CHECKUP

Write each time. [pages 88–89]

1. 

2. 

3. 

Match. [pages 88–89]

4. fifteen minutes after six
5. thirty minutes after six
6. fifteen minutes to six

a. quarter to six
b. half past six
c. quarter past six

Answer the questions. [pages 90–91]

7. How many seconds are in 1 minute?
8. How many minutes are in 1 hour?
9. How many hours are in 1 day?
10. Is 8:30 in the morning 8:30 A.M. or 8:30 P.M.?

How many minutes from [pages 92–93]

11. 3:45 to 4:30?

12. 7:08 to 8:00?

13. 2:06 to 2:43?

How much money? [pages 94–97]

14. 

15. 

16. 

CHAPTER PROJECT

TARGET BOARD

1. Bend a paper clip to look like this:

scoring tip

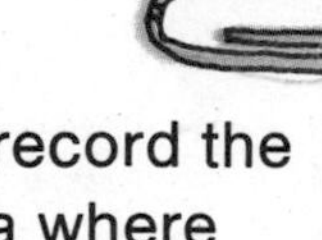

2. Hold the clip about 10 centimeters above the target board shown.

3. Drop the clip and record the number in the area where the scoring tip lands.

4. Repeat steps 2 and 3 to find out what sums are possible when you drop the clip three times.

Possible sums for clip dropped three times:
4 + 4 + 4 = 12
4 + 4 + 9 = 17

5. Suppose you can drop the clip as many times as you want to get a sum.
   a. List at least 20 numbers that can be the sum.
   b. How many numbers can you find that cannot be the sum?

CHAPTER REVIEW

Match.

1. quarter past one
2. half past one
3. quarter to one

a. 

b. 

c. 

True or false?

4. There are 60 seconds in 1 minute.
5. There are 60 minutes in 1 hour.
6. There are 12 hours in 1 day.
7. 4:45 P.M. is in the afternoon.

How many minutes?

8. from  to 

9. from to 

How much money?

10. 

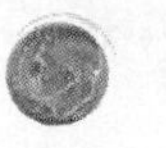

11. 

12. 

CHAPTER CHALLENGE

Computers, our armed forces, and some European countries use a different method of telling time. They use a "**24-hour**" clock, or **24-hour** time. **A.M.** and **P.M.** are not used. Study these examples.

	8:30 A.M.	12:00 noon	4:15 P.M.

24-hour time:	0830 This is generally read as "oh-eight-thirty."	1200 "twelve hundred"	1615 "sixteen fifteen"

Give the 24-hour time.

1. 5:20 A.M.
2. 6:00 A.M.
3. 9:57 A.M.
4. 10:30 A.M.
5. 11:45 A.M.
6. 12:00 noon
7. 1:00 P.M.
8. 2:00 P.M.
9. 4:00 P.M.
10. 6:25 P.M.
11. 8:45 P.M.
12. 10:40 P.M.

Write each time using A.M. or P.M.

13. 0100
14. 0430
15. 0950
16. 1140
17. 1330
18. 1545
19. 1900
20. 2130
21. 2245
22. 2315

MAJOR CHECKUP
STANDARDIZED FORMAT

Choose the correct letter.

1. In $17 - 9 = 8$, which number is the difference?

a. 17
b. 9
c. 8
d. none of these

2. The standard numeral for twenty-nine thousand five hundred fifty is

a. 29,550
b. 29,055
c. 29,505
d. none of these

3. 38,656 rounded to the nearest ten is

a. 38,700
b. 38,650
c. 38,660
d. 38,600

4. Add.

$$\begin{array}{r} 789 \\ +256 \\ \hline \end{array}$$

a. 935
b. 1045
c. 945
d. none of these

5. Add.

$$\begin{array}{r} 76 \\ 93 \\ 45 \\ +28 \\ \hline \end{array}$$

a. 242
b. 222
c. 243
d. 241

6. Subtract.

$$\begin{array}{r} 521 \\ -386 \\ \hline \end{array}$$

a. 145
b. 135
c. 265
d. 235

7. Subtract.

$$\begin{array}{r} 6204 \\ -3587 \\ \hline \end{array}$$

a. 2627
b. 2717
c. 3627
d. 2617

8. What time is shown?

a. quarter to five
b. quarter after five
c. quarter to six
d. quarter after six

9. How much time from 7:30 to 8:45?

a. 15 minutes
b. 45 minutes
c. 1 hour 15 minutes
d. none of these

10. How much money in all?
1 five-dollar bill
2 one-dollar bills
2 quarters
3 dimes

a. $6.80 b. $7.55
c. $7.70 d. none of these

11. Dick had 703 stamps. He sold 48 of them. How many did he have then?

a. 741 b. 751
c. 654 d. none of these

12. Dorothy had $20. She bought a game for $8 and a book for $5. How much money did she spend?

a. $7 b. $33
c. $13 d. none of these

Multiplication by 1-Digit Factors

5

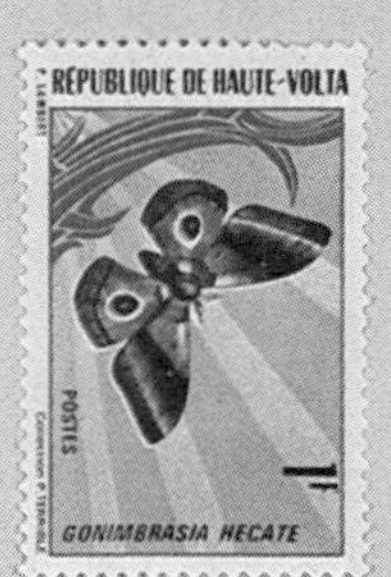

Multiplying 2, 3, and 4

The numbers you multiply are called **factors.** The answer is called the **product.**

$$3 \times 2 = 6$$

Read as "Three times two equals six."

EXERCISES

Give each product.

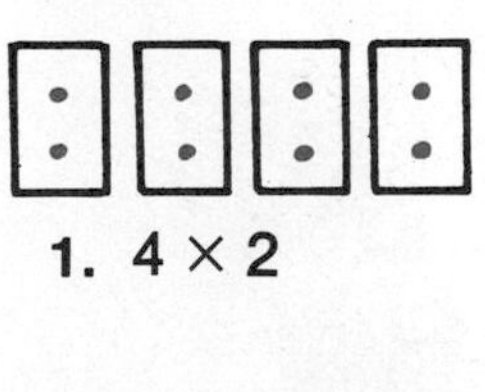

1. 4×2

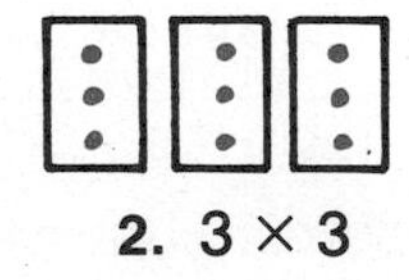

2. 3×3

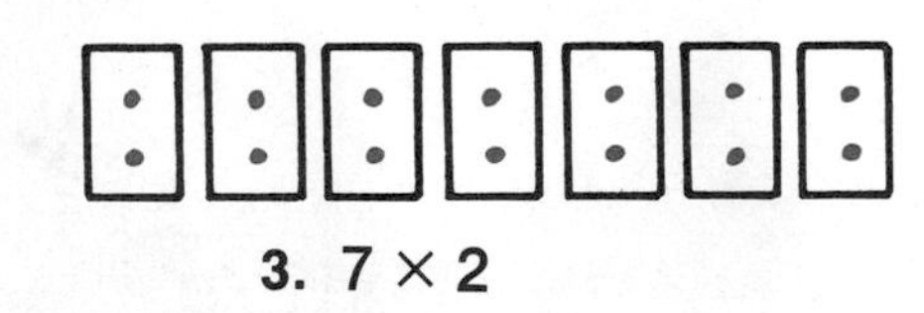

3. 7×2

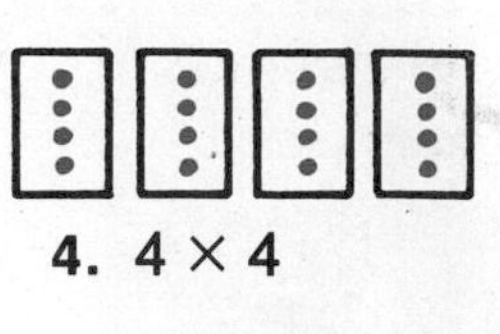

4. 4×4

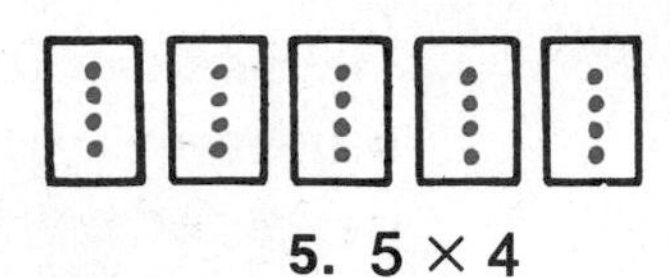

5. 5×4

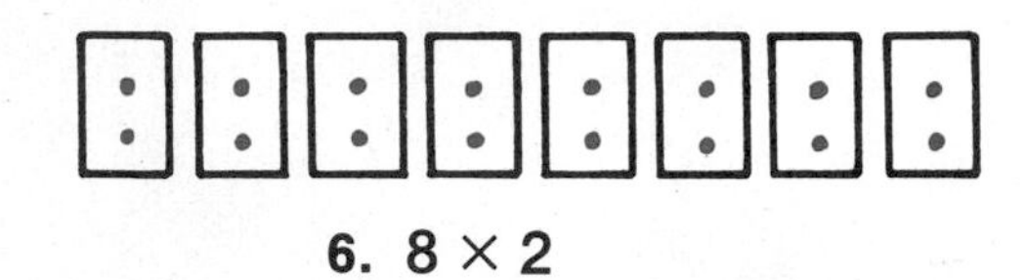

6. 8×2

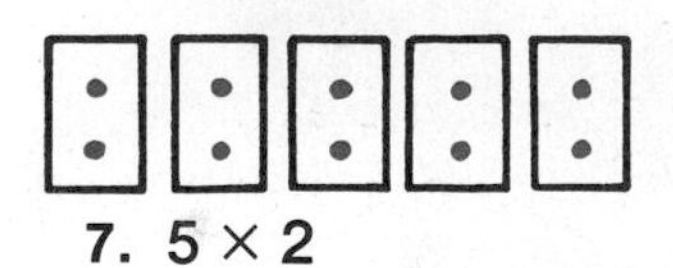

7. 5×2

8. 1×3

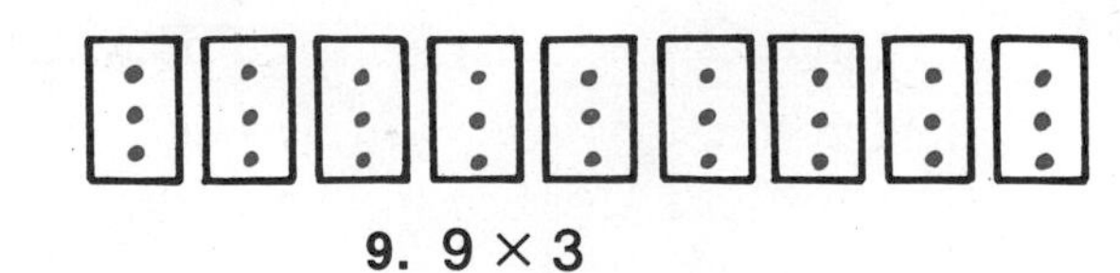

9. 9×3

Give each product.

10. 4×2	11. 0×2	12. 2×3	13. 5×2
14. 5×4	15. 1×3	16. 2×2	17. 9×3
18. 3×3	19. 4×3	20. 9×2	21. 2×4
22. 1×2	23. 8×4	24. 4×4	25. 8×2
26. 9×4	27. 6×3	28. 6×2	29. 6×4
30. 7×3	31. 7×2	32. 8×3	33. 5×3
34. 3×2	35. 1×4	36. 7×4	37. 3×4

Solve.

38.

How much will 8 cost?

39.

How much will 5 cost?

40. There were 7 tables with 2 books on each table. How many books were there?

41. There were 7 books on one table and 2 books on another. How many books were there?

42. There were 8 packages of erasers with 4 erasers in each package. How many erasers were there in all?

43. There were 8 red erasers and 4 green erasers. How many more red erasers were there?

44. There were 9 cartons of milk. If 3 cartons were sold, how many cartons of milk are left?

45. There were 9 cartons of milk. If there were 3 straws for each carton, how many straws were there all together?

Multiplying 0, 1, and 5

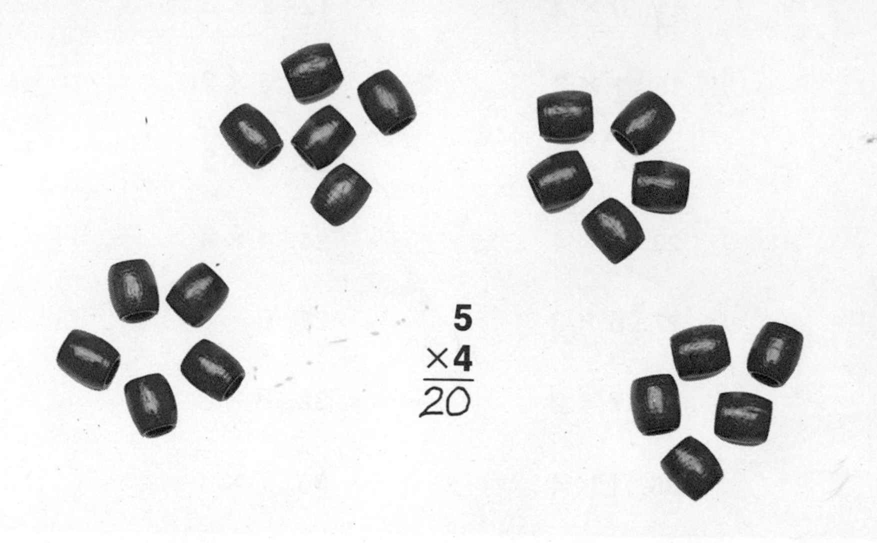

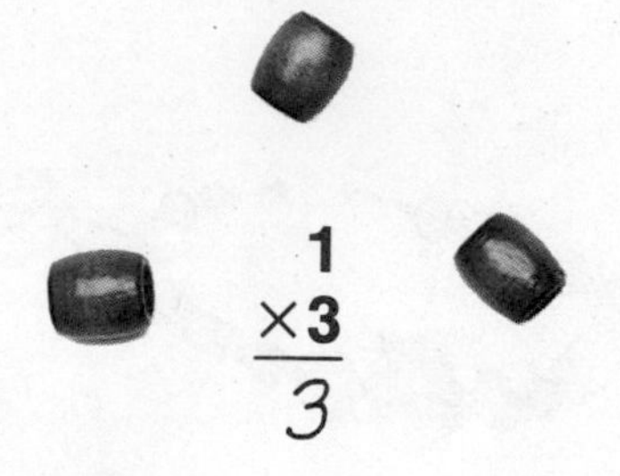

0
×3
0

EXERCISES

Multiply.

1. 1 ×3	**2.** 0 ×6	**3.** 1 ×9	**4.** 0 ×1	**5.** 5 ×7	**6.** 1 ×6
7. 1 ×5	**8.** 5 ×6	**9.** 1 ×1	**10.** 5 ×0	**11.** 0 ×5	**12.** 0 ×3
13. 5 ×2	**14.** 2 ×5	**15.** 1 ×7	**16.** 1 ×4	**17.** 0 ×8	**18.** 1 ×8
19. 0 ×7	**20.** 5 ×5	**21.** 0 ×9	**22.** 5 ×9	**23.** 0 ×4	**24.** 5 ×8

Give each product.

25. 1×7

26. 3×6

27. 5×3

28. 3×4

29. 2×3

30. 4×9

31. 3×2

32. 5×5

33. 4×6

34. 1×5

35. 2×5

36. 5×4

37. 5×6

38. 4×3

39. 1×6

40. 2×6

41. 4×2

42. 0×4

43. 3×7

44. 5×7

45. 2×7

46. 3×5

47. 5×8

48. 4×8

49. 0×8

50. 3×3

51. 1×8

52. 4×1

53. 2×9

54. 5×1

55. 4×4

56. 5×9

57. 4×7

58. 1×1

59. 1×4

60. 3×1

Solve.

61. Joan made this bracelet for her sister. Each red bead cost 5¢. Each yellow bead cost 4¢.

a. How much did the red beads cost?

b. How much did the yellow beads cost?

c. What was the total cost of the beads?

★ 62. Joan made a necklace to go with the bracelet. She paid 69¢ for the 15 beads she used to make the necklace. How many yellow beads did she use?

Practicing multiplication facts

You can use this page over and over again to practice multiplication facts. Cover a column with a piece of paper. First give the product. Then slide the paper down to check your answer.

2	3	4	5
2 ×0 0	3 ×0 0	4 ×0 0	5 ×0 0
2 ×1 2	3 ×1 3	4 ×1 4	5 ×1 5
2 ×2 4	3 ×2 6	4 ×2 8	5 ×2 10
2 ×3 6	3 ×3 9	4 ×3 12	5 ×3 15
2 ×4 8	3 ×4 12	4 ×4 16	5 ×4 20
2 ×5 10	3 ×5 15	4 ×5 20	5 ×5 25
2 ×6 12	3 ×6 18	4 ×6 24	5 ×6 30
2 ×7 14	3 ×7 21	4 ×7 28	5 ×7 35
2 ×8 16	3 ×8 24	4 ×8 32	5 ×8 40
2 ×9 18	3 ×9 27	4 ×9 36	5 ×9 45

1. Number your paper 1 through 60.
2. Keep track of the seconds that it takes you to write down all the products.
3. Use the flash cards on the facing page to check your answers.
4. To find your score, add the number of seconds and the number of mistakes. Keep trying for a better (lower) score.

1. 1×6	2. 4×5	3. 0×3	4. 2×9
5. 4×4	6. 2×6	7. 5×9	8. 0×7
9. 4×0	10. 2×1	11. 3×3	12. 5×5

13. 1×0	14. 5×3	15. 1×2	16. 4×3	17. 0×0	18. 4×9	19. 1×4	20. 5×8
21. 2×2	22. 0×4	23. 5×4	24. 1×3	25. 3×6	26. 0×9	27. 2×8	28. 5×6
29. 3×2	30. 3×7	31. 2×5	32. 3×8	33. 1×5	34. 5×1	35. 2×4	36. 0×1
37. 4×6	38. 1×1	39. 0×6	40. 5×0	41. 0×2	42. 4×1	43. 2×3	44. 2×7
45. 5×2	46. 4×2	47. 2×0	48. 4×7	49. 1×9	50. 3×4	51. 0×5	52. 3×9
53. 3×1	54. 0×8	55. 3×5	56. 5×7	57. 1×7	58. 4×8	59. 3×0	60. 1×8

Multiplication properties

Here are some properties that make it easier to remember the multiplication facts.

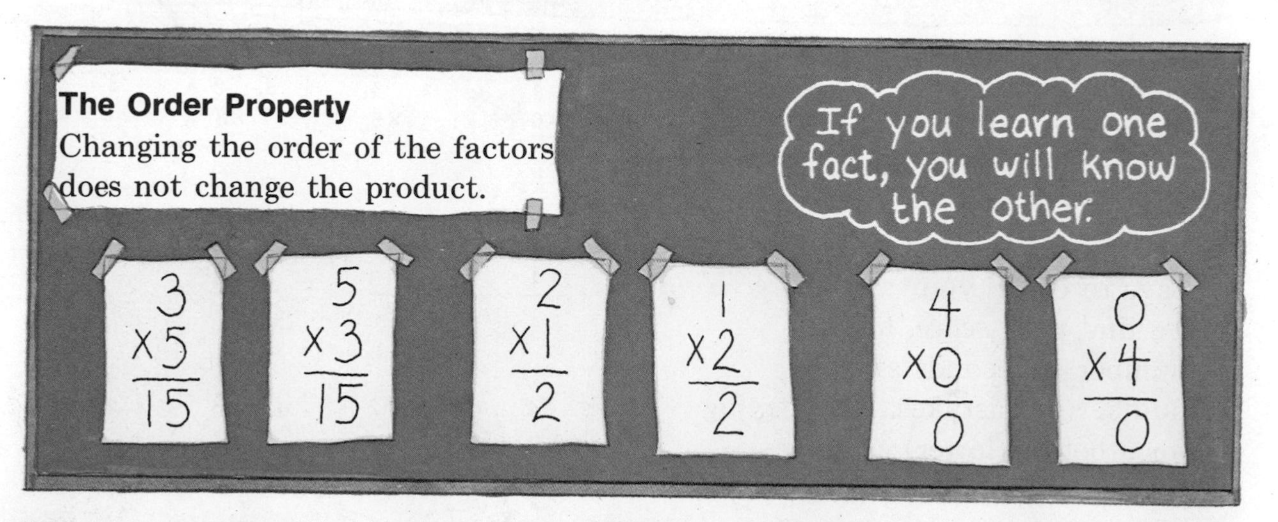

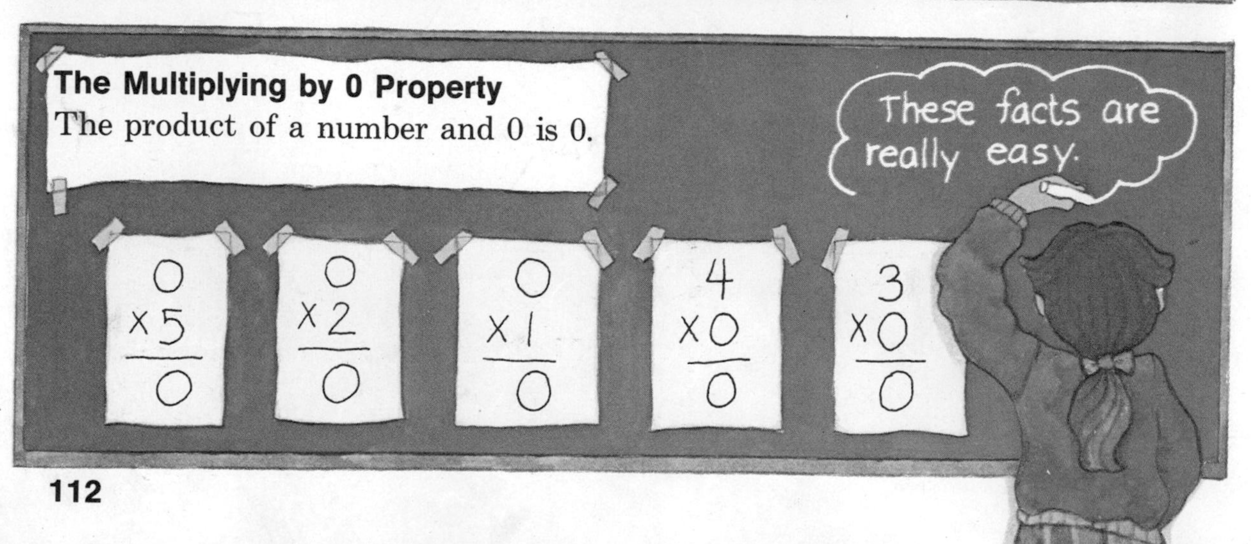

EXERCISES

Complete. ***Hint:*** **Use the order property.**

1. If $\begin{array}{r} 4 \\ \times 6 \\ \hline 24 \end{array}$ then $\begin{array}{r} 6 \\ \times 4 \\ \hline ? \end{array}$
2. If $\begin{array}{r} 5 \\ \times 7 \\ \hline 35 \end{array}$ then $\begin{array}{r} 7 \\ \times 5 \\ \hline ? \end{array}$
3. If $\begin{array}{r} 2 \\ \times 8 \\ \hline 16 \end{array}$ then $\begin{array}{r} 8 \\ \times 2 \\ \hline ? \end{array}$
4. If $\begin{array}{r} 3 \\ \times 9 \\ \hline 27 \end{array}$ then $\begin{array}{r} 9 \\ \times 3 \\ \hline ? \end{array}$
5. If $\begin{array}{r} 1 \\ \times 6 \\ \hline 6 \end{array}$ then $\begin{array}{r} 6 \\ \times 1 \\ \hline ? \end{array}$
6. If $\begin{array}{r} 3 \\ \times 7 \\ \hline 21 \end{array}$ then $\begin{array}{r} 7 \\ \times 3 \\ \hline ? \end{array}$

The Grouping Property
Changing the grouping of the factors does not change the product.

Multiply inside the grouping symbols first.

$(4 \times 2) \times 3 = ?$
$8 \times 3 = 24$

$4 \times (2 \times 3) = ?$
$4 \times 6 = 24$

Compute.

7. a. $(2 \times 3) \times 3$
   b. $2 \times (3 \times 3)$
8. a. $(4 \times 1) \times 5$
   b. $4 \times (1 \times 5)$
9. a. $(2 \times 2) \times 4$
   b. $2 \times (2 \times 4)$
10. a. $(5 + 3) + 6$
    b. $5 + (3 + 6)$
11. a. $(9 - 3) - 2$
    b. $9 - (3 - 2)$
12. a. $(12 - 8) - 3$
    b. $12 - (8 - 3)$

KEEPING SKILLS SHARP

1. $\begin{array}{r} 76 \\ +21 \\ \hline \end{array}$
2. $\begin{array}{r} 158 \\ +129 \\ \hline \end{array}$
3. $\begin{array}{r} 3826 \\ +2590 \\ \hline \end{array}$
4. $\begin{array}{r} 46{,}398 \\ +2{,}746 \\ \hline \end{array}$
5. $\begin{array}{r} 539{,}856 \\ +427{,}407 \\ \hline \end{array}$
6. $\begin{array}{r} 98 \\ -41 \\ \hline \end{array}$
7. $\begin{array}{r} 721 \\ -503 \\ \hline \end{array}$
8. $\begin{array}{r} 5601 \\ -429 \\ \hline \end{array}$
9. $\begin{array}{r} 82{,}642 \\ -9{,}758 \\ \hline \end{array}$
10. $\begin{array}{r} 634{,}182 \\ -387{,}096 \\ \hline \end{array}$

Multiplying 6 and 7

Use the facts you know to learn other facts.

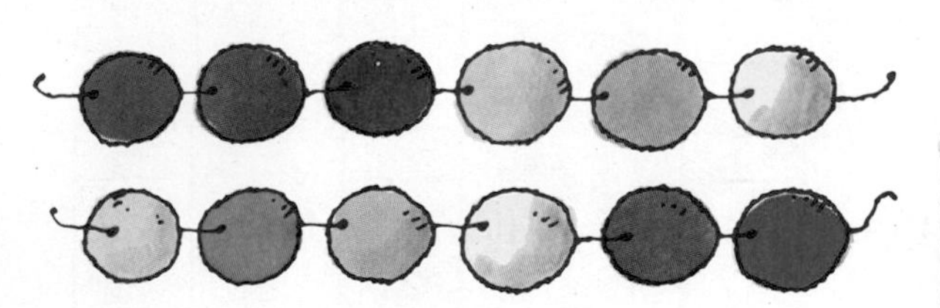

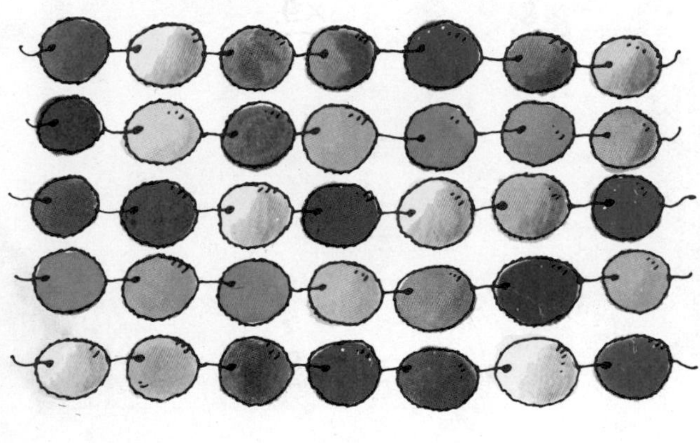

EXERCISES

Use the first fact to find the other product.

1. 

2. $7 \times 2 = 14$ $\quad$ 7×3

3. $7 \times 4 = 28$ $\quad$ 7×5

4. $6 \times 8 = 48$ $\quad$ 6×7

5. $7 \times 6 = 42$ $\quad$ 7×7

6. 

7. $5 \times 9 = 45$ $\quad$ 5×8

8. $6 \times 3 = 18$ $\quad$ 6×4

9. $7 \times 8 = 56$ $\quad$ 7×9

Multiply.

10. 4×5	11. 5×5	12. 7×5	13. 4×4	14. 6×1	15. 6×7
16. 3×6	17. 6×6	18. 3×7	19. 7×3	20. 5×6	21. 3×9
22. 7×6	23. 4×6	24. 6×2	25. 4×7	26. 3×8	27. 7×2
28. 5×4	29. 6×3	30. 5×7	31. 7×8	32. 4×9	33. 6×5
34. 6×8	35. 7×4	36. 5×2	37. 5×9	38. 7×1	39. 5×3
40. 6×9	41. 4×8	42. 6×4	43. 7×7	44. 5×8	45. 7×9

Copy and complete.

46.

weeks	1	2	3	4	5	6	7	8	9
days	7	?	?	?	?	?	?	?	?

47.

gallons	1	2	3	4	5
quarts	4	?	?	?	?

Challenge!

Guess my number.

48. Add my two numbers and you get 11. If you multiply them, you get 28.

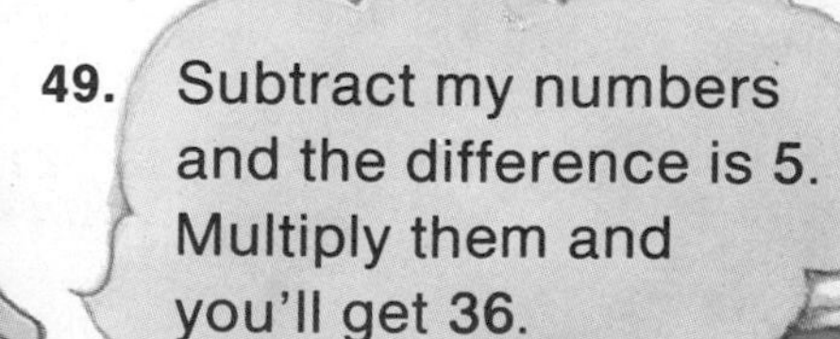

49. Subtract my numbers and the difference is 5. Multiply them and you'll get 36.

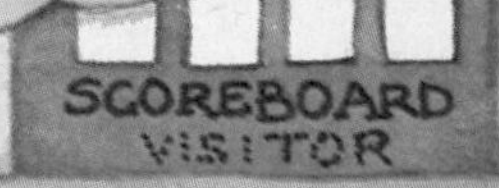

Multiplying 8 and 9

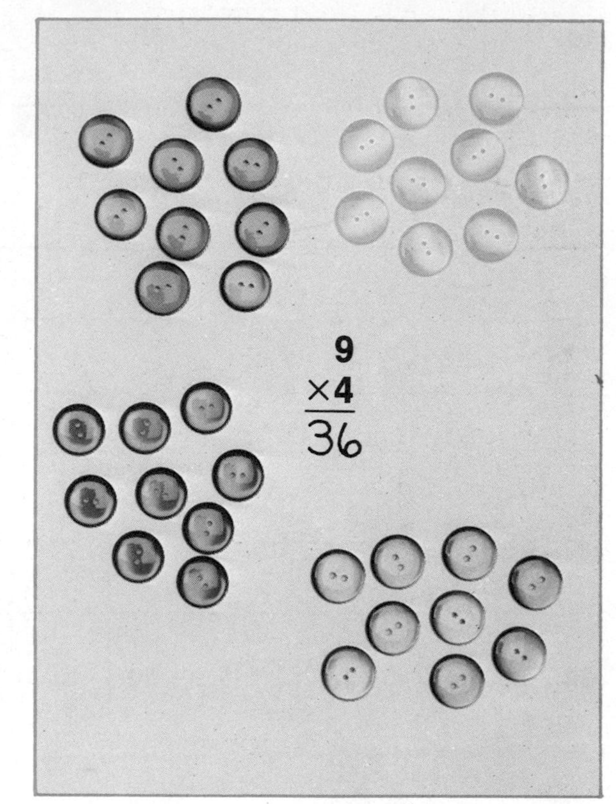

EXERCISES

Multiply.

1. $\begin{array}{r}8\\ \times 2\\ \hline\end{array}$	2. $\begin{array}{r}9\\ \times 1\\ \hline\end{array}$	3. $\begin{array}{r}8\\ \times 9\\ \hline\end{array}$	4. $\begin{array}{r}8\\ \times 1\\ \hline\end{array}$	5. $\begin{array}{r}7\\ \times 9\\ \hline\end{array}$	6. $\begin{array}{r}9\\ \times 4\\ \hline\end{array}$
7. $\begin{array}{r}9\\ \times 6\\ \hline\end{array}$	8. $\begin{array}{r}7\\ \times 5\\ \hline\end{array}$	9. $\begin{array}{r}8\\ \times 3\\ \hline\end{array}$	10. $\begin{array}{r}6\\ \times 7\\ \hline\end{array}$	11. $\begin{array}{r}9\\ \times 8\\ \hline\end{array}$	12. $\begin{array}{r}6\\ \times 6\\ \hline\end{array}$
13. $\begin{array}{r}8\\ \times 8\\ \hline\end{array}$	14. $\begin{array}{r}8\\ \times 4\\ \hline\end{array}$	15. $\begin{array}{r}7\\ \times 4\\ \hline\end{array}$	16. $\begin{array}{r}9\\ \times 2\\ \hline\end{array}$	17. $\begin{array}{r}6\\ \times 5\\ \hline\end{array}$	18. $\begin{array}{r}8\\ \times 7\\ \hline\end{array}$
19. $\begin{array}{r}6\\ \times 8\\ \hline\end{array}$	20. $\begin{array}{r}9\\ \times 3\\ \hline\end{array}$	21. $\begin{array}{r}7\\ \times 3\\ \hline\end{array}$	22. $\begin{array}{r}8\\ \times 5\\ \hline\end{array}$	23. $\begin{array}{r}7\\ \times 6\\ \hline\end{array}$	24. $\begin{array}{r}7\\ \times 8\\ \hline\end{array}$
25. $\begin{array}{r}9\\ \times 5\\ \hline\end{array}$	26. $\begin{array}{r}6\\ \times 9\\ \hline\end{array}$	27. $\begin{array}{r}8\\ \times 6\\ \hline\end{array}$	28. $\begin{array}{r}9\\ \times 9\\ \hline\end{array}$	29. $\begin{array}{r}7\\ \times 7\\ \hline\end{array}$	30. $\begin{array}{r}9\\ \times 7\\ \hline\end{array}$

31. Copy and complete this multiplication table. (Multiply the "red numbers.")

×	0	1	2	3	4	5	6	7	8	9
0	0						0			
1										
2										
3					12					
4									32	
5										
6			12							
7						35				
8										72
9								63		

Keep your multiplication table to review the multiplication facts.

Solve.

32. There were 8 bottles in one carton and 7 bottles in another. How many bottles were there in all?

33. There were 8 bottles in each of 7 cartons. How many bottles were there in all?

34. There were 9 pencils in a large box. There were 8 pencils in a small box. How many more pencils were there in the large box?

35. There were 9 pencils in each large box. There were 8 large boxes. How many pencils were there in the large boxes?

Practicing multiplication facts

You can use this page over and over again to practice multiplication facts. Cover a column with a piece of paper. First give the product. Then slide the paper down to check your answer.

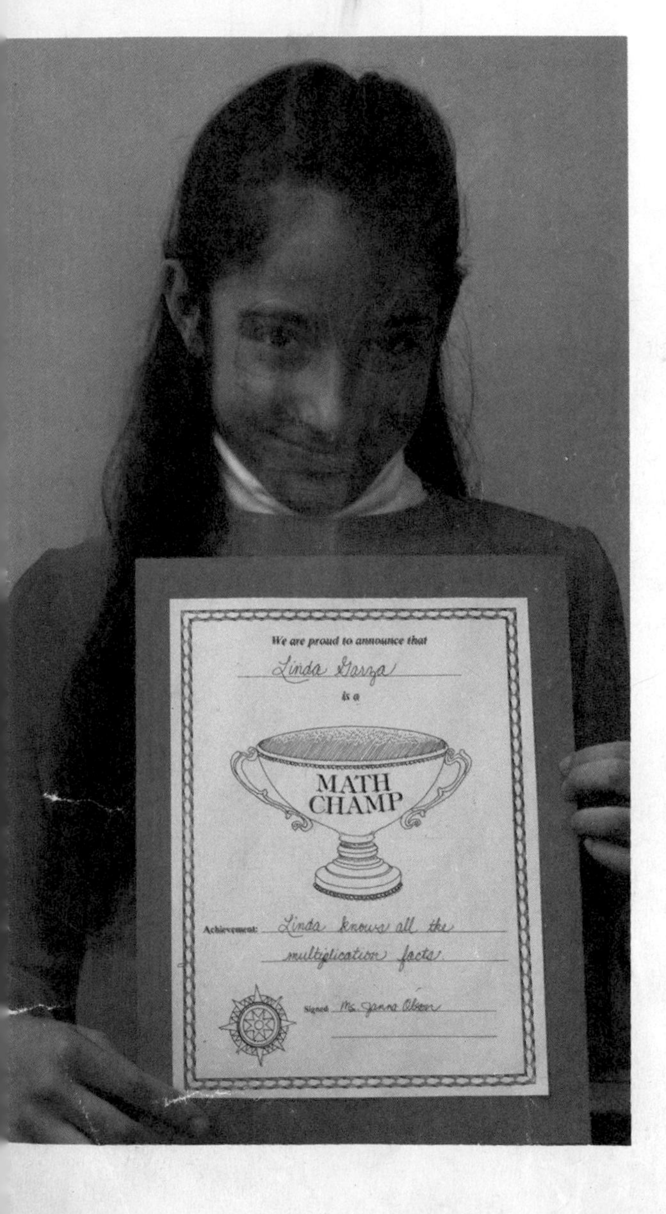

6	7	8	9
6 ×0 0	7 ×0 0	8 ×0 0	9 ×0 0
6 ×1 6	7 ×1 7	8 ×1 8	9 ×1 9
6 ×2 12	7 ×2 14	8 ×2 16	9 ×2 18
6 ×3 18	7 ×3 21	8 ×3 24	9 ×3 27
6 ×4 24	7 ×4 28	8 ×4 32	9 ×4 36
6 ×5 30	7 ×5 35	8 ×5 40	9 ×5 45
6 ×6 36	7 ×6 42	8 ×6 48	9 ×6 54
6 ×7 42	7 ×7 49	8 ×7 56	9 ×7 63
6 ×8 48	7 ×8 56	8 ×8 64	9 ×8 72
6 ×9 54	7 ×9 63	8 ×9 72	9 ×9 81

Speed drill

1. Number your paper 1 through 100.
2. Keep track of the seconds that it takes to write down the products.
3. Use pages 110 and 118 to check your answers.
4. To find your score, add the number of seconds and the number of mistakes. Keep trying for a better (lower) score.

1. 9×2	2. 2×1	3. 7×5	4. 0×2
5. 2×8	6. 1×5	7. 0×8	8. 2×5
9. 1×0	10. 0×3	11. 1×4	12. 8×2

13. 5×4	14. 4×8	15. 0×0	16. 0×1	17. 3×8	18. 7×7	19. 5×9	20. 8×5
21. 5×0	22. 6×1	23. 2×0	24. 9×8	25. 0×6	26. 9×3	27. 1×3	28. 9×1
29. 2×9	30. 6×4	31. 4×6	32. 8×4	33. 8×0	34. 2×6	35. 4×7	36. 3×4
37. 5×8	38. 9×5	39. 7×9	40. 7×0	41. 8×8	42. 8×6	43. 3×6	44. 9×0
45. 1×6	46. 4×1	47. 2×2	48. 0×9	49. 3×3	50. 5×3	51. 2×3	52. 5×2
53. 9×7	54. 2×4	55. 1×1	56. 8×7	57. 1×9	58. 6×3	59. 8×3	60. 4×9
61. 0×7	62. 1×8	63. 3×9	64. 4×0	65. 9×6	66. 1×2	67. 5×5	68. 7×8
69. 2×7	70. 6×0	71. 3×1	72. 9×9	73. 1×7	74. 4×4	75. 3×7	76. 7×1
77. 5×6	78. 4×5	79. 7×6	80. 6×7	81. 7×2	82. 3×2	83. 4×2	84. 6×2
85. 0×4	86. 3×5	87. 5×1	88. 6×9	89. 6×6	90. 8×1	91. 6×5	92. 0×5
93. 4×3	94. 5×7	95. 6×8	96. 8×9	97. 7×4	98. 7×3	99. 3×0	100. 9×4

Problem solving

The people that work at Manley Boyce's service station provide many services to the community. They sell gas, oil, tires, and other items for cars. They work on car engines to keep them running smoothly. Service stations also give car owners tips on safety and saving energy.

Solve.

1. The station is open from 6:00 A.M. until 10:00 P.M. How many hours is the station open?

2. It takes 25 gallons of gasoline to fill a large tank and 17 gallons to fill a small tank. How many more gallons does it take to fill the large tank?

Sometimes people use credit cards to pay for items and repairs at service stations.

Fill in the totals on these credit-card slips.

3. Credit-Card Sale
583 2806

Paul Beckett

Paul Beckett

Manley Boyce
44 Great Road
Acton, MA 01720

Gas	$19	60
Oil	1	75
Battery	72	65
Tax	3	68
Total		

4. Credit-Card Sale
982 8244

Doris Kroner

Doris Kroner

Manley Boyce
44 Great Road
Acton, MA 01720

Gas	$15	30
Oil	5	25
Tire repair	3	75
Tax		19
Total		

Mr. Boyce needs more workers during the busy hours. He made this picture graph to help him find his busy time. It shows the number of cars that bought gas during certain times on an average day.

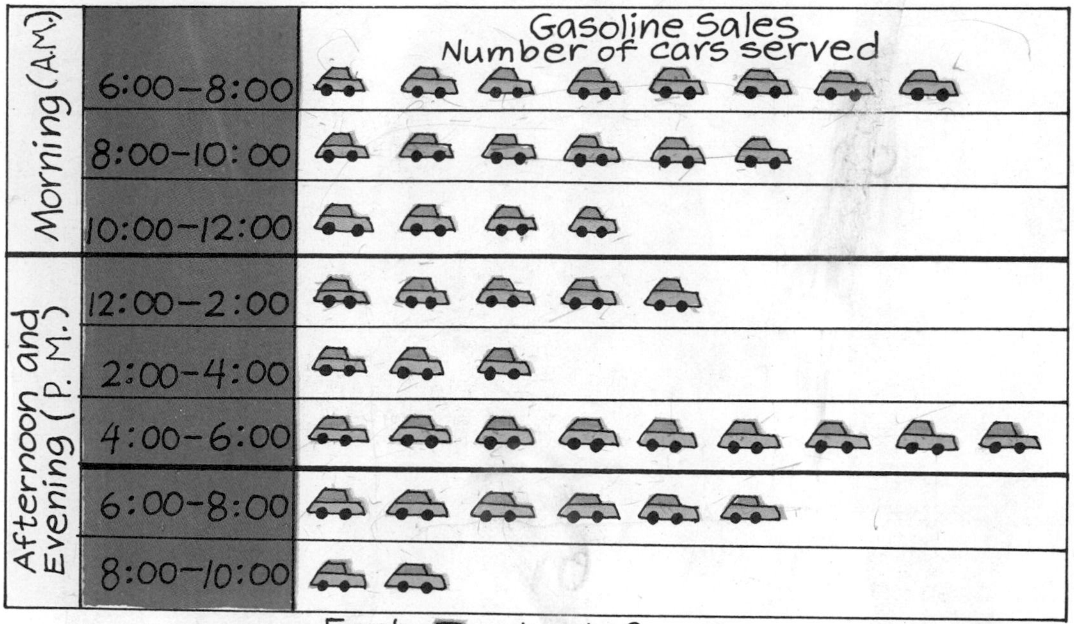

Gasoline Sales
Number of cars served

Morning (A.M.)	6:00–8:00
	8:00–10:00
	10:00–12:00
Afternoon and Evening (P.M.)	12:00–2:00
	2:00–4:00
	4:00–6:00
	6:00–8:00
	8:00–10:00

Each [car] stands for 5 cars.

5. What two-hour time period is the station's busiest time?

6. What two-hour time period is the station's slowest time?

7. How many cars were served during the first two hours the station was open?

8. How many cars were served during the first four hours the station was open?

9. Were there more cars during 8:00–10:00 A.M. or 8:00–10:00 P.M.? How many more?

10. How many cars were served during the evening (last four) hours of the day?

11. How many cars were served during the morning (A.M.) hours?

12. How many cars were served during the afternoon and evening (P.M.) hours?

13. At two times during the day, the same number of cars were served. What were they?

14. In what four-hour time period were 45 cars served?

★ 15. In what four-hour time period were 75 cars served?

★ 16. What was the total number of cars that were served that day?

Missing factors

EXERCISES

Give each missing factor.

1. $8 \times \underline{?} = 16$
2. $9 \times \underline{?} = 27$
3. $6 \times \underline{?} = 0$
4. $7 \times \underline{?} = 21$
5. $5 \times \underline{?} = 30$
6. $5 \times \underline{?} = 25$
7. $5 \times \underline{?} = 45$
8. $6 \times \underline{?} = 18$
9. $8 \times \underline{?} = 32$
10. $8 \times \underline{?} = 56$
11. $4 \times \underline{?} = 36$
12. $6 \times \underline{?} = 36$
13. $9 \times \underline{?} = 72$
14. $8 \times \underline{?} = 40$
15. $8 \times \underline{?} = 64$
16. $8 \times \underline{?} = 48$
17. $9 \times \underline{?} = 81$
18. $7 \times \underline{?} = 49$
19. $7 \times \underline{?} = 63$
20. $6 \times \underline{?} = 42$
21. $6 \times \underline{?} = 54$
22. $5 \times \underline{?} = 35$
23. $7 \times \underline{?} = 28$
24. $3 \times \underline{?} = 24$

Give each missing factor.

25. $4 \times \underline{?} = 12$
26. $6 \times \underline{?} = 48$
27. $9 \times \underline{?} = 63$
28. $\underline{?} \times 5 = 20$
29. $\underline{?} \times 9 = 18$
30. $\underline{?} \times 5 = 15$
31. $7 \times \underline{?} = 56$
32. $5 \times \underline{?} = 40$
33. $9 \times \underline{?} = 54$
34. $\underline{?} \times 2 = 14$
35. $\underline{?} \times 7 = 35$
36. $\underline{?} \times 4 = 24$
37. $9 \times \underline{?} = 45$
38. $7 \times \underline{?} = 42$
39. $4 \times \underline{?} = 32$
40. $\underline{?} \times 2 = 12$
41. $\underline{?} \times 4 = 0$
42. $\underline{?} \times 4 = 16$
43. $3 \times \underline{?} = 27$
44. $4 \times \underline{?} = 28$
45. $8 \times \underline{?} = 72$

Solve.

46. A baseball card costs 4¢. How many cards can be bought for 32¢?

47. Three tennis balls fill a can. How many cans can be filled with 25 balls? How many balls will be left over?

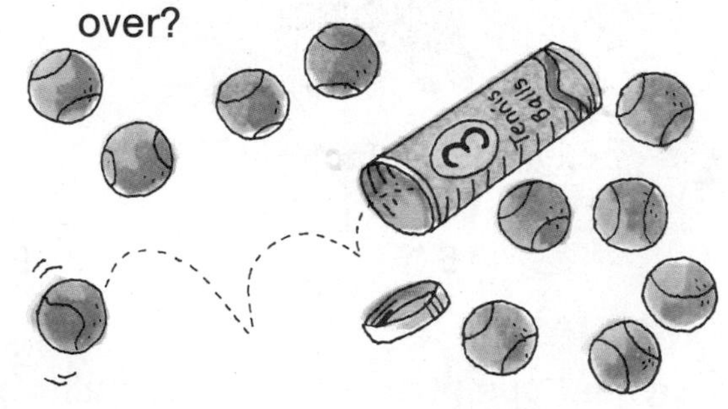

Challenge!

Copy and complete.

48.

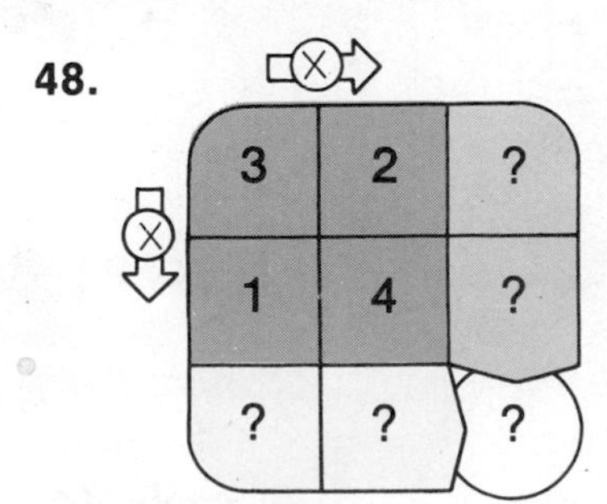

49.

3	?	9
?	2	?
6	?	?

50.

3	?	?
?	?	5
3	?	30

Multiples

Here are some **multiples** of 2 and 3.

Multiples of 2

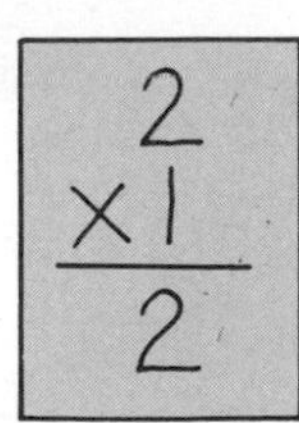

$\begin{array}{r} 2 \\ \times 1 \\ \hline 2 \end{array}$	$\begin{array}{r} 2 \\ \times 2 \\ \hline 4 \end{array}$	$\begin{array}{r} 2 \\ \times 3 \\ \hline 6 \end{array}$	$\begin{array}{r} 2 \\ \times 4 \\ \hline 8 \end{array}$	$\begin{array}{r} 2 \\ \times 5 \\ \hline 10 \end{array}$	$\begin{array}{r} 2 \\ \times 6 \\ \hline 12 \end{array}$

Multiples of 3

$\begin{array}{r} 3 \\ \times 1 \\ \hline 3 \end{array}$	$\begin{array}{r} 3 \\ \times 2 \\ \hline 6 \end{array}$	$\begin{array}{r} 3 \\ \times 3 \\ \hline 9 \end{array}$	$\begin{array}{r} 3 \\ \times 4 \\ \hline 12 \end{array}$	$\begin{array}{r} 3 \\ \times 5 \\ \hline 15 \end{array}$	$\begin{array}{r} 3 \\ \times 6 \\ \hline 18 \end{array}$

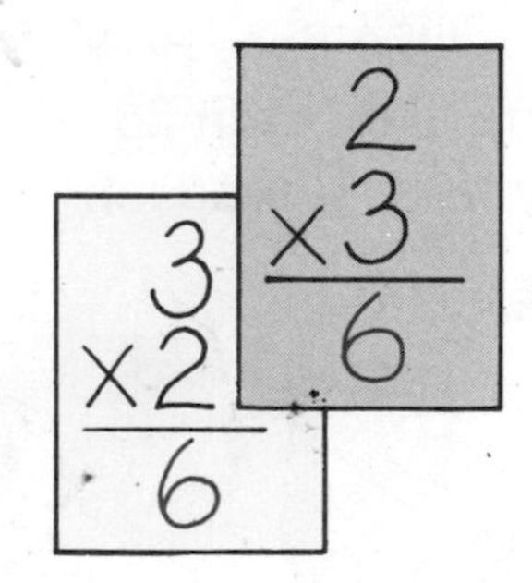

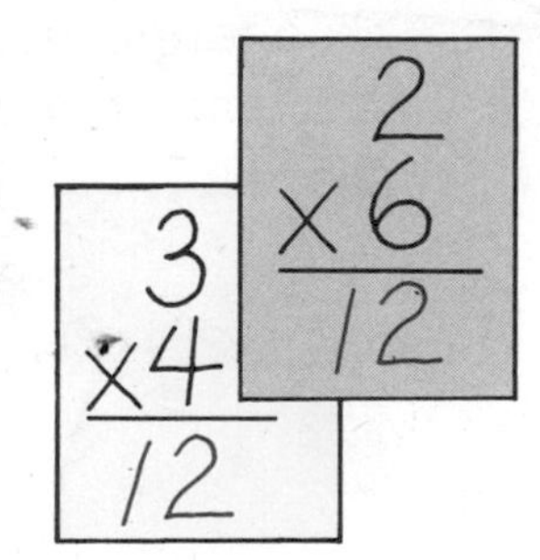

6 and 12 are multiples of both 2 and 3. They are called **common multiples.**

6 is called the **least common multiple** of 2 and 3.

EXERCISES

Copy and complete each list of multiples.

1. 3, 6, 9, ?, ?, ?, ?, ?, ?

2. 4, 8, 12, ?, ?, ?, ?, ?, ?

3. 6, 12, 18, ?, ?, ?, ?, ?, ?

4. 8, 16, 24, ?, ?, ?, ?, ?, ?

5. 5, 10, 15, ?, ?, ?, ?, ?, ?

6. 9, 18, 27, ?, ?, ?, ?, ?, ?

7. 7, 14, 21, ?, ?, ?, ?, ?, ?

Use your lists of multiples to find the first two common multiples of

8. 4 and 6
9. 2 and 4
10. 4 and 8
11. 3 and 6
12. 2 and 5
13. 2 and 6

Use your lists of multiples to find the least common multiple of

14. 3 and 5
15. 4 and 5
16. 4 and 6
17. 3 and 8
18. 4 and 8
19. 6 and 9
20. 3 and 9
21. 6 and 7

0 and multiples of 2 are called **even numbers.** The other whole numbers are called **odd numbers.**

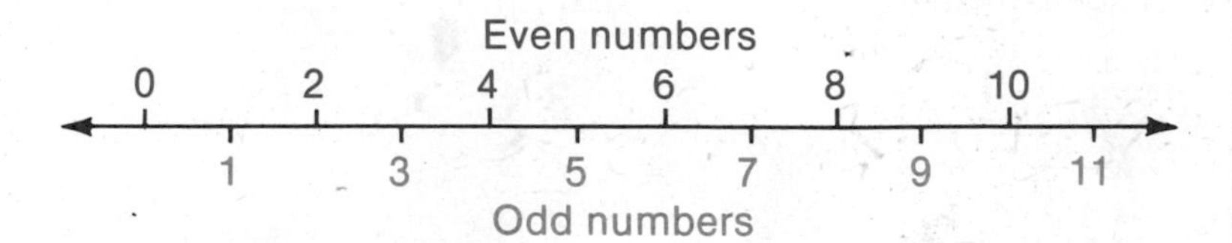

Even or odd?

22. 6
23. 7
24. 9
25. 12
26. 18
27. 13
28. 16
29. 19
30. 23
31. 26
32. 41
33. 54
34. 70
35. 75
36. 83
★ 37. 168
★ 38. 241
★ 39. 350

KEEPING SKILLS SHARP

< or >?

1. 87 ● 78
2. 56 ● 65
3. 309 ● 390
4. 542 ● 544
5. 682 ● 79
6. 742 ● 651
7. 6905 ● 6991
8. 5934 ● 5926
9. 8342 ● 8432
10. 783 ● 2873
11. 6914 ● 6710
12. 3999 ● 4000
13. 56,938 ● 6884
14. 34,291 ● 35,384
15. 72,916 ● 75,388
16. 26,593 ● 26,504
17. 458,371 ● 453,871
18. 100,000 ● 99,999

Multiplying a 2-digit number

The examples show how to multiply a 2-digit number by a 1-digit number.

Jennifer and David each bought 43 energy stamps. How many stamps did they buy in all?

EXAMPLE 1.

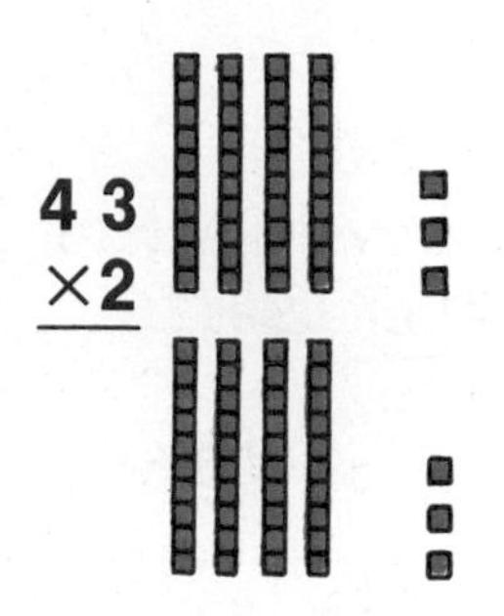

Step 1.

Multiply to find how many ones.

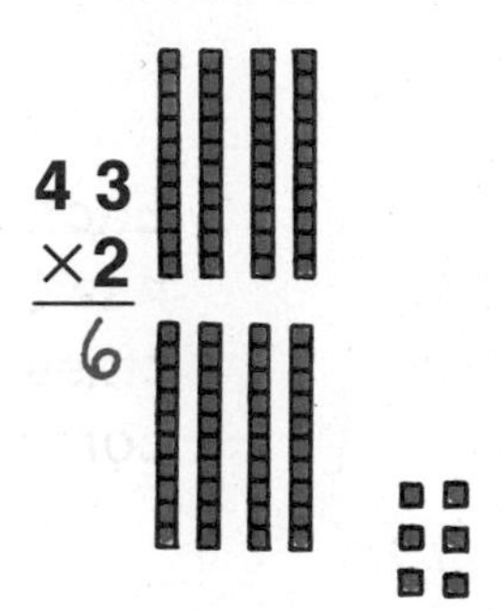

Step 2.

Multiply to find how many tens.

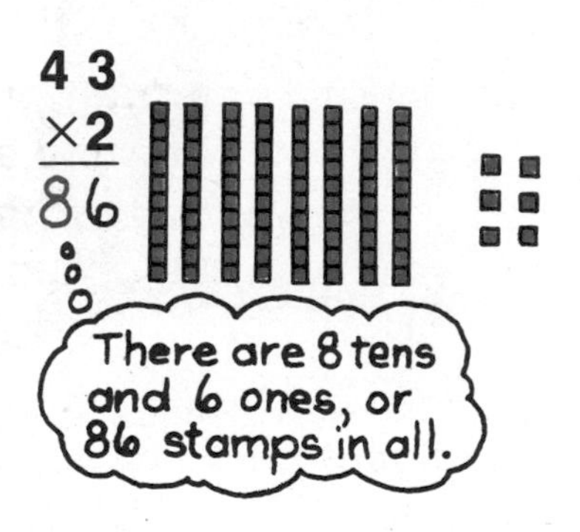

EXAMPLE 2.

31
×3

Step 1.

Multiply the ones.

31
×3
3

Step 2.

Multiply the tens.

31
×3

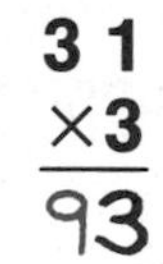

EXERCISES

Multiply.

1. 12 ×2	2. 21 ×3	3. 11 ×3	4. 20 ×4	5. 11 ×9	6. 12 ×3
7. 22 ×2	8. 10 ×7	9. 31 ×3	10. 10 ×9	11. 11 ×8	12. 30 ×2

Give each product.

13. 32×3	14. 23×2	15. 43×2	16. 20×4	17. 22×2
18. 10×6	19. 12×3	20. 10×8	21. 32×2	22. 21×4
23. 34×2	24. 11×5	25. 22×3	26. 20×3	27. 12×2
28. 12×4	29. 11×6	30. 22×4	31. 42×2	32. 23×3

Solve.

33. David's computer program had 43 steps. He found a 'bug' and wrote 3 new steps to fix the bug. How many steps did he have in all?

34. Each classroom in David's school has 3 computers. There are 32 classrooms. How many computers are in David's school?

35. Jennifer can save 12 programs on each computer disk. How many programs can she save on 4 disks?

36. David has 9 more disks than Jennifer. If Jennifer has 17 disks, how many disks does David have?

Challenge!

Find the missing number.

37.

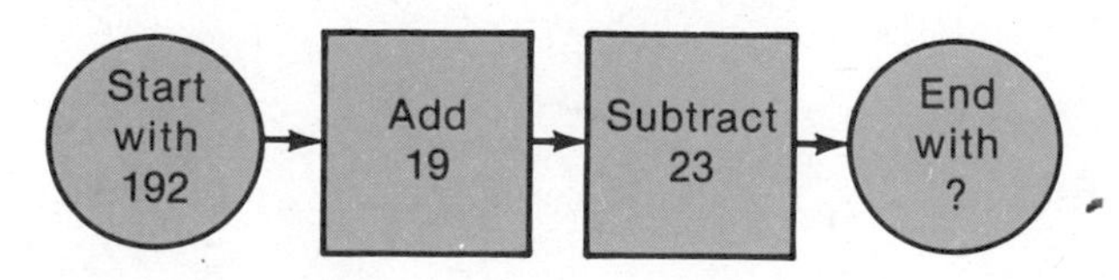

More about multiplying

EXAMPLE.

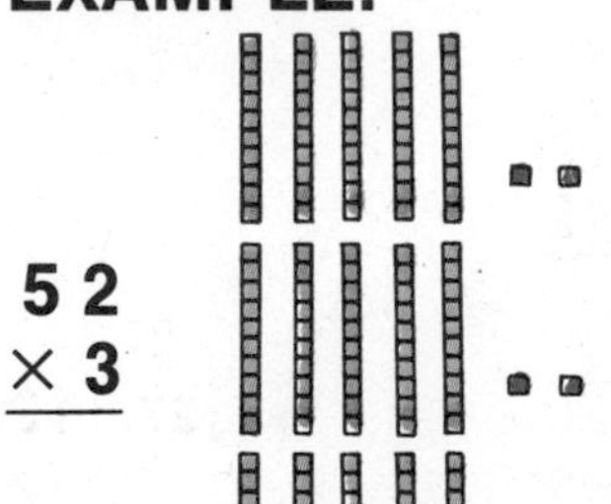

$$\begin{array}{r} 52 \\ \times\ 3 \\ \hline \end{array}$$

Step 1. Multiply to find how many ones.

$$\begin{array}{r} 52 \\ \times\ 3 \\ \hline 6 \end{array}$$

Step 2. Multiply to find how many tens.

15 tens, or 1 hundred and 5 tens.

$$\begin{array}{r} 52 \\ \times\ 3 \\ \hline 156 \end{array}$$

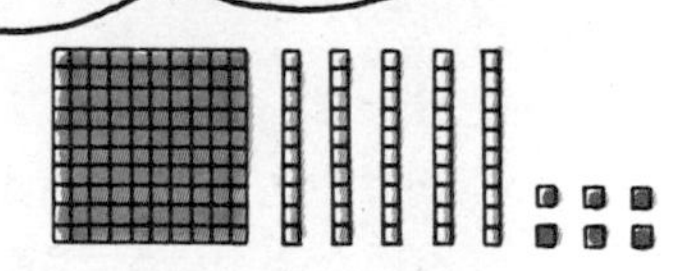

There are 156 blocks in all.

EXERCISES

Multiply.

1. $\begin{array}{r} 42 \\ \times 3 \\ \hline \end{array}$
2. $\begin{array}{r} 21 \\ \times 5 \\ \hline \end{array}$
3. $\begin{array}{r} 52 \\ \times 4 \\ \hline \end{array}$
4. $\begin{array}{r} 43 \\ \times 3 \\ \hline \end{array}$
5. $\begin{array}{r} 64 \\ \times 2 \\ \hline \end{array}$
6. $\begin{array}{r} 41 \\ \times 4 \\ \hline \end{array}$
7. $\begin{array}{r} 50 \\ \times 3 \\ \hline \end{array}$
8. $\begin{array}{r} 54 \\ \times 2 \\ \hline \end{array}$
9. $\begin{array}{r} 62 \\ \times 4 \\ \hline \end{array}$
10. $\begin{array}{r} 31 \\ \times 5 \\ \hline \end{array}$

11. 31 ×7	12. 64 ×2	13. 21 ×5	14. 50 ×8	15. 40 ×6
16. 53 ×2	17. 83 ×3	18. 60 ×5	19. 73 ×2	20. 43 ×3
21. 60 ×6	22. 50 ×5	23. 40 ×8	24. 80 ×7	25. 70 ×4

Give each product.

26. 52 × 4 27. 40 × 7 28. 63 × 3 29. 72 × 2 30. 31 × 8

Solve.

31. What would the refund be for 32 bottles?
32. What would the refund be for 21 cans?
33. What would the refund be for 12 bottles and 12 cans?
34. What would the refund be for 31 bottles and 23 cans?
35. John returned 11 bottles and Ruth returned 32 cans. How much more money did Ruth get?
★ 36. Sarah returned a total of 5 bottles and cans. The total refund was 19¢. How many of each did she return?

KEEPING SKILLS SHARP

1. 58 +21	2. 72 −19	3. 85 +28	4. 761 −385	5. 297 +378
6. 1859 −376	7. 2485 +999	8. 5018 −3596	9. 8421 +3678	10. 5214 −3786

Multiplying with regrouping

EXAMPLE.

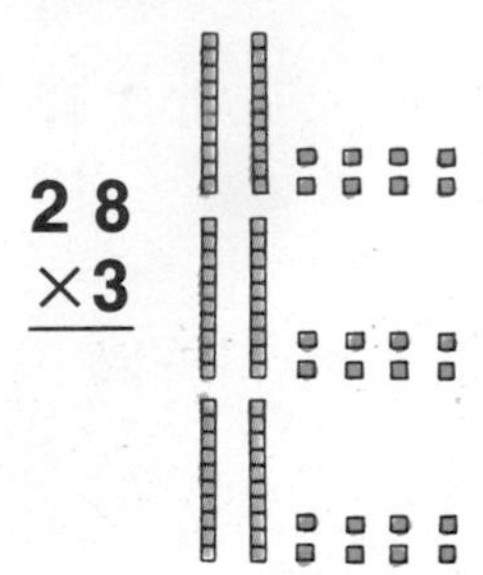

Step 1.

Multiply the ones and regroup.

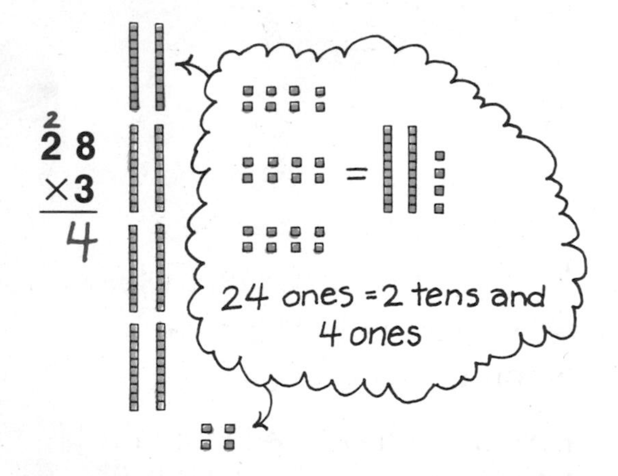

Step 2.

Multiply the tens and add.

$$\begin{array}{r} \overset{2}{2}8 \\ \times 3 \\ \hline 84 \end{array}$$

There are 84 blocks in all.

Rounding may be used to help estimate a product. An estimate can help you decide whether you made a mistake.

$$\begin{array}{r} 28 \\ \times 3 \\ \hline \end{array} \quad \text{Round to} \quad \begin{array}{r} 30 \\ \times 3 \\ \hline 90 \end{array}$$

EXERCISES

Multiply. Then estimate the product to see if your answer makes sense.

1. $\begin{array}{r} 46 \\ \times 2 \\ \hline \end{array}$
2. $\begin{array}{r} 81 \\ \times 4 \\ \hline \end{array}$
3. $\begin{array}{r} 37 \\ \times 5 \\ \hline \end{array}$
4. $\begin{array}{r} 92 \\ \times 7 \\ \hline \end{array}$
5. $\begin{array}{r} 54 \\ \times 5 \\ \hline \end{array}$
6. $\begin{array}{r} 61 \\ \times 6 \\ \hline \end{array}$
7. $\begin{array}{r} 38 \\ \times 4 \\ \hline \end{array}$
8. $\begin{array}{r} 31 \\ \times 9 \\ \hline \end{array}$
9. $\begin{array}{r} 75 \\ \times 6 \\ \hline \end{array}$
10. $\begin{array}{r} 86 \\ \times 7 \\ \hline \end{array}$
11. $\begin{array}{r} 78 \\ \times 3 \\ \hline \end{array}$
12. $\begin{array}{r} 65 \\ \times 7 \\ \hline \end{array}$
13. $\begin{array}{r} 73 \\ \times 6 \\ \hline \end{array}$
14. $\begin{array}{r} 48 \\ \times 3 \\ \hline \end{array}$
15. $\begin{array}{r} 68 \\ \times 7 \\ \hline \end{array}$
16. $\begin{array}{r} 80 \\ \times 7 \\ \hline \end{array}$
17. $\begin{array}{r} 53 \\ \times 9 \\ \hline \end{array}$
18. $\begin{array}{r} 43 \\ \times 2 \\ \hline \end{array}$
19. $\begin{array}{r} 70 \\ \times 4 \\ \hline \end{array}$
20. $\begin{array}{r} 46 \\ \times 8 \\ \hline \end{array}$
21. $\begin{array}{r} \$.63 \\ \times 8 \\ \hline \end{array}$
22. $\begin{array}{r} \$.56 \\ \times 2 \\ \hline \end{array}$
23. $\begin{array}{r} \$.62 \\ \times 4 \\ \hline \end{array}$
24. $\begin{array}{r} \$.54 \\ \times 9 \\ \hline \end{array}$
25. $\begin{array}{r} \$.83 \\ \times 5 \\ \hline \end{array}$

Solve.

26.   

How many cents?

27. How many hours are there in 4 days?
28. How many minutes are there in 8 hours?
29. How many days are there in 24 weeks?

Challenge!

Use the shortcut to find each product.

30. 54×3

31. 63×5
32. 48×4
33. 72×6
34. 87×3

Problem solving

1. Study and understand.

2. Plan and do.

3. Answer and check.

Julie and Matt set up a vegetable stand.

1. They had 72 tomatoes in one basket and 59 in another. How many tomatoes were there in all?
2. One day they sold 39 green peppers and 28 red peppers. How many peppers were sold?
3. A customer bought 3 dozen tomatoes. Matt took them out of the basket that contained 72 tomatoes. How many were left in the basket?
4. One week they sold 75 tomatoes a day. How many tomatoes did they sell that week?
5. Mrs. Yazzie bought 3 pounds of peas for 65¢ a pound. How much did the peas cost?
6. Mrs. Yazzie paid with a $5 bill. How much change did she get? [See exercise 5.]

7. Mr. Martinez bought 4 dozen ears of corn for $.95 a dozen and 2 pounds of tomatoes for 75¢ a pound. How much did the corn cost in all?

8. Miss Peck bought a dozen ears of corn for $.95, 2 pounds of peas for $.65 each, and 3 peppers for $.13 each. How much did she owe?

9. Miss Peck gave Julie $3.00. How much change did she get? [See exercise 8.]

★ 10. Matt and Julie sold bunches of beets in two sizes. The large bunches had 9 beets; the small ones had 5 beets. If they had 15 large bunches and 21 small bunches, how many beets were there in all?

CHAPTER CHECKUP

Multiply. [pages 106–119, 122–125]

1. 3×7
2. 5×5
3. 6×8
4. 4×6
5. 8×9
6. 7×6
7. 9×6
8. 4×9
9. 6×3
10. 8×5
11. 2×9
12. 9×3
13. 8×4
14. 4×4
15. 9×9
16. 7×4
17. 8×2
18. 9×7

Multiply. [pages 126–131]

19. 23×2
20. 30×3
21. 12×4
22. 20×4
23. 11×6
24. 42×2
25. 54×2
26. 40×3
27. 31×6
28. 62×4
29. 73×3
30. 81×5
31. 74×8
32. 83×5
33. 49×5
34. 62×7
35. 78×6
36. 59×8

Solve. [pages 120–121, 132–133]

37. There were 6 books on each of 4 shelves. How many books were there?
38. There were 9 pencils in one box and 7 pencils in another. How many pencils were there?
39. There were 24 crayons on a big table and 3 crayons on a small table. How many crayons were there?
40. There were 8 cards with 32 tacks on each card. How many tacks were there?

CHAPTER PROJECT

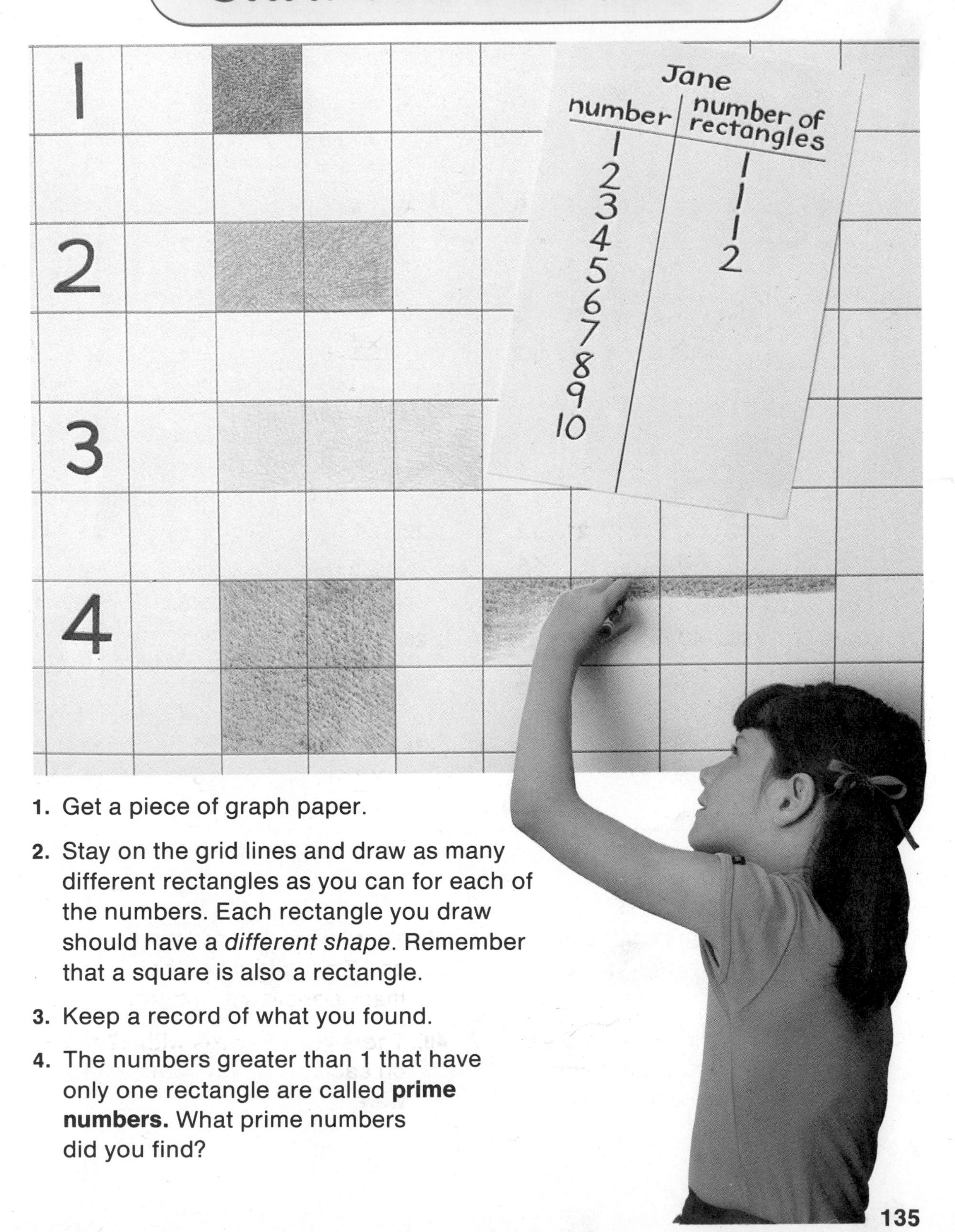

1. Get a piece of graph paper.
2. Stay on the grid lines and draw as many different rectangles as you can for each of the numbers. Each rectangle you draw should have a *different shape*. Remember that a square is also a rectangle.
3. Keep a record of what you found.
4. The numbers greater than 1 that have only one rectangle are called **prime numbers.** What prime numbers did you find?

CHAPTER REVIEW

6
×3
18

Multiply.

1. 4×2	2. 6×3	3. 8×4	4. 7×3	5. 6×5
6. 4×9	7. 5×8	8. 7×7	9. 7×8	10. 5×7
11. 8×8	12. 7×9	13. 8×9	14. 6×9	15. 7×6

14
×2
28

Multiply.

16. 10×5	17. 13×2	18. 12×4	19. 23×3	20. 22×4
21. 31×6	22. 52×2	23. 70×3	24. 61×2	25. 53×3

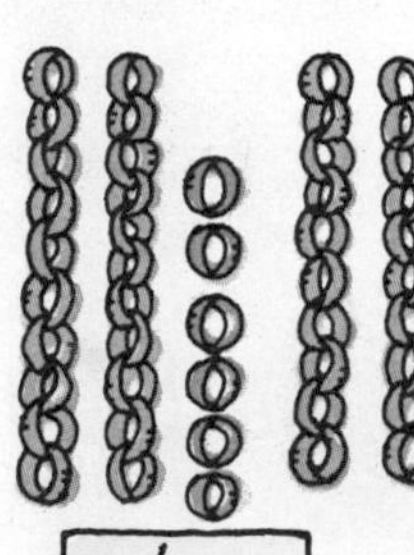

1
26
×2
52

Multiply.

26. 28×2	27. 34×3	28. 46×8	29. 57×7	30. 48×9
31. 63×7	32. 59×8	33. 85×5	34. 93×6	35. 74×4

CHAPTER CHALLENGE

Here is a **factor tree.**

First, 30 was written as 5×6. Then, 6 was written as 3×2.

Copy and complete each factor tree.
Do not use 1 as a factor.

1.

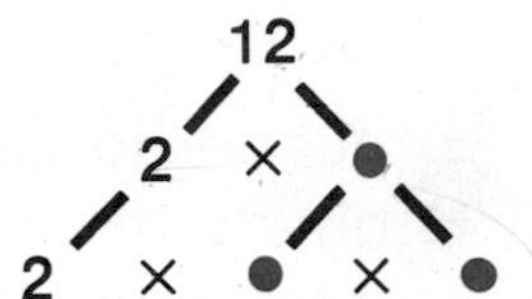

2.

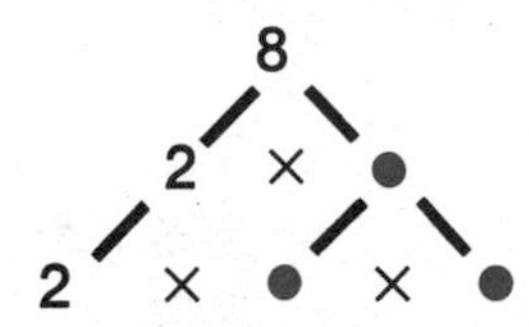

3.

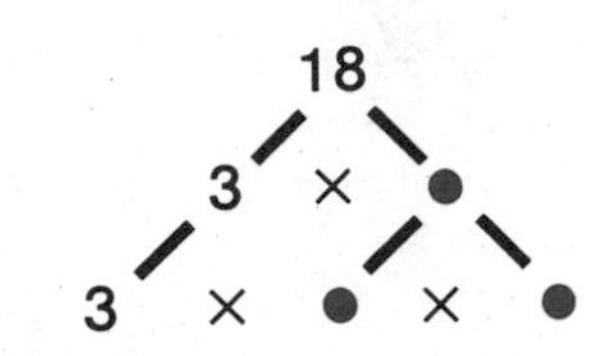

4.

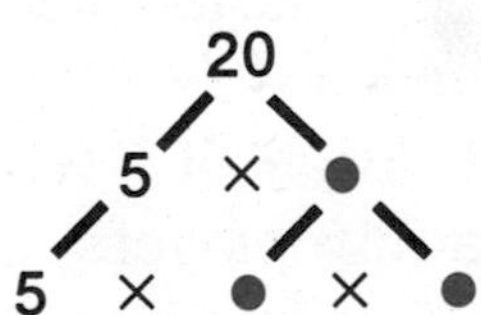

5.

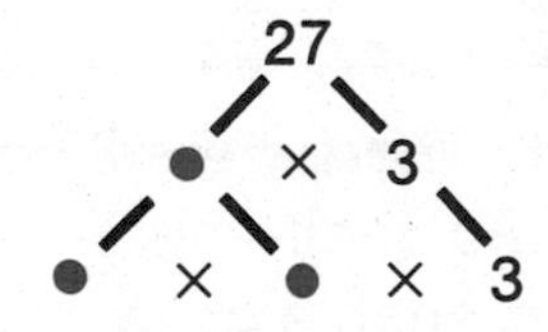

6.

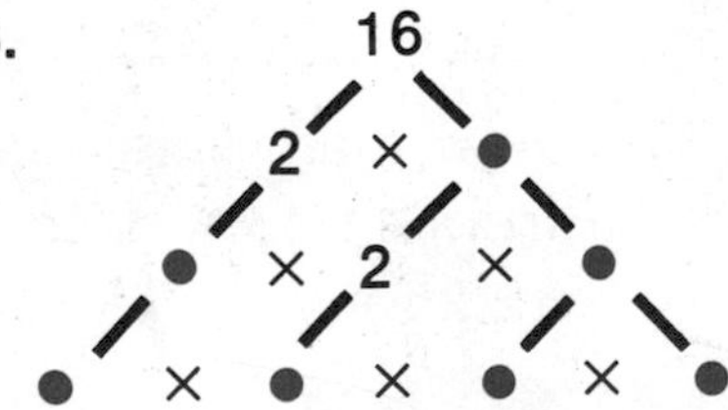

Make a factor tree for each of these numbers. Make each tree as large as possible. Do not use 1 as a factor.

7. 14 **8.** 30 **9.** 54 **10.** 32 **11.** 48 **12.** 64

MAJOR CHECKUP
STANDARDIZED FORMAT

2. a b c d 5. a b c d 6. a b c d
8. 9. 10. 11. 12.

Choose the correct letter.

1. 708 rounded to the nearest ten is
   - a. 700
   - b. 710
   - c. 800
   - d. none of these

2. 448 rounded to the nearest hundred is
   - a. 450
   - b. 400
   - c. 500
   - d. none of these

3. Which digit in 793,465 is in the hundred thousands place?
   - a. 7
   - b. 3
   - c. 9
   - d. 4

4. The standard numeral for five hundred five thousand, seventy-seven is
   - a. 55,077
   - b. 550,077
   - c. 505,707
   - d. none of these

5. Add.
$$\begin{array}{r} \$9.63 \\ +2.57 \\ \hline \end{array}$$
   - a. $12.20
   - b. $11.10
   - c. $11.20
   - d. none of these

6. Add.
$$\begin{array}{r} 593 \\ 269 \\ +742 \\ \hline \end{array}$$
   - a. 1494
   - b. 1604
   - c. 1504
   - d. none of these

7. Subtract.
$$\begin{array}{r} 5362 \\ -1785 \\ \hline \end{array}$$
   - a. 3577
   - b. 4423
   - c. 4687
   - d. none of these

8. To check this subtraction, you can add
$$\begin{array}{r} 603 \\ -259 \\ \hline 344 \end{array}$$
   - a. 344 and 603
   - b. 259 and 603
   - c. 344 and 259
   - d. none of these

9. What time is it?

   - a. quarter to three
   - b. quarter past three
   - c. quarter to ten
   - d. quarter past ten

10. How many minutes from 11:45 A.M. to 12:25 P.M.?
   - a. 40
   - b. 30
   - c. 50
   - d. 20

11. There were 8 books on a shelf and 5 books on a table. How many books were there?
   - a. 40
   - b. 3
   - c. 12
   - d. none of these

12. Each of 7 teams had 18 players. Each of 6 teams had 24 players. How many players were there?
   - a. 270
   - b. 126
   - c. 144
   - d. none of these

Division by 1-Digit Divisors

6

Dividing by 2, 3, and 4

12 apples have been divided equally into 3 baskets. You can write two division equations to show this.

baskets

$12 \div 4 = \underline{3}$

Read as "Twelve divided by four equals three."

$12 \div 3 = \underline{4}$

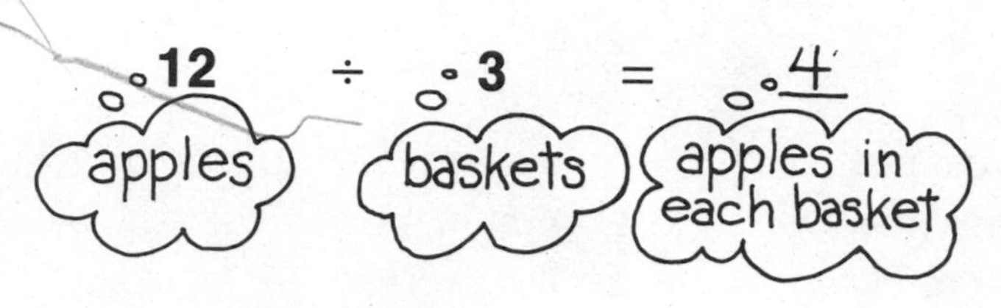

When you divide one number by another number, the answer is called the **quotient.**

Multiplication and division are related. You can find a quotient by finding a missing factor.

$16 \div 2 = \underline{?}$

The quotient is 8.

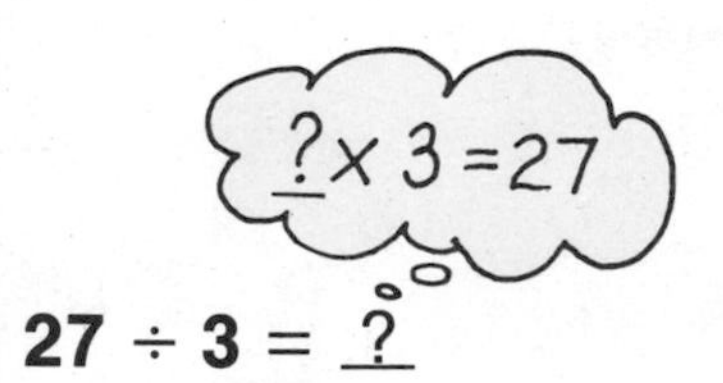

$27 \div 3 = \underline{?}$

The quotient is 9.

EXERCISES

Divide.

1. $20 \div 4$
2. $12 \div 2$
3. $16 \div 4$
4. $16 \div 2$
5. $24 \div 3$
6. $28 \div 4$
7. $21 \div 3$
8. $27 \div 3$
9. $14 \div 2$
10. $10 \div 2$
11. $18 \div 3$
12. $32 \div 4$

Give each quotient.

13. $14 \div 2$
14. $15 \div 3$
15. $6 \div 3$
16. $6 \div 2$
17. $12 \div 4$
18. $12 \div 3$
19. $18 \div 3$
20. $10 \div 2$
21. $8 \div 4$
22. $27 \div 3$
23. $36 \div 4$
24. $3 \div 3$
25. $27 \div 3$
26. $24 \div 4$
27. $9 \div 3$
28. $2 \div 2$
29. $4 \div 2$
30. $8 \div 2$
31. $24 \div 3$
32. $18 \div 2$
33. $4 \div 4$
34. $20 \div 4$
35. $16 \div 4$
36. $16 \div 2$

Solve.

37. There are 15 oranges. If you put 3 oranges in each package, how many packages will you have?

38. There are 28 baskets with 4 apples in each basket. How many apples are there?

39. There are 20 pears in a large box. If you divide the pears equally into 4 smaller boxes, then how many pears will be in each box?

40. There were 36 plums in each box. If there were 4 boxes, then how many plums were there in all?

Challenge!

Find the missing input or output.

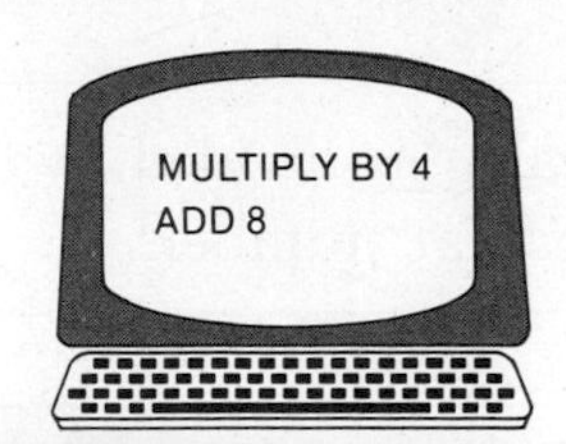

	Input	Output
	9	44
41.	7	?
42.	4	?
43.	?	40

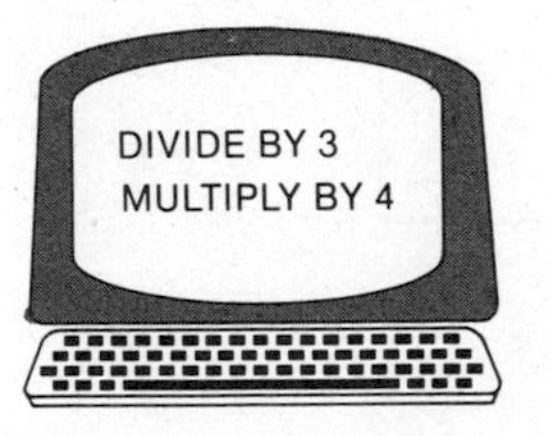

	Input	Output
	12	16
44.	9	?
45.	27	?
46.	?	4

More about division

Here is another way to write about division.

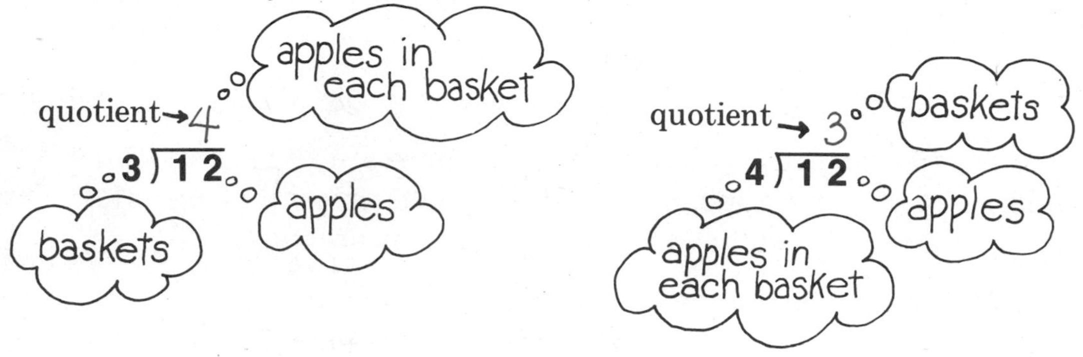

Here are some properties that make it easier to remember some division facts.

$3\overline{)0}$?	?×3=0	$3\overline{)0}$ 0	When 0 is divided by any number, the quotient is always 0.
$1\overline{)7}$?	?×1=7	$1\overline{)7}$ 7	When any number is divided by 1, the quotient is that number.
$4\overline{)4}$?	?×4=4	$4\overline{)4}$ 1	When any number is divided by itself, the quotient is always 1.
$0\overline{)6}$?	?×0=6 Impossible! Any number times 0 is 0, not 6.	~~$0\overline{)6}$~~	We don't divide by 0!

EXERCISES

Divide.

Do for homework Tommarrw

1. $3\overline{)18}$ 2. $3\overline{)21}$ 3. $1\overline{)6}$ 4. $3\overline{)24}$ 5. $4\overline{)20}$

6. $3\overline{)15}$ 7. $4\overline{)24}$ 8. $4\overline{)32}$ 9. $3\overline{)27}$ 10. $3\overline{)12}$

11. $4\overline{)28}$ 12. $1\overline{)7}$ 13. $3\overline{)0}$ 14. $1\overline{)4}$ 15. $4\overline{)12}$

16. $1\overline{)5}$ 17. $4\overline{)0}$ 18. $1\overline{)9}$ 19. $4\overline{)16}$ 20. $4\overline{)36}$

Give each quotient.

21. $2\overline{)0}$ 22. $4\overline{)8}$ 23. $3\overline{)6}$ 24. $4\overline{)4}$ 25. $3\overline{)21}$

26. $3\overline{)18}$ 27. $2\overline{)14}$ 28. $2\overline{)10}$ 29. $2\overline{)4}$ 30. $4\overline{)0}$

31. $3\overline{)24}$ 32. $1\overline{)3}$ 33. $4\overline{)24}$ 34. $4\overline{)20}$ 35. $3\overline{)9}$

36. $2\overline{)12}$ 37. $4\overline{)28}$ 38. $3\overline{)3}$ 39. $1\overline{)9}$ 40. $4\overline{)32}$

41. $3\overline{)0}$ 42. $2\overline{)16}$ 43. $3\overline{)12}$ 44. $4\overline{)36}$ 45. $2\overline{)6}$

46. $4\overline{)12}$ 47. $2\overline{)18}$ 48. $2\overline{)8}$ 49. $3\overline{)15}$ 50. $4\overline{)4}$

Challenge!

Find the missing input or output.

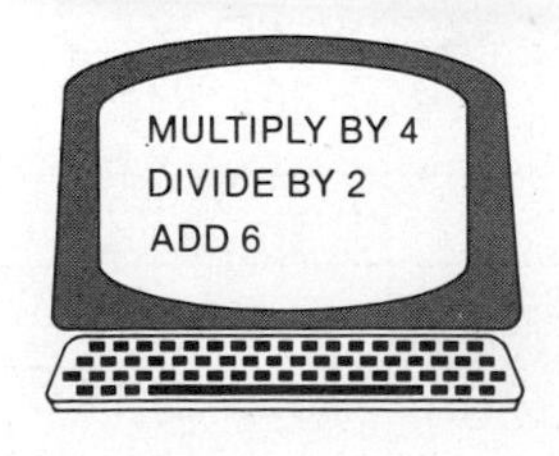

	Input	Output
	1	8
51.	4	?
52.	7	?
53.	?	18

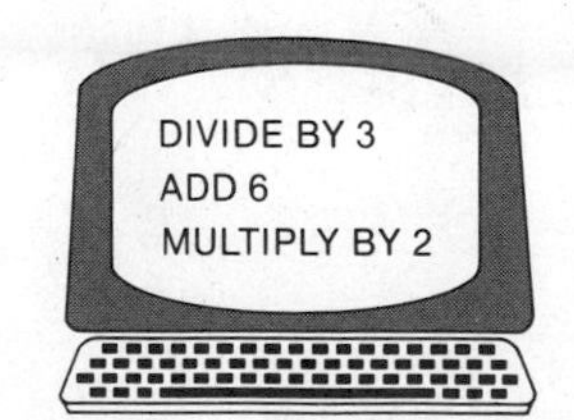

	Input	Output
	9	18
54.	3	?
55.	6	?
56.	?	20

Dividing by 5 and 6

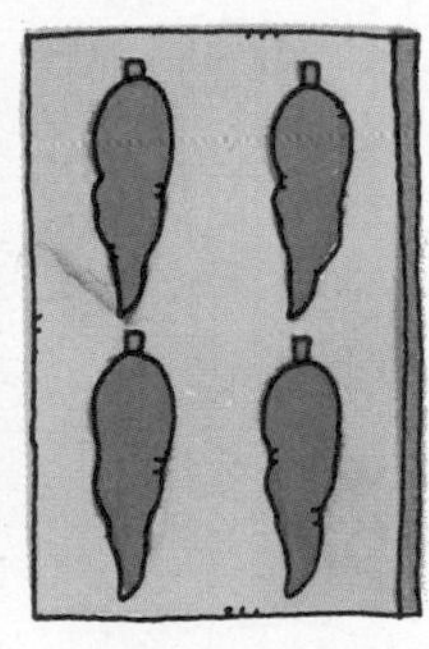
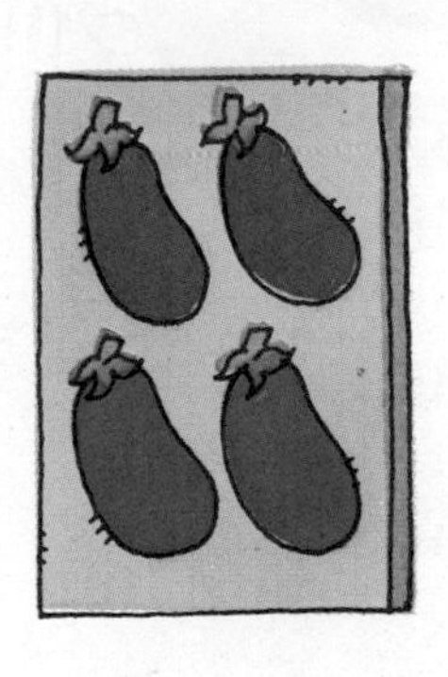
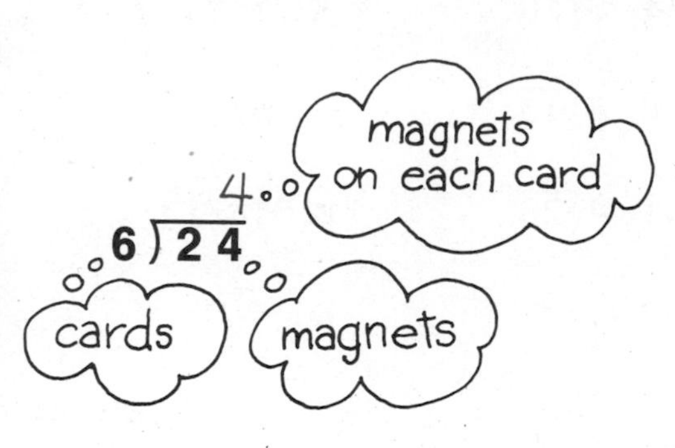

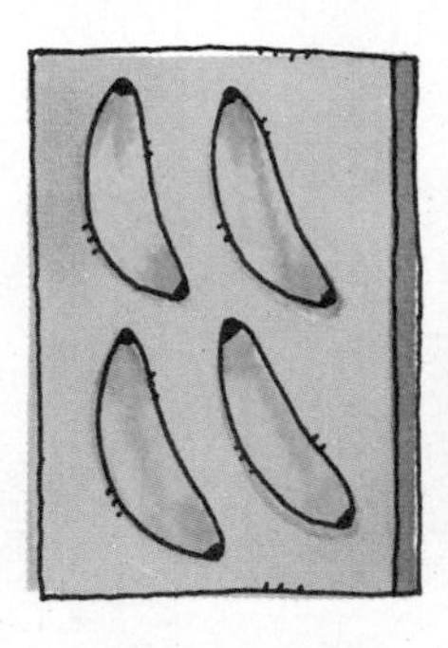
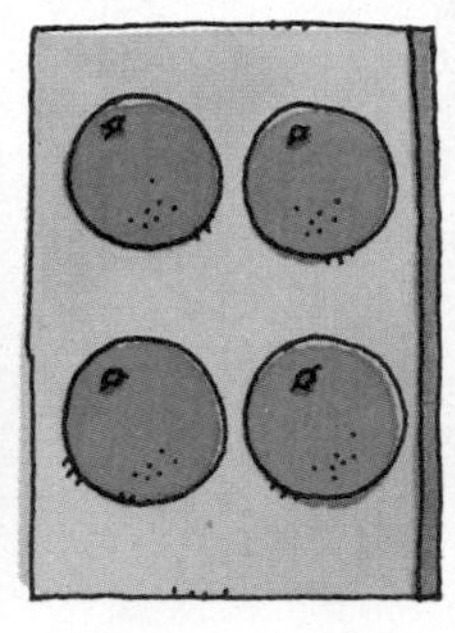

cards
6
4)24
magnets
magnets on each card

EXERCISES

Divide.

1. $5\overline{)20}$	**2.** $4\overline{)12}$	**3.** $3\overline{)24}$	**4.** $3\overline{)21}$	**5.** $4\overline{)16}$
6. $5\overline{)25}$	**7.** $2\overline{)14}$	**8.** $2\overline{)18}$	**9.** $2\overline{)12}$	**10.** $6\overline{)18}$
11. $4\overline{)20}$	**12.** $4\overline{)8}$	**13.** $6\overline{)24}$	**14.** $4\overline{)24}$	**15.** $5\overline{)5}$
16. $3\overline{)15}$	**17.** $3\overline{)18}$	**18.** $5\overline{)15}$	**19.** $5\overline{)30}$	**20.** $4\overline{)36}$
21. $6\overline{)36}$	**22.** $4\overline{)28}$	**23.** $5\overline{)10}$	**24.** $3\overline{)27}$	**25.** $6\overline{)42}$
26. $4\overline{)32}$	**27.** $5\overline{)0}$	**28.** $6\overline{)12}$	**29.** $2\overline{)10}$	**30.** $5\overline{)45}$
31. $5\overline{)35¢}$	**32.** $6\overline{)48¢}$	**33.** $3\overline{)12¢}$	**34.** $6\overline{)6¢}$	**35.** $2\overline{)16¢}$
36. $6\overline{)30¢}$	**37.** $5\overline{)40¢}$	**38.** $6\overline{)0¢}$	**39.** $6\overline{)54¢}$	**40.** $3\overline{)9¢}$

Solve.

41.   

How many stamps can you buy?

42.     

How many stamps can you buy?

43. 

How many stamps can you buy?

44. 

How many stamps can you buy?

45. How many 6¢ stamps can you buy with 24¢?

46. You have a quarter and a dime. How many 5¢ stamps can you buy?

47. Which costs more, nine 3¢ stamps or six 4¢ stamps?

48. You have a half-dollar and a penny. You buy seven 5¢ stamps. How much money will you have left?

Challenge!

Guess my number.

49. If you divide my number by 3, you get 7.

50. Divide my number by 6, add 3, and you get 10.

Dividing by 7, 8, and 9

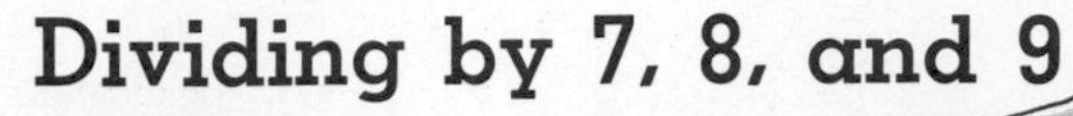

6 — markers in each box

9)54 — boxes; markers

9 — boxes

6)54 — markers in each box; markers

EXERCISES

Divide.

1. 3)27	**2.** 6)24	**3.** 3)24	**4.** 8)8	**5.** 7)21
6. 9)18	**7.** 8)24	**8.** 7)42	**9.** 6)30	**10.** 8)32
11. 7)14	**12.** 7)28	**13.** 5)35	**14.** 9)27	**15.** 7)35
16. 8)16	**17.** 9)36	**18.** 6)36	**19.** 4)36	**20.** 8)48
21. 4)32	**22.** 8)40	**23.** 7)56	**24.** 9)0	**25.** 5)40
26. 6)18	**27.** 7)49	**28.** 9)54	**29.** 8)64	**30.** 6)42
31. 8)56	**32.** 9)63	**33.** 6)54	**34.** 5)45	**35.** 9)81
36. 9)45	**37.** 6)48	**38.** 8)72	**39.** 9)72	**40.** 7)63
41. 4)32	**42.** 3)21	**43.** 4)16	**44.** 8)8	**45.** 9)81
46. 5)30	**47.** 4)20	**48.** 2)16	**49.** 9)9	**50.** 9)72

Give each quotient.

51. $4\overline{)28}$ 52. $5\overline{)20}$ 53. $3\overline{)15}$ 54. $8\overline{)64}$ 55. $7\overline{)56}$

56. $5\overline{)25}$ 57. $3\overline{)18}$ 58. $5\overline{)15}$ 59. $9\overline{)63}$ 60. $6\overline{)36}$

61. $4\overline{)24}$ 62. $2\overline{)18}$ 63. $2\overline{)14}$ 64. $7\overline{)49}$ 65. $6\overline{)42}$

Solve.

66. Susan had $2.25. She spent $1.69 for the markers. How much money did she have left?

67. There were 30 students in Susan's class. Twenty-one of them came to the party. How many did not come to the party?

68. 21 students were seated at 3 tables. The same number of students were at each table. How many students were at each table?

69. Susan and her friends brought 54 cookies to the party. They put 9 cookies on each dish. How many dishes did they need?

KEEPING SKILLS SHARP

1. $\begin{array}{r} 58 \\ -24 \\ \hline \end{array}$ 2. $\begin{array}{r} 60 \\ -37 \\ \hline \end{array}$ 3. $\begin{array}{r} 379 \\ -146 \\ \hline \end{array}$ 4. $\begin{array}{r} 572 \\ -249 \\ \hline \end{array}$ 5. $\begin{array}{r} 625 \\ -467 \\ \hline \end{array}$

6. $\begin{array}{r} 302 \\ -186 \\ \hline \end{array}$ 7. $\begin{array}{r} 500 \\ -374 \\ \hline \end{array}$ 8. $\begin{array}{r} 8371 \\ -2593 \\ \hline \end{array}$ 9. $\begin{array}{r} 6025 \\ -3748 \\ \hline \end{array}$ 10. $\begin{array}{r} 9211 \\ -6987 \\ \hline \end{array}$

Addition, subtraction, multiplication, and division

Work inside the grouping symbols first.

$(12 \div 4) + 2 = 5$

$12 \div (4 + 2) = 2$

EXERCISES

Compute.

1. $(18 \div 6) + 3$
2. $18 \div (6 + 3)$
3. $(18 \div 6) \div 3$
4. $(16 \div 4) + 4$
5. $16 \div (4 + 4)$
6. $(24 \div 8) - 2$
7. $24 \div (8 - 2)$
8. $9 + (6 \div 3)$
9. $(9 + 6) \div 3$
10. $(15 - 6) \times 2$
11. $15 - (6 \times 2)$
12. $(36 \div 4) \times 8$
13. $(48 \div 8) + 7$
14. $(42 \div 6) \times 6$
15. $(9 + 8) - 8$
16. $(54 \div 9) \times 9$
17. $(8 + 7) - 7$
18. $(5 \times 9) \div 9$

Challenge!

Find the end number.

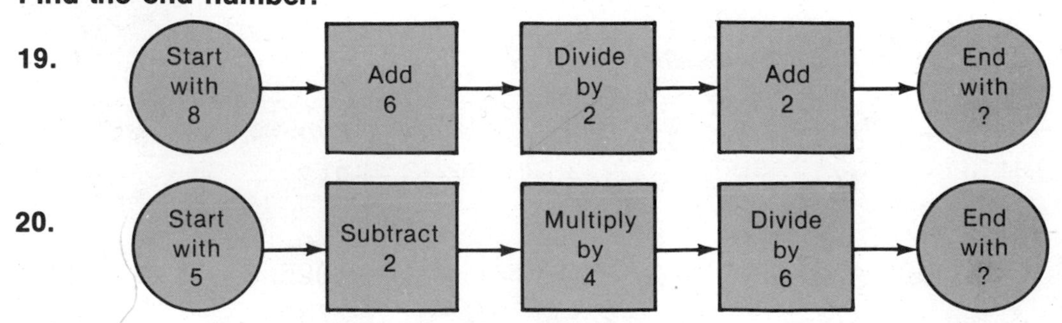

You can add, subtract, multiply, or divide. Can you use the three numbers to build the target number?

1. 2 3 6 — target number 4

2. 9 6 7 — target number 8

3. target number

4. target number

Play the game.

1. Make digit cards for the numbers 0 through 9.
2. Choose a leader.
3. Divide the class into two teams, team A and team B.
4. Without looking, the leader picks a target-number card and three other cards.
5. If team A can build the target number using the other three numbers, it scores 1 point.
6. The leader chooses new cards, and team B plays.
7. The first team to get 12 points wins the game.

Division with remainder

Some divisions do not come out evenly.

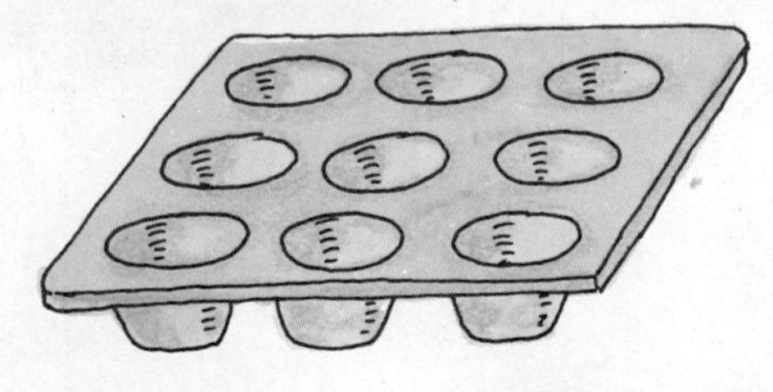

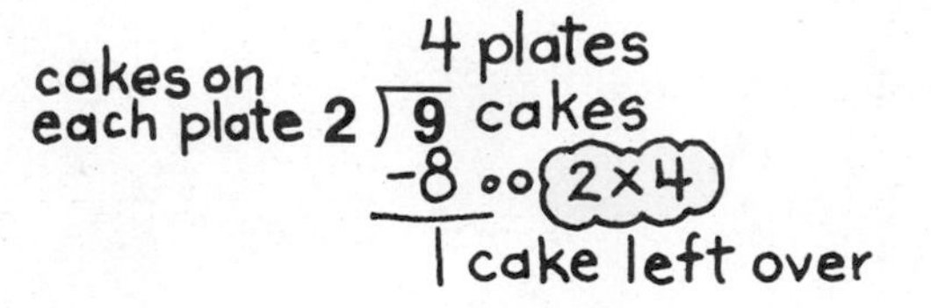

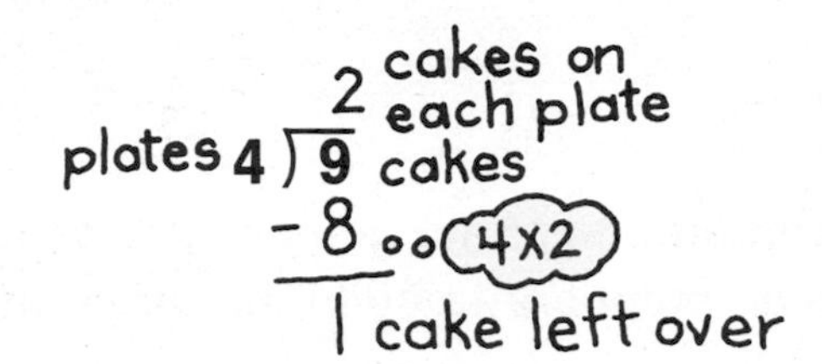

The 1 that is left over is called the **remainder.**

Here is how to write the answer to a division-with-remainder problem.

$$\begin{array}{r} 2\ \text{R}1 \\ 4\overline{)9} \\ -8 \\ \hline 1 \end{array}$$

divisor (4)
remainder (1)

The remainder is always less than the divisor.

EXERCISES

Copy and complete.

1. $4\overline{)15}$ — 3 R ▢; −12; ▢
2. $3\overline{)13}$ — 4 R ▢; −▢; ▢
3. $4\overline{)34}$ — 8 R ▢; −▢; ▢
4. $6\overline{)29}$ — ▢ R ▢; −▢; ▢
5. $5\overline{)49}$ — ▢ R ▢; −▢; ▢
6. $7\overline{)45}$ — ▢ R ▢; −▢; ▢
7. $8\overline{)43}$ — ▢ R ▢; −▢; ▢
8. $9\overline{)50}$ — ▢ R ▢; −▢; ▢
9. $8\overline{)60}$ — ▢ R ▢; −▢; ▢
10. $7\overline{)39}$ — ▢ R ▢; −▢; ▢

Give each quotient and remainder.

11. $5\overline{)29}$	12. $9\overline{)20}$	13. $2\overline{)17}$
14. $5\overline{)27}$	15. $6\overline{)50}$	16. $9\overline{)50}$
17. $7\overline{)43}$	18. $4\overline{)37}$	19. $8\overline{)25}$
20. $7\overline{)30}$	21. $8\overline{)46}$	22. $9\overline{)35}$
23. $6\overline{)45}$	24. $4\overline{)30}$	25. $8\overline{)67}$
26. $7\overline{)15}$	27. $6\overline{)39}$	28. $9\overline{)25}$
29. $7\overline{)45}$	30. $5\overline{)49}$	31. $3\overline{)20}$
32. $7\overline{)54}$	33. $8\overline{)35}$	34. $9\overline{)80}$

Solve.

35. Twenty-six students signed up for a relay race. If each team had 4 players on it, how many teams would there be for the race? How many students would be left over?

Challenge!

Guess our numbers.

KEEPING SKILLS SHARP

1. 12×3	2. 11×5
3. 21×4	4. 32×2
5. 10×6	6. 11×8
7. 23×4	8. 35×2
9. 26×3	10. 13×5
11. 38×3	12. 42×5
13. 52×7	14. 46×4
15. 70×6	16. 68×9
17. 83×8	18. 92×7
19. 75×6	20. 91×9

Problem solving—choosing the operation

1. Study and understand.
2. Plan and do.
3. Answer and check.

Add, subtract, multiply, or divide? Decide what to do.

1. Karen had □ German and ● United States stamps. How many did she have in all?

2. Jerry had ● stamps. He sold □ of them. How many did he have then?

3. Roberto had ⬡ stamps in his collection. He bought ● more stamps. How many did he have then?

4. Ellen bought □ stamps. She paid ⬡¢ for each stamp. What was the total price?

5. ● stamps were pasted on □ pages of an album. The same number of stamps was pasted on each page. How many was that?

6. In a collection of ⬡ stamps, ● were European stamps. How many stamps were not from Europe?

7. One of Roberto's albums has ● pages. If he can put ⬡ stamps on a page, how many stamps will the album hold?

8. Another album holds □ stamps per page. How many pages would ● stamps fill?

9. David had □ stamps in one album and ● stamps in another album. How many stamps did he have in all?

10. The oldest stamp in Jerry's collection was printed in ●. How many years ago was that?

11. Donna has filled ● albums with stamps. Each album has □ pages. How many stamps does she have if each page holds ⬡ stamps?

12. Eva spent ●¢ for □ stamps. Each of the stamps cost the same amount. What was the price of one stamp?

13. Pick your own numbers for exercises 9–12. Then solve the problems.

Problem solving

Solve.

1. Brad had 37 sections of track. He got 18 more for his birthday. How many sections did he have then?

2. Brad could buy one engine for $24.95 or another for $19.50. How much more did the first engine cost?

3. He decided that he wanted the $24.95 engine. If he had $18.38, how much more did he need?

4. Each section of straight track was 9 inches long. How many inches of track were there in 24 sections?

5. Brad bought a station house and a water tower for $8.72. The station house cost $4.85. What did the water tower cost?

7. Brad was thinking about buying a tank car for $6.97, a flatcar for $3.98, or a caboose for $3.50. He had only $8. He decided to buy the flatcar and the caboose. How much money did he have left?

9. Fireplugs cost 8¢ each. How many could Brad buy with 60¢? How much would he have left?

11. Brad sold a boxcar for $2.50. Then he bought 2 sections of track for $.43 each. How much did he have left?

6. In laying out his new railroad, Brad placed the station house and water tower 42 inches apart. How many 6-inch sections of track were needed to reach from the station house to the water tower?

8. He bought 2 flatcars for $2.15 each and a mail car for $3.25. What was the total cost?

10. Trees cost 9¢ each and bushes cost 6¢ each. How much would 5 trees and 6 bushes cost?

12. He could put 4 trucks on each of 3 large flatcars and 3 trucks on each of 5 small flatcars. How many trucks could his train carry?

Dividing a 2-digit number

Sue, Joe, and Mikey helped do the shopping. The clerk gave them 69 Bonus Stamps. They each had their own card for saving the stamps.

The example shows how they divided the stamps.

EXAMPLE 1. Divide 69 stamps equally among 3 people.

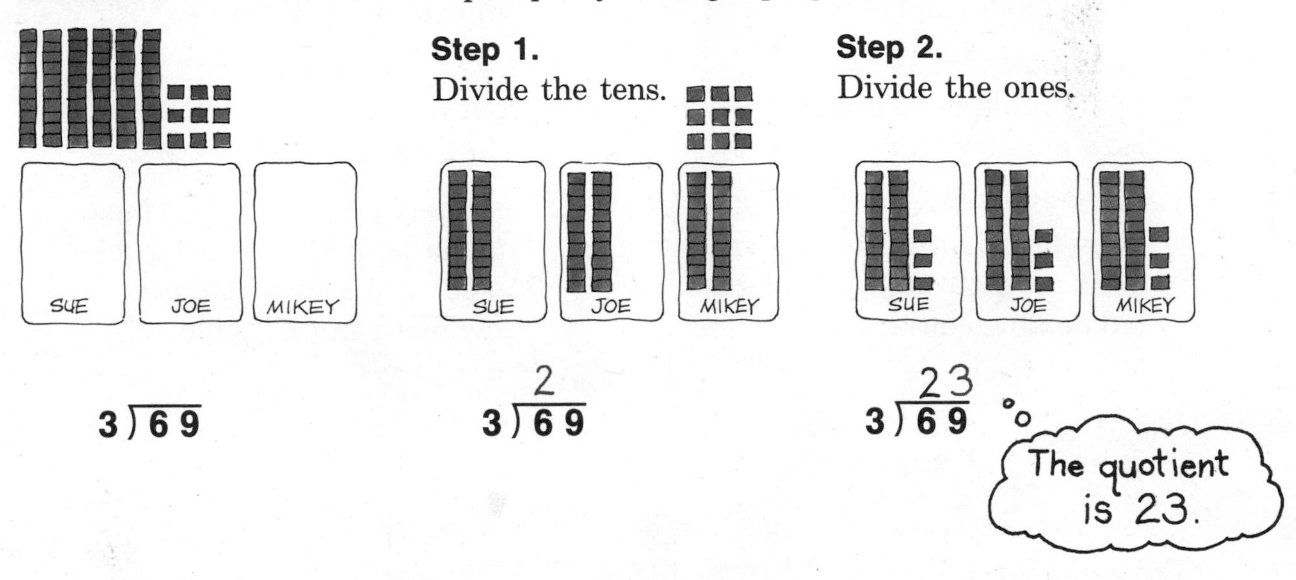

EXAMPLE 2. Divide 80 stamps equally between 2 people.

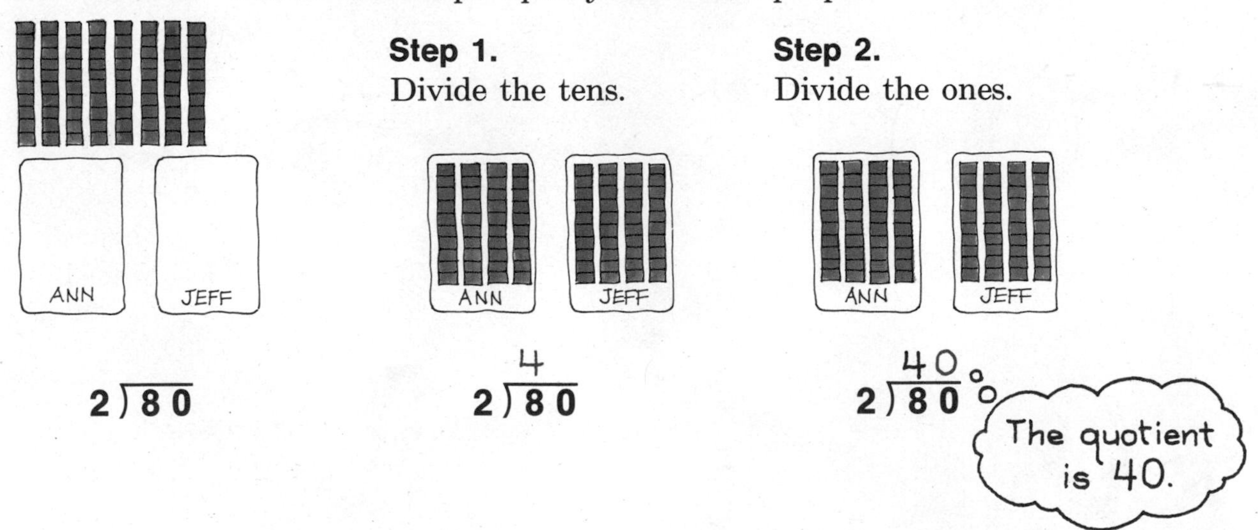

EXERCISES

Divide.

1. $2\overline{)42}$ **2.** $4\overline{)40}$ **3.** $2\overline{)68}$ **4.** $5\overline{)55}$ **5.** $2\overline{)62}$

6. $6\overline{)66}$ **7.** $3\overline{)33}$ **8.** $4\overline{)88}$ **9.** $2\overline{)86}$ **10.** $4\overline{)44}$

11. $2\overline{)80}$ **12.** $3\overline{)63}$ **13.** $2\overline{)44}$ **14.** $3\overline{)36}$ **15.** $2\overline{)88}$

16. $4\overline{)84}$ **17.** $4\overline{)48}$ **18.** $3\overline{)66}$ **19.** $2\overline{)84}$ **20.** $2\overline{)46}$

21. $3\overline{)30}$ **22.** $2\overline{)64}$ **23.** $3\overline{)39}$ **24.** $4\overline{)80}$ **25.** $3\overline{)99}$

26. What problem did Beth get wrong? What mistake did she make?

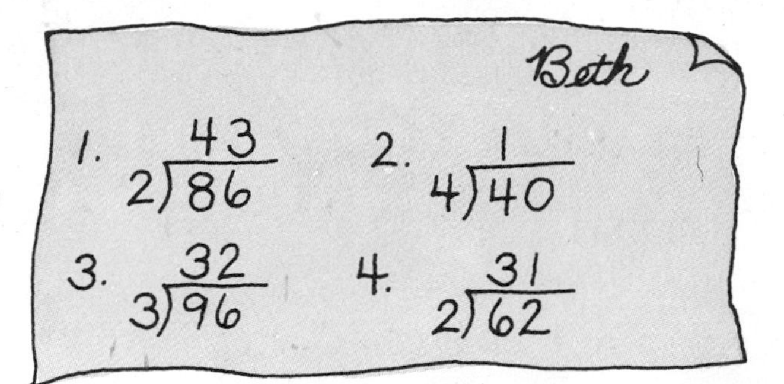

Beth

1. $\begin{array}{r} 43 \\ 2\overline{)86} \end{array}$ 2. $\begin{array}{r} 1 \\ 4\overline{)40} \end{array}$

3. $\begin{array}{r} 32 \\ 3\overline{)96} \end{array}$ 4. $\begin{array}{r} 31 \\ 2\overline{)62} \end{array}$

Solve.

27.

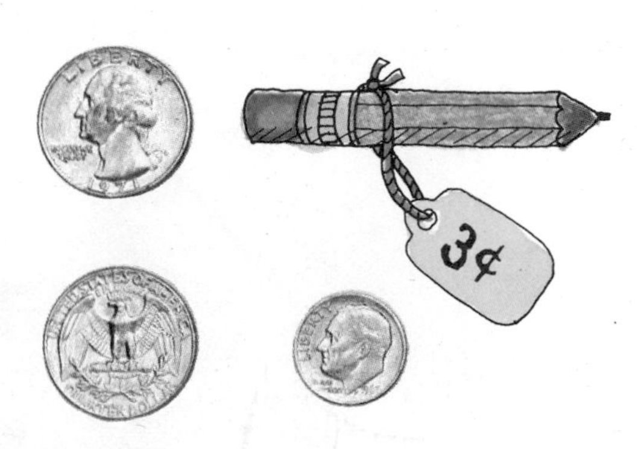

How many pencils can you buy?

28.

How many erasers can you buy?

29. Bill divided up 48 marbles. He put the same number of marbles in each of 2 boxes. How many marbles did he put in each box?

30. Sandy found 66 seashells. He shared them equally with his 2 sisters. How many seashells did each child get?

Division with regrouping

Sometimes you will need to regroup when dividing.

EXAMPLE.

Divide 81 stamps equally among 3 people.

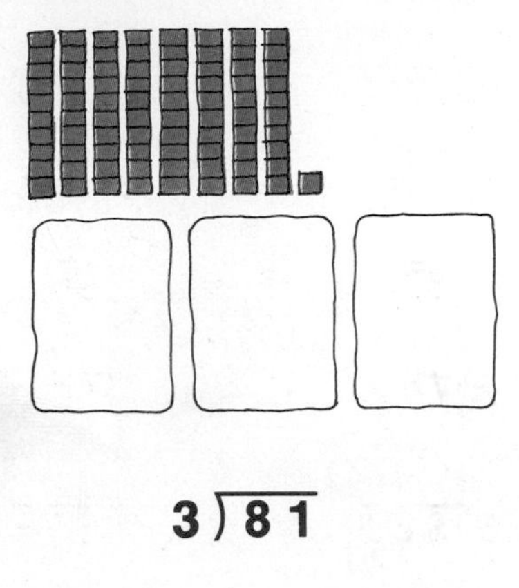

Step 1.
Divide the tens.
Subtract.

Step 2.
Regroup 2 tens to 20 ones.

Step 3.
Divide the ones.

EXERCISES

Copy and complete.

1. $\begin{array}{r} 15 \\ 3\overline{)45} \\ -3 \\ \hline 15 \\ -15 \\ \hline ? \end{array}$

2. $\begin{array}{r} 26 \\ 2\overline{)52} \\ -4 \\ \hline 12 \\ -?? \\ \hline ? \end{array}$

3. $\begin{array}{r} 1? \\ 4\overline{)72} \\ -4 \\ \hline 32 \\ -?? \\ \hline ? \end{array}$

4. $\begin{array}{r} 1? \\ 5\overline{)75} \\ -5 \\ \hline ?? \\ -?? \\ \hline ? \end{array}$

5. $\begin{array}{r} ?? \\ 7\overline{)84} \\ -? \\ \hline ?? \\ -?? \\ \hline ? \end{array}$

Divide.

6. $6\overline{)84}$
7. $2\overline{)64}$
8. $3\overline{)57}$
9. $7\overline{)91}$
10. $6\overline{)90}$
11. $3\overline{)48}$
12. $2\overline{)74}$
13. $5\overline{)70}$
14. $3\overline{)87}$
15. $8\overline{)96}$
16. $4\overline{)80}$
17. $3\overline{)42}$
18. $6\overline{)96}$
19. $5\overline{)65}$
20. $2\overline{)58}$
21. $2\overline{)\$.96}$
22. $5\overline{)\$.90}$
23. $7\overline{)\$.98}$
24. $5\overline{)\$.80}$
25. $6\overline{)\$.78}$

Solve.

7 days = 1 week

26. How many weeks are there in 84 days?

4 quarts = 1 gallon

27. How many quarts are there in 96 gallons?

3 feet = 1 yard

★ 28. How many yards are there in 77 feet? How many feet are left over?

2 cups = 1 pint

★ 29. How many pints are there in 43 cups? How many cups are left over?

Challenge!

30. Which path did the rider use?

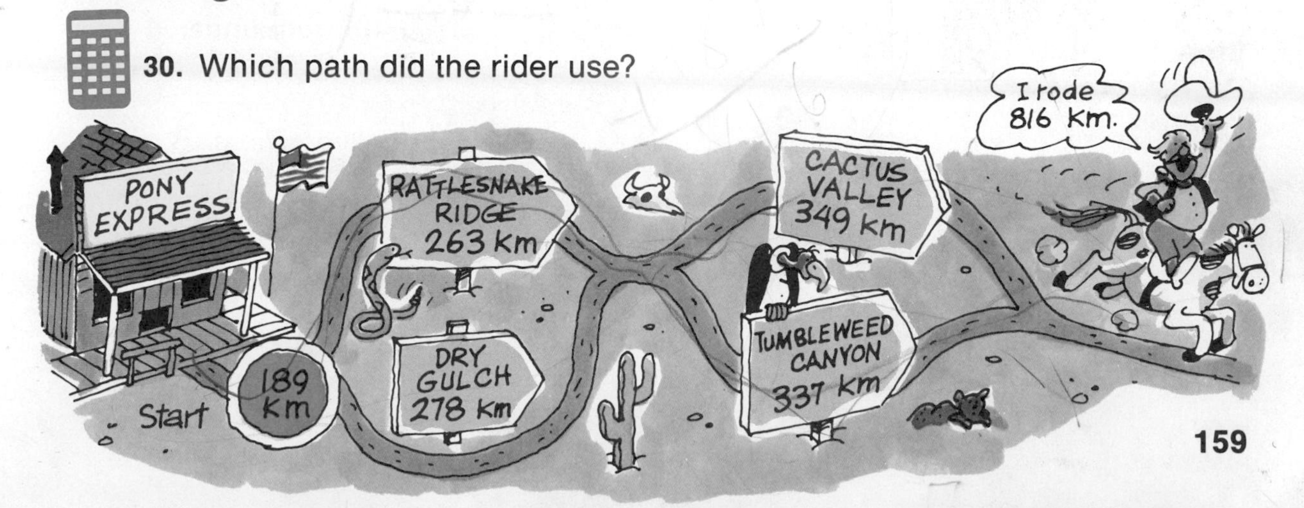

Division with remainder

Sometimes you can't divide a number evenly.

EXAMPLE.

Divide 58 stamps equally among 3 people.

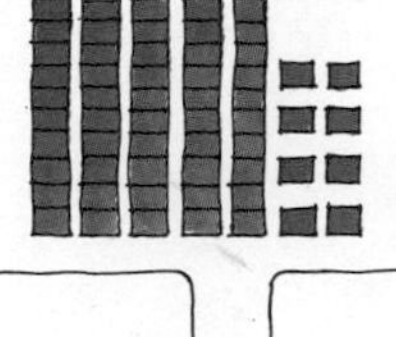

$3\overline{)58}$

Step 1.
Divide the tens.
Subtract.

$$\begin{array}{r} 1 \\ 3\overline{)58} \\ -3 \\ \hline 2 \end{array}$$

Step 2.
Regroup the tens.

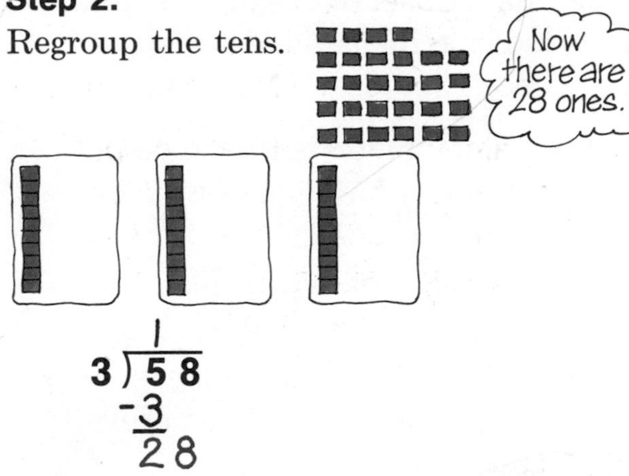

$$\begin{array}{r} 1 \\ 3\overline{)58} \\ -3 \\ \hline 28 \end{array}$$

Step 3.
Divide the ones.
Subtract.

There is 1 stamp left.

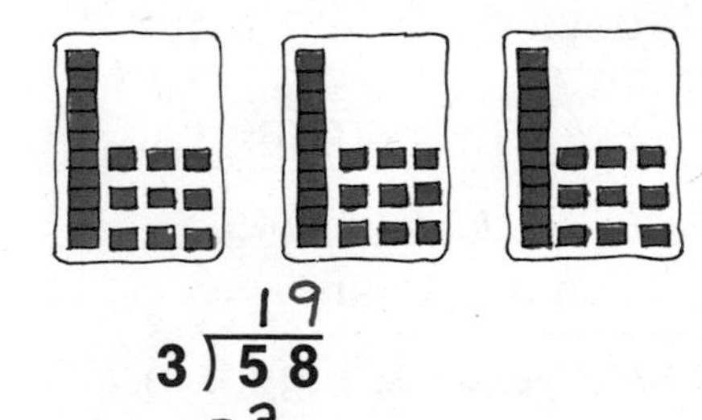

$$\begin{array}{r} 19 \\ 3\overline{)58} \\ -3 \\ \hline 28 \\ -27 \\ \hline 1 \end{array}$$

This number is called the remainder.

Each person got 19 stamps. There is 1 stamp left over.

Step 4.
Write the remainder.

$$\begin{array}{r} 19 \text{ R1} \\ 3\overline{)58}\phantom{\text{ R1}} \\ -3\phantom{\text{ R1}} \\ \hline 28\phantom{\text{ R1}} \\ -27\phantom{\text{ R1}} \\ \hline 1\phantom{\text{ R1}} \end{array}$$

EXERCISES

First divide. Then estimate the quotient to see if your answer makes sense.

1. $3\overline{)68}$	**2.** $5\overline{)58}$	**3.** $4\overline{)58}$	**4.** $3\overline{)38}$	**5.** $6\overline{)83}$
6. $4\overline{)72}$	**7.** $5\overline{)74}$	**8.** $7\overline{)79}$	**9.** $8\overline{)86}$	**10.** $5\overline{)68}$
11. $6\overline{)60}$	**12.** $3\overline{)78}$	**13.** $6\overline{)84}$	**14.** $3\overline{)71}$	**15.** $4\overline{)81}$
16. $4\overline{)69}$	**17.** $7\overline{)89}$	**18.** $5\overline{)72}$	**19.** $8\overline{)94}$	**20.** $6\overline{)75}$
21. $3\overline{)92}$	**22.** $5\overline{)73}$	**23.** $4\overline{)53}$	**24.** $3\overline{)95}$	**25.** $7\overline{)85}$
26. $5\overline{)97}$	**27.** $6\overline{)91}$	**28.** $3\overline{)68}$	**29.** $4\overline{)98}$	**30.** $4\overline{)92}$

Solve.

31. Julie sold 42 bottles. If she sold each bottle for 6¢, how much money did she get?

32. Mark collected 102 bottles. Sarah collected 46 bottles. How many more bottles did Mark collect?

33. Robert had 82 bottles. He put them in 6-bottle cartons. How many cartons did he fill? How many bottles were left over?

34. Elaine had 95 bottles. She put them in 8-bottle cartons. How many cartons did she fill? How many bottles were left over?

KEEPING SKILLS SHARP

1. 23×2	**2.** 12×4	**3.** 30×3	**4.** 41×4	**5.** 62×3
6. 57×6	**7.** 86×5	**8.** 68×8	**9.** 95×7	**10.** 87×9

Finding an average

The PTA of Clark Elementary School had a school carnival to raise money. The students sold tickets for one week. The table shown below was kept on the school bulletin board. It shows how many tickets each class sold every day.

Ticket Sales

	Monday	Tuesday	Wednesday	Thursday	Friday
Kindergarten	5	8	5	6	6
First Grade	7	5	6	3	9
Second Grade	7	6	5	3	9
Third Grade	8	8	9	10	10
Fourth Grade	9	8	9	7	12
Fifth Grade	11	8	14	17	15
Sixth Grade	9	7	8	12	9

The kindergarten class sold 30 tickets in all. Suppose that they had sold the same number of tickets each day. How many tickets would they have sold each day?

$$\begin{array}{r} 5 \\ 8 \\ 5 \\ 6 \\ +6 \\ \hline 30 \end{array}$$

$$5 \overline{)\,30\,}\ \ \text{quotient } 6$$

6 is the **average** number of tickets sold each day.

EXERCISES

1. Find the average number of tickets sold each day by the second grade.
2. Find the average number of tickets sold each day by the fifth grade.
3. Find the average number of tickets sold by each class on Monday.
4. Find the average number of tickets sold by each class on Friday.

5. At the carnival, David threw darts 3 times. His scores were 28, 23, and 21. What was his average score?

6. Alicia won the basketball free-throw contest for sixth graders. Four times she took 20 shots. She made 11, 12, 14, and 15 baskets. What was her average?

7. a. Brian won the baseball-throwing contest. In his 6 throws, he knocked over 8, 6, 3, 1, 5, and 7 bottles. What was his average?

   b. How many times did he knock down fewer bottles than his average? More bottles than his average?

8. a. Four students sold 64 chances and 2 students sold 20 chances on a radio. How many chances were sold?

   b. What was the average number of chances sold per student?

Problem solving—interpreting a graph

1. Jim takes a typing class after school. Class begins at 2:30 and ends at 3:18. How many minutes is that?
2. Jim takes a typing test each month. The graph shows Jim's improvement during the year. What was Jim's typing rate (words per minute) on the first test?
3. On what test was his rate 27 words per minute?
4. By how much did he improve his rate from the first test to the second test?

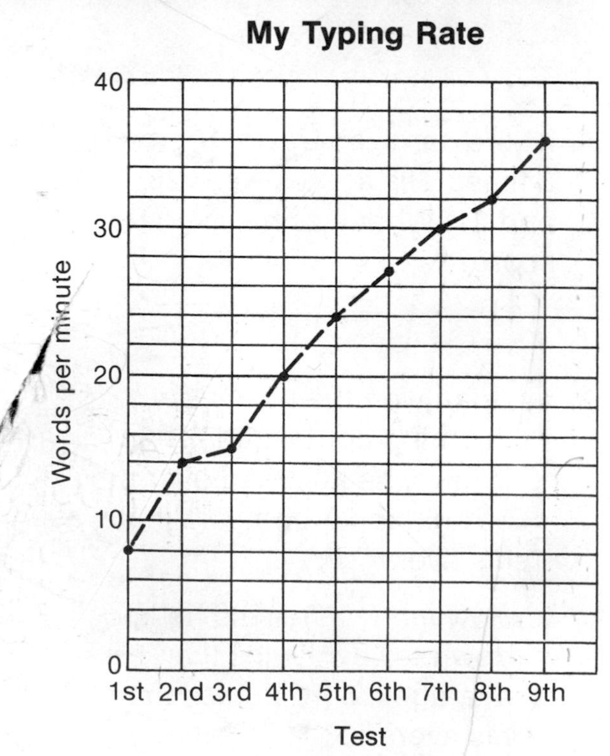

5. By how many words per minute did he improve from the first to the ninth test?

6. What was his rate on the fifth test?

7. Using the rate shown for the fifth test, how many words could he have typed in 8 minutes?

8. Using the rate shown for the ninth test, how many words could he have typed in 9 minutes?

9. One school week he practiced typing 25 minutes each day. How many minutes did he practice that week (5 days)?

10. During a 6-minute test, Jim typed 195 words. Was that more or less than 32 words per minute?

11. On her first test, Laura typed 63 words in 9 minutes. How many words per minute was that?

12. Laura typed 27 words per minute for 8 minutes, and Jim typed 31 words per minute for 8 minutes. How many more words did Jim type in the 8 minutes?

13. Rick started typing at 2:30 and stopped at 2:36. He typed 96 words. How many words did he average per minute?

14. Marianne started typing at 2:49 and stopped at 2:56. She typed 91 words. How many words did she average per minute?

15. Donna started typing at 2:57 and stopped at 3:04. She typed 98 words. How many words did she average per minute?

16. Alan typed an average of 9 words per minute. He started typing at 2:52 and typed 99 words. When did he stop typing?

CHAPTER CHECKUP

Divide. [pages 140–151]

1. $6\overline{)36}$	2. $3\overline{)27}$	3. $6\overline{)30}$	4. $8\overline{)72}$	5. $7\overline{)49}$
6. $8\overline{)24}$	7. $6\overline{)54}$	8. $5\overline{)35}$	9. $8\overline{)56}$	10. $9\overline{)81}$
11. $9\overline{)63}$	12. $9\overline{)45}$	13. $6\overline{)48}$	14. $5\overline{)40}$	15. $6\overline{)42}$
16. $5\overline{)25}$	17. $4\overline{)16}$	18. $2\overline{)18}$	19. $8\overline{)72}$	20. $7\overline{)56}$

Divide. [pages 156–163]

21. $2\overline{)46}$	22. $4\overline{)84}$	23. $3\overline{)96}$	24. $5\overline{)55}$	25. $2\overline{)80}$
26. $4\overline{)92}$	27. $6\overline{)90}$	28. $3\overline{)54}$	29. $7\overline{)91}$	30. $5\overline{)85}$
31. $2\overline{)79}$	32. $3\overline{)82}$	33. $4\overline{)73}$	34. $2\overline{)95}$	35. $6\overline{)80}$
36. $4\overline{)79}$	37. $8\overline{)94}$	38. $5\overline{)76}$	39. $9\overline{)83}$	40. $7\overline{)85}$

Solve. [pages 152–155, 164–165]

41. Eight people were going on a bicycle trip. They planned to ride a bicycle trail in 4 hours. If the trail is 24 kilometers long, then how many kilometers per hour would they average?

42. Together, the 8 people planned to take 96 kilograms of gear. They wanted to divide the gear equally. How much should each person carry?

43. Each of the 8 people paid $1.50 for a box lunch. What was the total cost for the lunches?

44. They bicycled 15 kilometers of the 24 kilometers before lunch. How many kilometers do they have to go after lunch?

CHAPTER PROJECT

1. Hold a centimeter ruler flat on a table and measure the length of your hand span. Be sure to spread your fingers as wide apart as possible.
2. Make a table of the hand-span lengths of yourself and three classmates.
3. Find the average length of a hand span.
4. Find the average length of the hand spans of the people in this table.

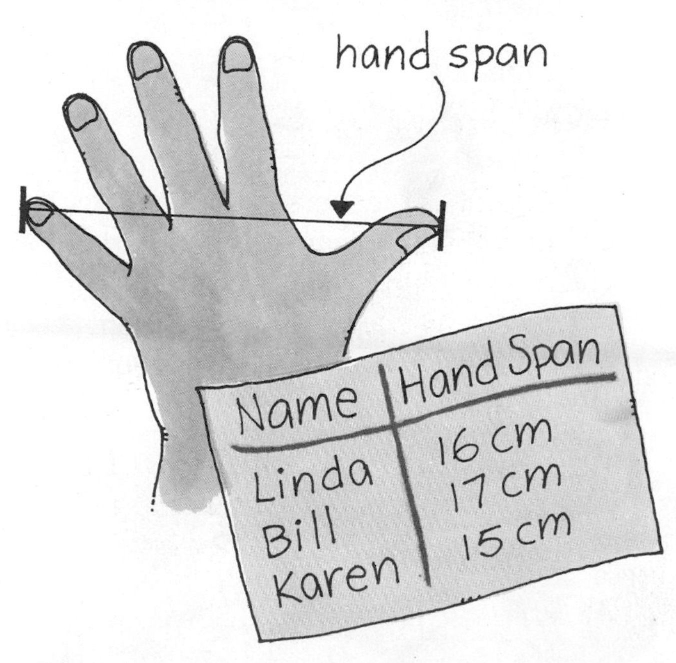

CHAPTER REVIEW

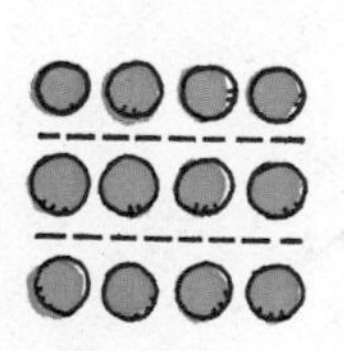

$$3\overline{)12}\quad 4$$

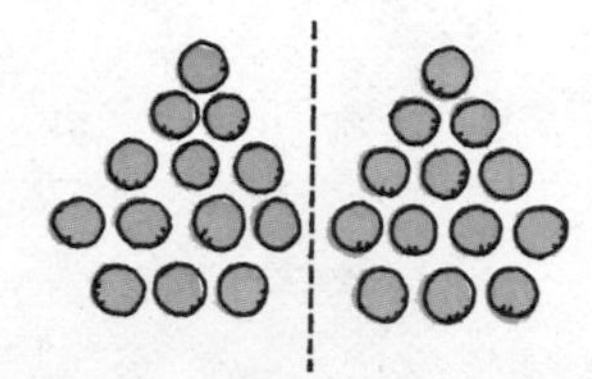

$$2\overline{)26}\quad 13$$

Divide tens.
Regroup.
Divide ones.

$$\begin{array}{r} 18 \\ 4\overline{)72} \\ -4 \\ \hline 32 \\ -32 \\ \hline 0 \end{array}$$

Divide tens.
Regroup.
Divide ones.
Write remainder.

$$\begin{array}{r} 17\text{R}3 \\ 5\overline{)88} \\ -5 \\ \hline 38 \\ -35 \\ \hline 3 \end{array}$$

Divide.

1. $4\overline{)32}$	**2.** $5\overline{)30}$	**3.** $8\overline{)48}$
4. $6\overline{)42}$	**5.** $7\overline{)49}$	**6.** $3\overline{)27}$
7. $9\overline{)63}$	**8.** $5\overline{)45}$	**9.** $7\overline{)56}$
10. $2\overline{)48}$	**11.** $3\overline{)39}$	**12.** $4\overline{)80}$
13. $5\overline{)50}$	**14.** $2\overline{)86}$	**15.** $3\overline{)96}$
16. $6\overline{)66}$	**17.** $4\overline{)84}$	**18.** $2\overline{)68}$
19. $3\overline{)42}$	**20.** $6\overline{)84}$	**21.** $5\overline{)75}$
22. $2\overline{)78}$	**23.** $4\overline{)68}$	**24.** $6\overline{)96}$
25. $4\overline{)92}$	**26.** $5\overline{)85}$	**27.** $3\overline{)87}$
28. $4\overline{)75}$	**29.** $3\overline{)49}$	**30.** $6\overline{)80}$
31. $5\overline{)93}$	**32.** $6\overline{)85}$	**33.** $2\overline{)63}$
34. $3\overline{)71}$	**35.** $5\overline{)88}$	**36.** $4\overline{)79}$

CHAPTER CHALLENGE

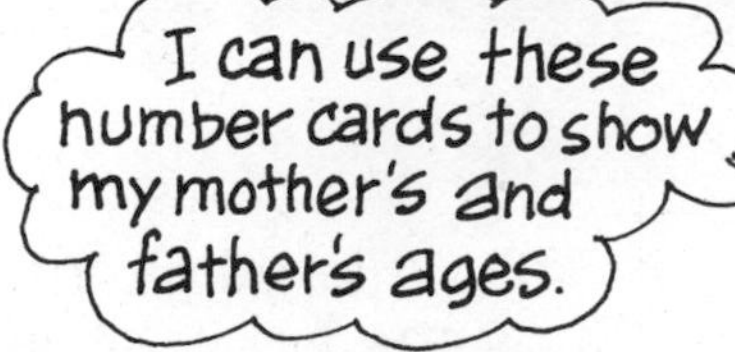

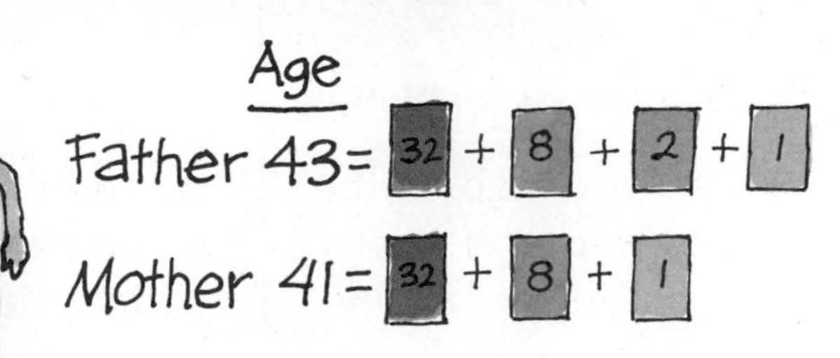

1. Can you tell what cards I used to show the rest of my family's ages?

	Age	
Barbara	20	= ? + ?
Mike	13	= ? + ? + ?
Joan	9	= ? + ?

2. Make a list of your family's ages. Show the cards you would use.

3. What is the oldest age you could show with this set of cards?

4. A country auction used color-code tags to show how old each item was. How old is each of these items?

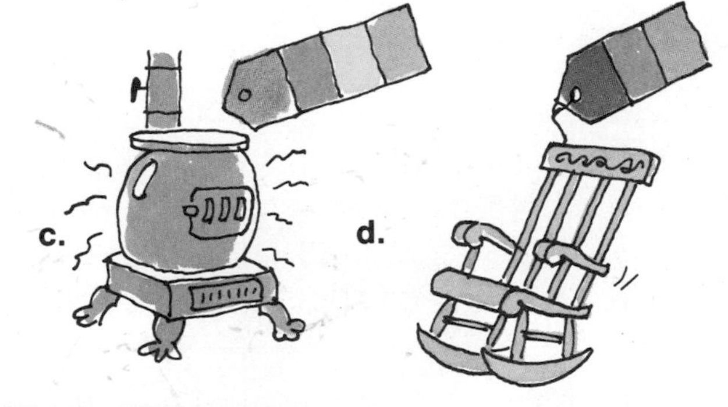

5. Which item is the oldest?

6. The auction was given a quilt that was 83 years old. They made this tag to put on the quilt:

What number does the white card stand for?

Choose the correct letter.

1. 736 rounded to the nearest ten is
   a. 700
   b. 730
   c. 740
   d. 800

2. The standard numeral for forty thousand nine hundred six is
   a. 4,096
   b. 40,960
   c. 40,096
   d. 40,906

3. Which digit in 596,814 is in the hundred thousands place?
   a. 5
   b. 9
   c. 6
   d. 8

4. Which number is greater than 58,163?
   a. 58,171
   b. 6999
   c. 57,238
   d. none of these

5. Add.
   $$\begin{array}{r} 5387 \\ +2694 \\ \hline \end{array}$$
   a. 7971
   b. 7981
   c. 7071
   d. none of these

6. Subtract.
   $$\begin{array}{r} 5285 \\ -3436 \\ \hline \end{array}$$
   a. 2251
   b. 2859
   c. 1849
   d. none of these

7. 

   What time will it be in 45 minutes?
   a. 9:30 b. 9:35
   c. 10:30 d. 10:35

8. How much money?
   1 ten-dollar bill
   1 five-dollar bill
   3 one-dollar bills
   2 quarters
   4 nickels
   a. $18.70 b. $18.90
   c. $16.90 d. $16.70

9. Multiply.
   $$\begin{array}{r} 76 \\ \times 4 \\ \hline \end{array}$$
   a. 304
   b. 284
   c. 2824
   d. none of these

10. Divide.
    $3\overline{)89}$
    a. 23
    b. 29 R1
    c. 29 R2
    d. none of these

11. Ann bought a book for $3.87. She gave the clerk $5. How much change did she get?
    a. $8.87 b. $1.13
    c. $1.23 d. $2.23

12. Pencils sell for 3¢ each. How many pencils can you buy for 60¢?
    a. 63 b. 180
    c. 20 d. none of these

Fractions

7

Fractions and regions

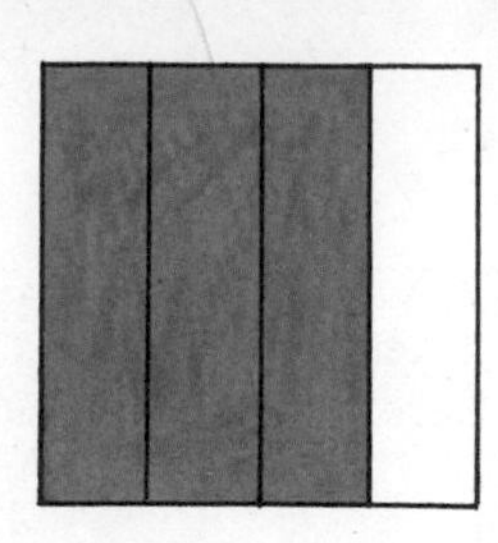

3 parts are blue
4 equal parts

numerator ⟶ 3
denominator ⟶ 4

$\frac{3}{4}$ is blue

Read as "three fourths."

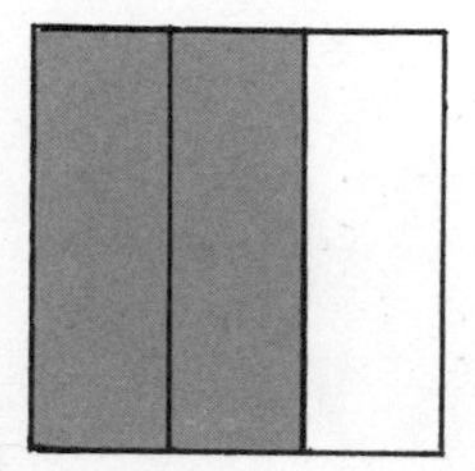

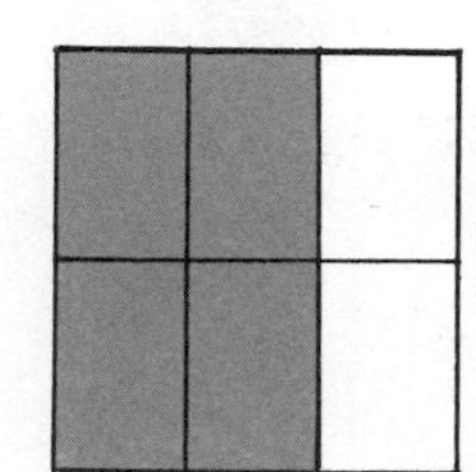

$$\frac{2}{3} = \frac{4}{6}$$

The same amount is shaded in both squares. So the fractions are **equivalent.**

EXERCISES

What fraction is colored?

1. 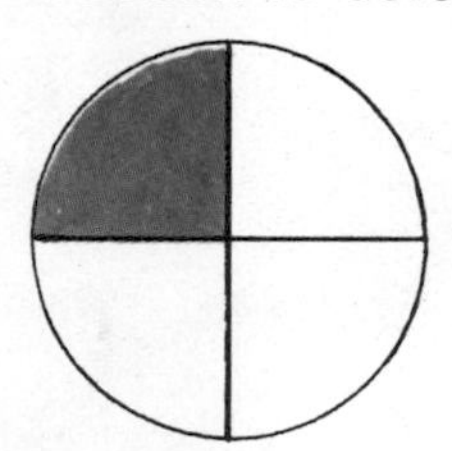

2. 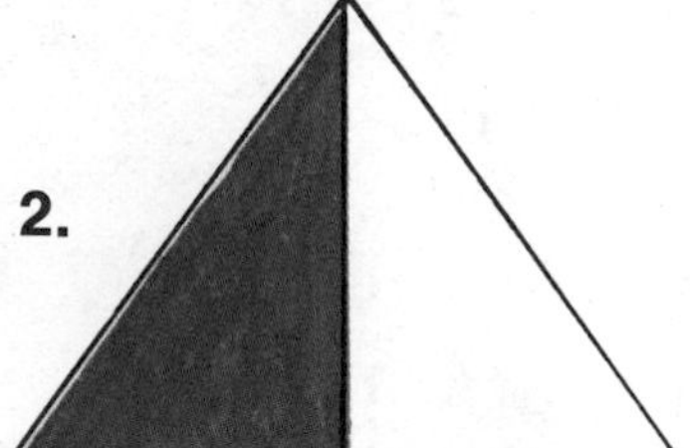

3. 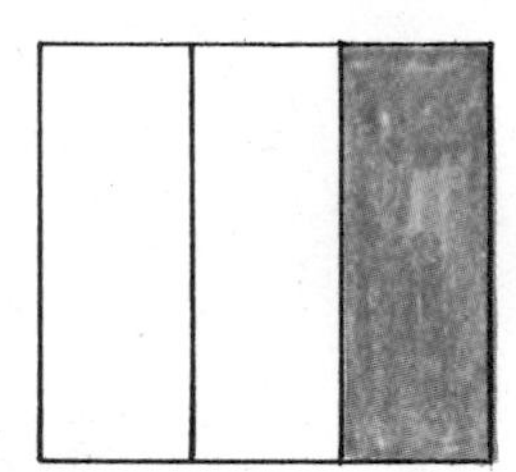

4. 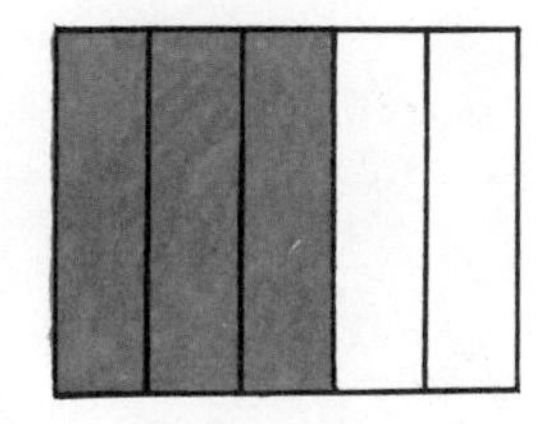

5. 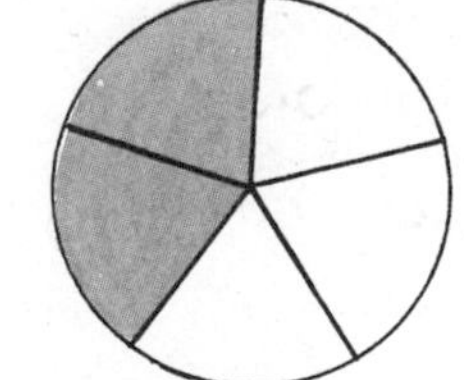

6. 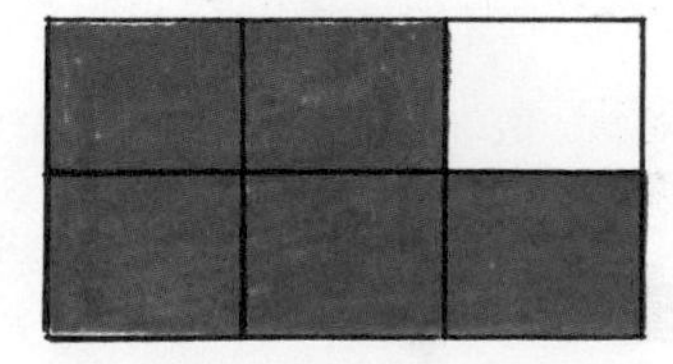

7. 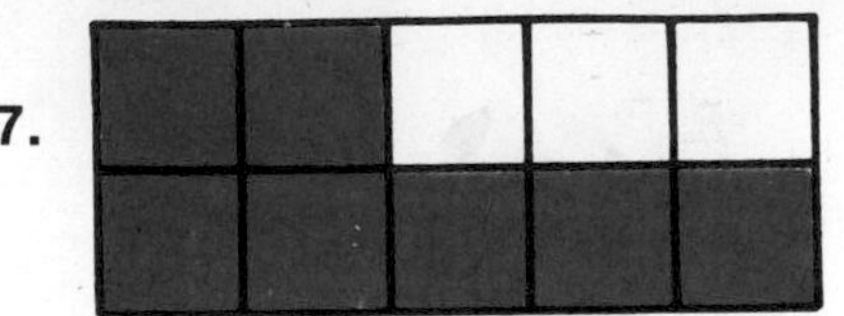

8. 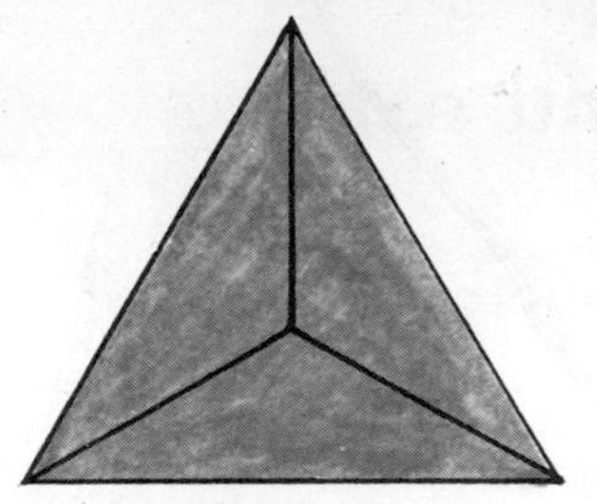

9. 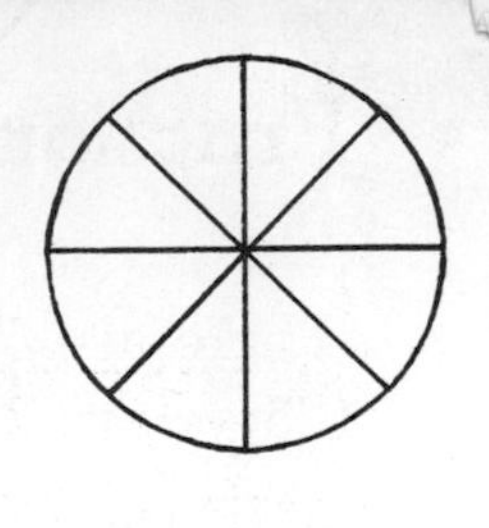

Give the equivalent fractions that are pictured.

10. 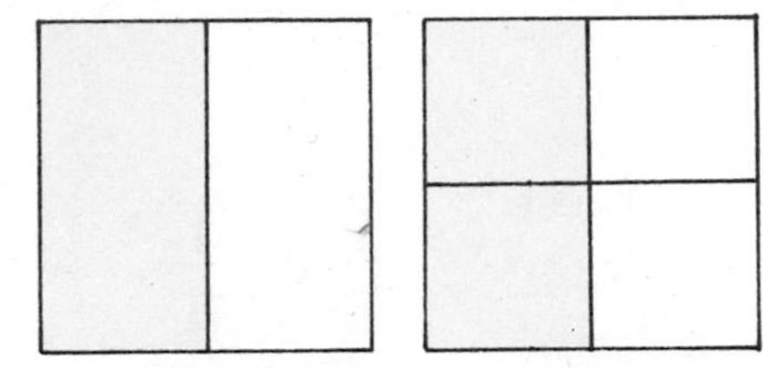

11. 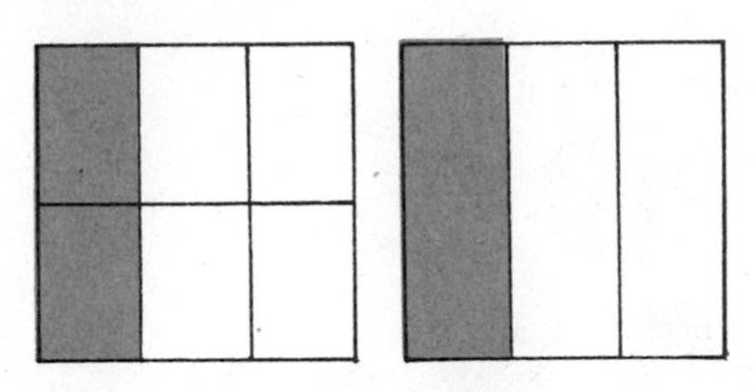

12. 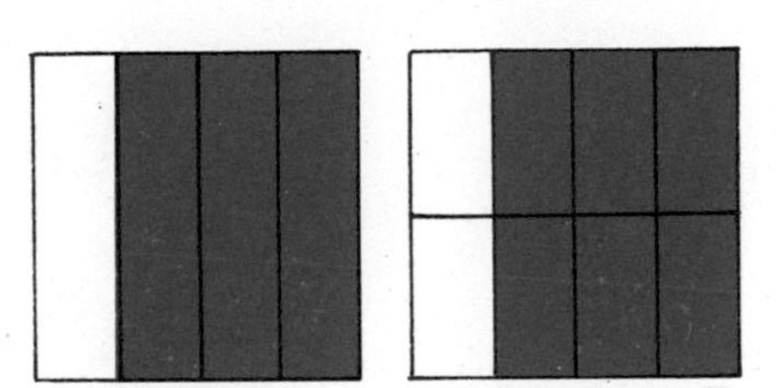

13. 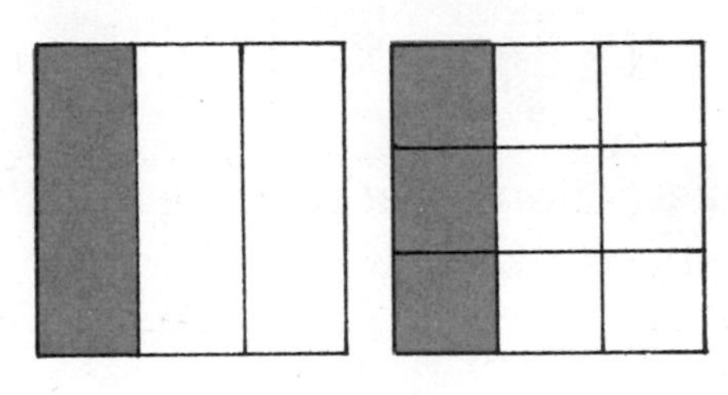

14. 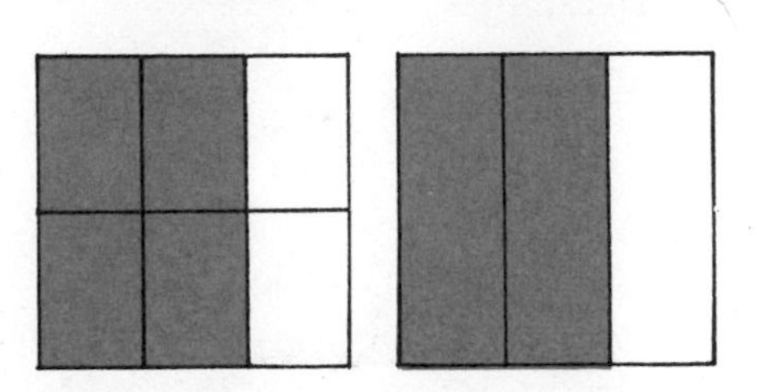

15. 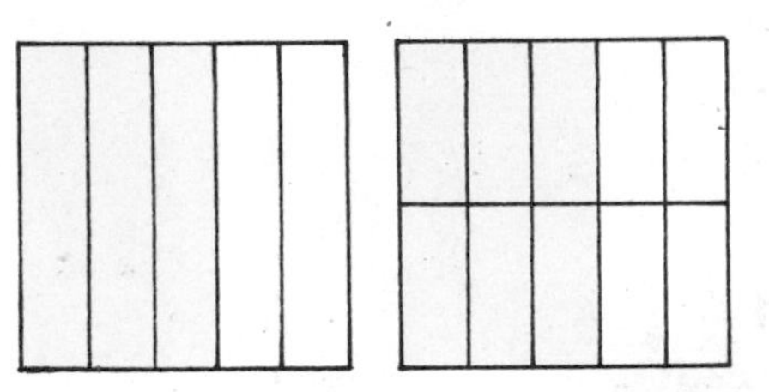

Complete. If you need to, draw pictures.

16. $\frac{1}{2} = \frac{?}{4}$ **17.** $\frac{1}{2} = \frac{?}{6}$ **18.** $\frac{1}{2} = \frac{?}{8}$ **19.** $\frac{1}{2} = \frac{?}{10}$

20. $\frac{1}{3} = \frac{?}{6}$ **21.** $\frac{1}{3} = \frac{?}{9}$ **22.** $\frac{1}{4} = \frac{?}{8}$ **23.** $\frac{1}{4} = \frac{?}{12}$

Challenge!

Use the clues to find the mystery fractions.

24. The numerator is 3. The denominator is 5.

25. The denominator is 3 more than the numerator. The sum of the numerator and the denominator is 11.

Fractions and sets

There are 3 red umbrellas.
There are 8 umbrellas in all.
$\frac{3}{8}$ of the umbrellas are red.

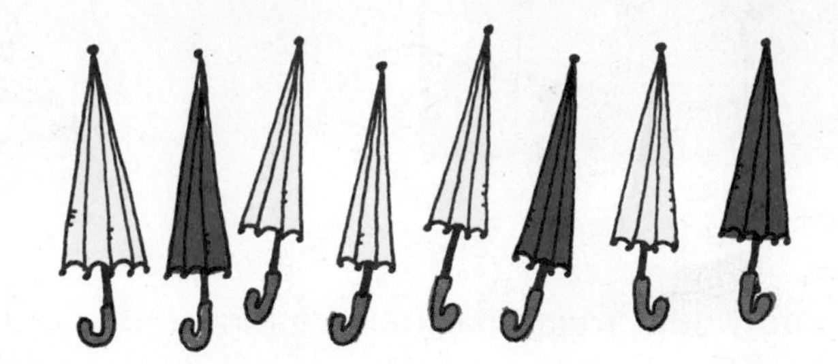

We can use either of these equivalent fractions to tell what fraction of the balls are green.

$\frac{2}{6}$ or $\frac{1}{3}$

Give 2 fractions that tell what fraction of the balls are red.

EXERCISES

1.

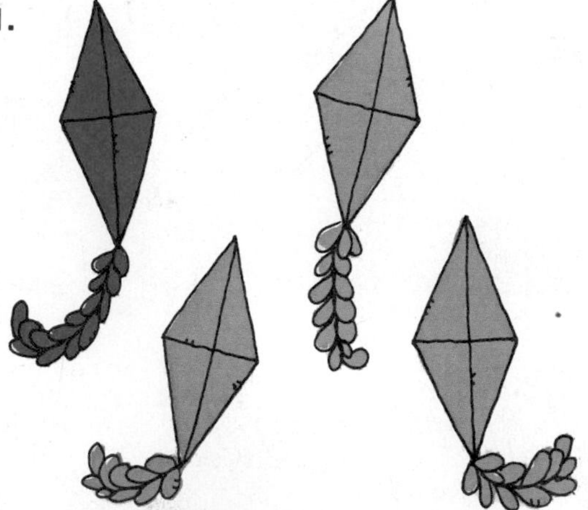

a. What fraction of the kites are red?

b. What fraction of the kites are blue?

2.

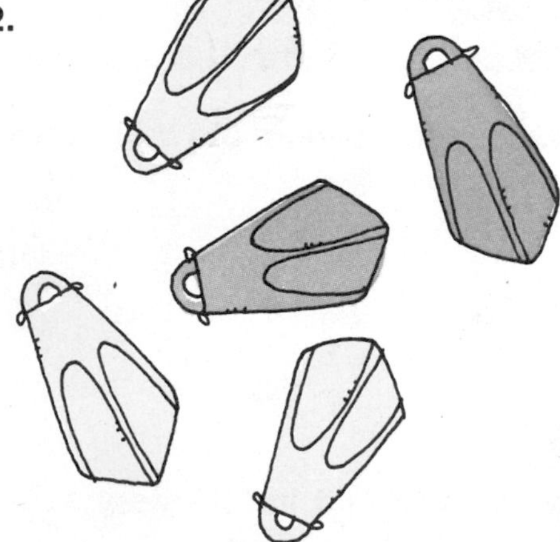

a. What fraction of the fins are green?

b. What fraction of the fins are yellow?

3.

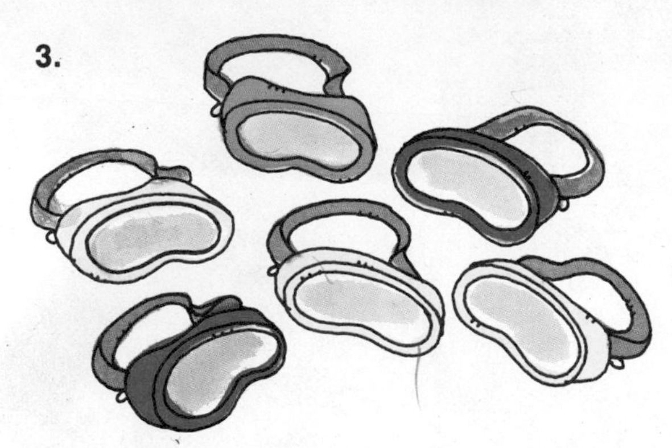

What fraction are

a. red?
b. yellow?
c. blue?
d. not blue?
e. not yellow?

4.

What fraction are

a. green?
b. red?
c. blue?
d. not blue?
e. not red?

5. What fraction are blue?

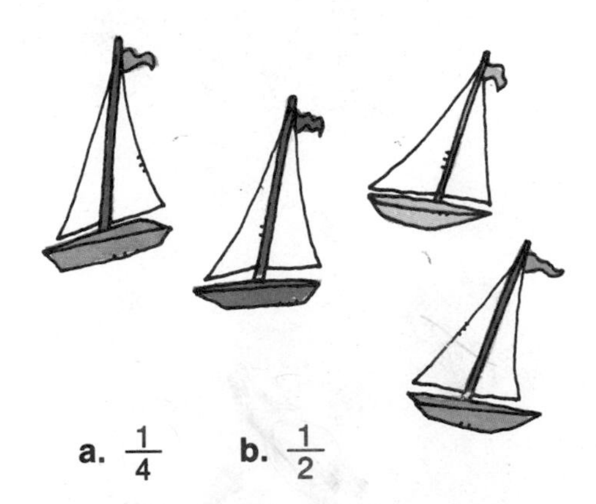

a. $\frac{1}{4}$ b. $\frac{1}{2}$

6. What fraction are green?

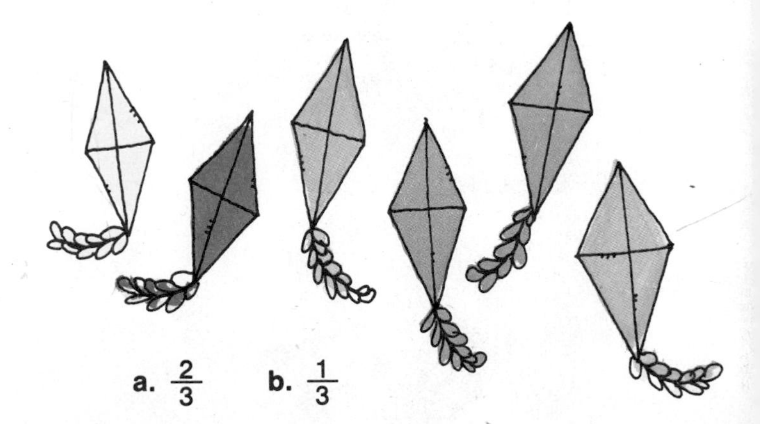

a. $\frac{2}{3}$ b. $\frac{1}{3}$

7. What fraction are red?

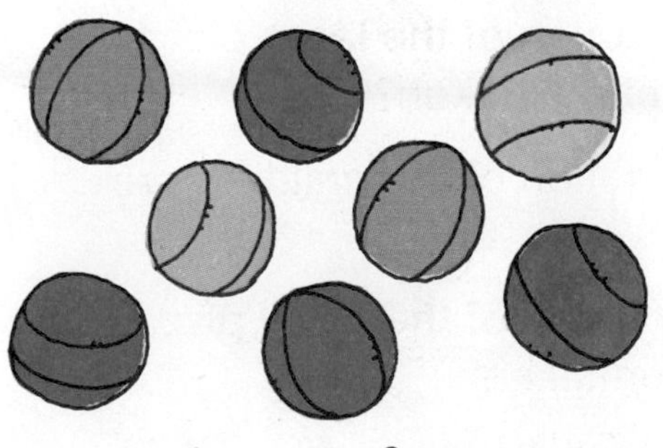

a. $\frac{1}{2}$ b. $\frac{2}{8}$

8. What fraction are not blue?

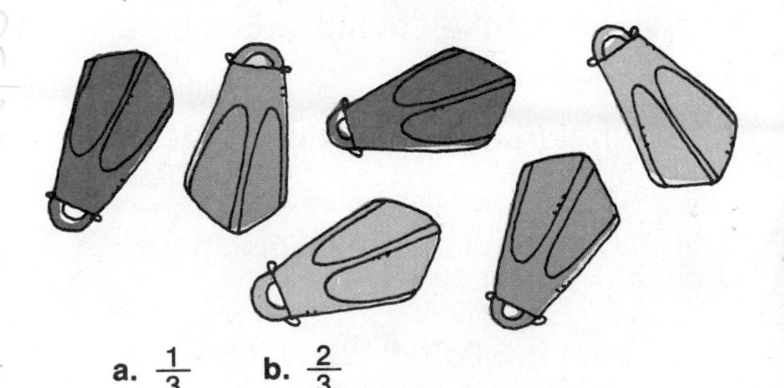

a. $\frac{1}{3}$ b. $\frac{2}{3}$

Problem solving

Use the picture to solve.

1. What fraction of the large pizza has been taken?
2. What fraction of the large pizza has not been taken?
3. What fraction of the children are boys?
4. What fraction of the children are girls?
5. What fraction of the small pizza has not been taken?
6. What fraction of the small pizza has been taken?
7. What fraction of the children are drinking root beer?
8. What fraction of the children are drinking milk?

9. What fraction of the boys are drinking milk?
10. What fraction of the girls are not drinking milk?
11. When two more pieces of the small pizza are taken, what fraction of the pizza will be left? Give two answers.
12. When three more pieces of the large pizza are taken, what fraction of the pizza will be left? Give two answers.

KEEPING SKILLS SHARP

1. $\begin{array}{r} 5 \\ \times 4 \\ \hline \end{array}$
2. $\begin{array}{r} 8 \\ \times 3 \\ \hline \end{array}$
3. $\begin{array}{r} 5 \\ \times 6 \\ \hline \end{array}$
4. $\begin{array}{r} 4 \\ \times 5 \\ \hline \end{array}$
5. $\begin{array}{r} 6 \\ \times 4 \\ \hline \end{array}$
6. $\begin{array}{r} 4 \\ \times 8 \\ \hline \end{array}$
7. $\begin{array}{r} 3 \\ \times 7 \\ \hline \end{array}$
8. $\begin{array}{r} 7 \\ \times 4 \\ \hline \end{array}$
9. $\begin{array}{r} 15 \\ \times 5 \\ \hline \end{array}$
10. $\begin{array}{r} 16 \\ \times 6 \\ \hline \end{array}$
11. $\begin{array}{r} 23 \\ \times 8 \\ \hline \end{array}$
12. $\begin{array}{r} 24 \\ \times 9 \\ \hline \end{array}$
13. $\begin{array}{r} 37 \\ \times 5 \\ \hline \end{array}$
14. $\begin{array}{r} 42 \\ \times 9 \\ \hline \end{array}$
15. $\begin{array}{r} 56 \\ \times 7 \\ \hline \end{array}$
16. $\begin{array}{r} 87 \\ \times 8 \\ \hline \end{array}$
17. $\begin{array}{r} 76 \\ \times 8 \\ \hline \end{array}$
18. $\begin{array}{r} 93 \\ \times 8 \\ \hline \end{array}$

Fraction of a number

To find $\frac{1}{2}$ of a number, divide the number by 2.

$\frac{1}{2}$ of 8 = ?

$\frac{1}{2}$ of 8 = 4

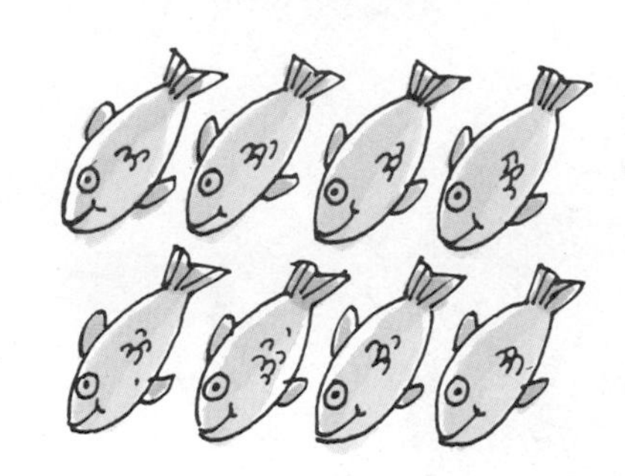

To find $\frac{1}{3}$ of a number, divide it by 3.

$\frac{1}{3}$ of 15 = ?

$\frac{1}{3}$ of 15 = 5

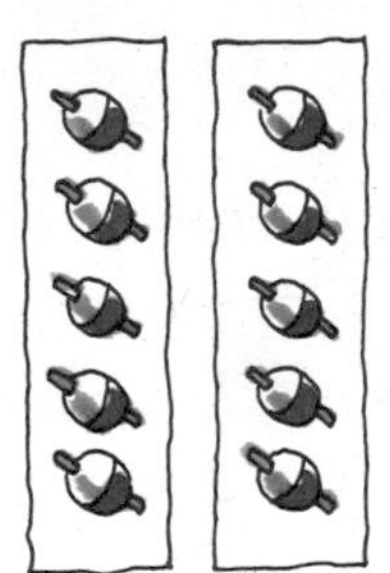
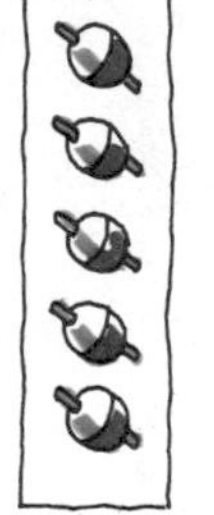

EXERCISES

Complete.

1. $\frac{1}{2}$ of 12 = ?

2. $\frac{1}{2}$ of 10 = ?

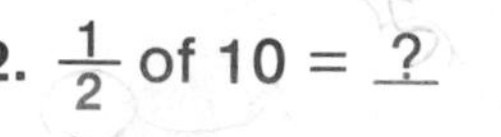

3. $\frac{1}{3}$ of 9 = ?

4. $\frac{1}{3}$ of 12 = ?

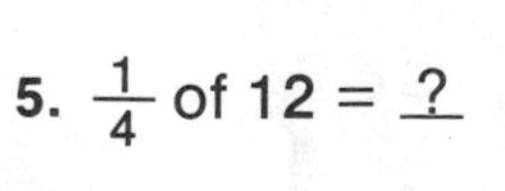

5. $\frac{1}{4}$ of 12 = ?

6. $\frac{1}{4}$ of 20 = ?

Complete.

7. $\frac{1}{2}$ of 4 = ?
8. $\frac{1}{3}$ of 6 = ?
9. $\frac{1}{5}$ of 25 = ?
10. $\frac{1}{6}$ of 18 = ?
11. $\frac{1}{4}$ of 8 = ?
12. $\frac{1}{2}$ of 8 = ?
13. $\frac{1}{4}$ of 28 = ?
14. $\frac{1}{5}$ of 30 = ?
15. $\frac{1}{3}$ of 21 = ?
16. $\frac{1}{6}$ of 24 = ?
17. $\frac{1}{2}$ of 6 = ?
18. $\frac{1}{8}$ of 24 = ?

Find the number of fishhooks each person has.

1 dozen = 12

19. Sonya: $\frac{1}{2}$ of a dozen.
20. Alex: $\frac{1}{3}$ of a dozen.
21. Maria: $\frac{1}{4}$ of a dozen.
22. Barry: $\frac{1}{6}$ of a dozen.

Solve.

23. Sonya caught 12 fish.
    Her cat ate $\frac{1}{6}$ of them.
    How many fish did her cat eat?
24. Alex caught 14 fish.
    $\frac{1}{2}$ of them were over 8 inches long. How many fish were over 8 inches long?
25. Maria caught 24 fish.
    She gave $\frac{1}{4}$ of them away.
    How many did she give away?
    How many did she keep?
26. Barry caught 20 fish.
    He gave $\frac{1}{5}$ of them away.
    How many did he keep?

Challenge!

Make up a problem.

Fraction of a number

To find a fraction of a number: Divide the number by the denominator, then multiply by the numerator.

$\frac{2}{3}$ of 12 = ?

Step 1. Divide 12 by 3 to find $\frac{1}{3}$. $12 \div 3 = 4$

Step 2. Multiply the quotient by 2 to find $\frac{2}{3}$. $4 \times 2 = 8$

$\frac{2}{3}$ of 12 = 8

$\frac{3}{5}$ of 15 = ?

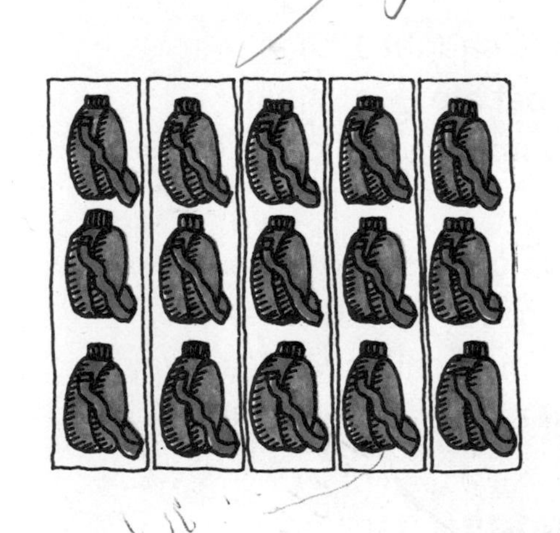

Step 1. Divide 15 by 5 to find $\frac{1}{5}$.

Step 2. Multiply the quotient by 3 to find $\frac{3}{5}$.

$\frac{3}{5}$ of 15 = 9

EXERCISES

Complete.

1. $\frac{2}{3}$ of 9 = ?

2. $\frac{2}{3}$ of 15 = ?

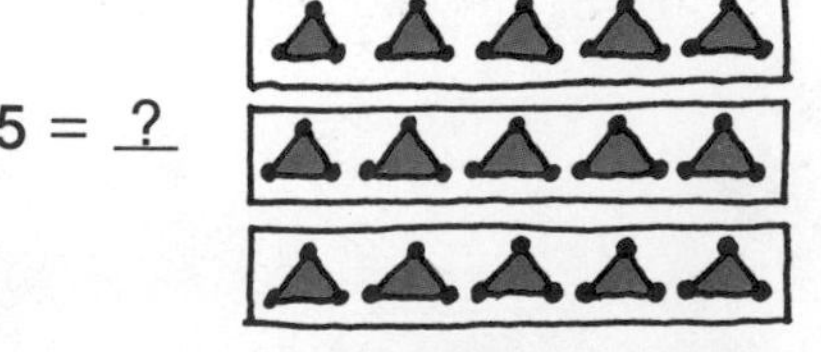

3. $\frac{3}{4}$ of 8 = ?

4. $\frac{3}{4}$ of 16 = ?

Divide by 3.
Multiply by 2.

Divide by 3.
Multiply by 3.

5. $\frac{1}{3}$ of 18 = ?

6. $\frac{2}{3}$ of 18 = ?

7. $\frac{3}{3}$ of 18 = ?

8. $\frac{1}{4}$ of 20 = ?

9. $\frac{2}{4}$ of 20 = ?

10. $\frac{3}{4}$ of 20 = ?

11. $\frac{1}{5}$ of 30 = ?

12. $\frac{2}{5}$ of 30 = ?

13. $\frac{3}{5}$ of 30 = ?

14. $\frac{1}{6}$ of 42 = ?

15. $\frac{1}{8}$ of 40 = ?

16. $\frac{3}{8}$ of 24 = ?

17. $\frac{3}{8}$ of 8 = ?

18. $\frac{5}{6}$ of 36 = ?

19. $\frac{5}{8}$ of 32 = ?

20. $\frac{5}{6}$ of 30 = ?

21. $\frac{7}{8}$ of 16 = ?

22. $\frac{5}{9}$ of 18 = ?

Find the sale price.

23. $\frac{2}{3}$ of marked price

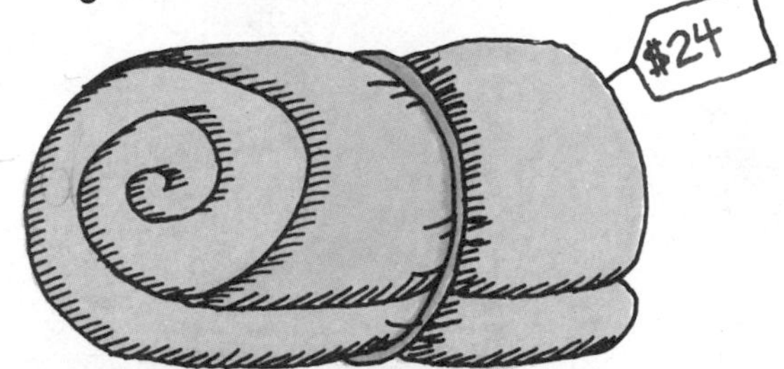

24. $\frac{3}{4}$ of marked price

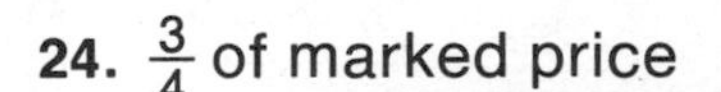

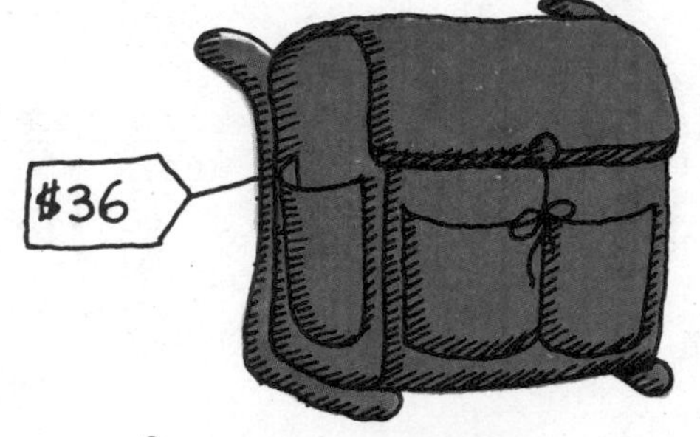

25. $\frac{3}{4}$ of marked price

26. $\frac{2}{3}$ of marked price

Solve.

27. Sally went on 15-kilometer hike. At the end of 3 hours, she had hiked $\frac{2}{3}$ of the way. How many kilometers had she hiked?

28. Brendan had $12. He spent $\frac{2}{3}$ of it for a camping knife. Later, he bought a canteen for $2.79. How much money did he have left?

Problem solving

EXAMPLE.
What is the sale price of the kickstand?

Step 1. Find how much off.

$\frac{1}{3}$ of \$6 = \$2

Step 2. Subtract to find sale price.

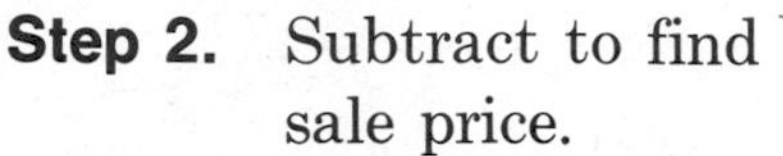

\$6 − \$2 = \$4

Solve.

1. Sarah wanted to buy the bicycle. She had \$62.75. How much more money did she need?

2. The store owner offered Juan \$37.50 for his old bicycle. How much more money would Juan need to buy the new bicycle?

3. Before the sale, what was the total price of a bell and a reflector?

4. Before the sale, how much more did the bell cost than the handle grips?

5. **a.** What is the regular price of the lock?
   b. How many dollars off on sale?
   c. What is the sale price?

6. **a.** What is the regular price of the pump?
   b. How many dollars off on sale?
   c. What is the sale price?

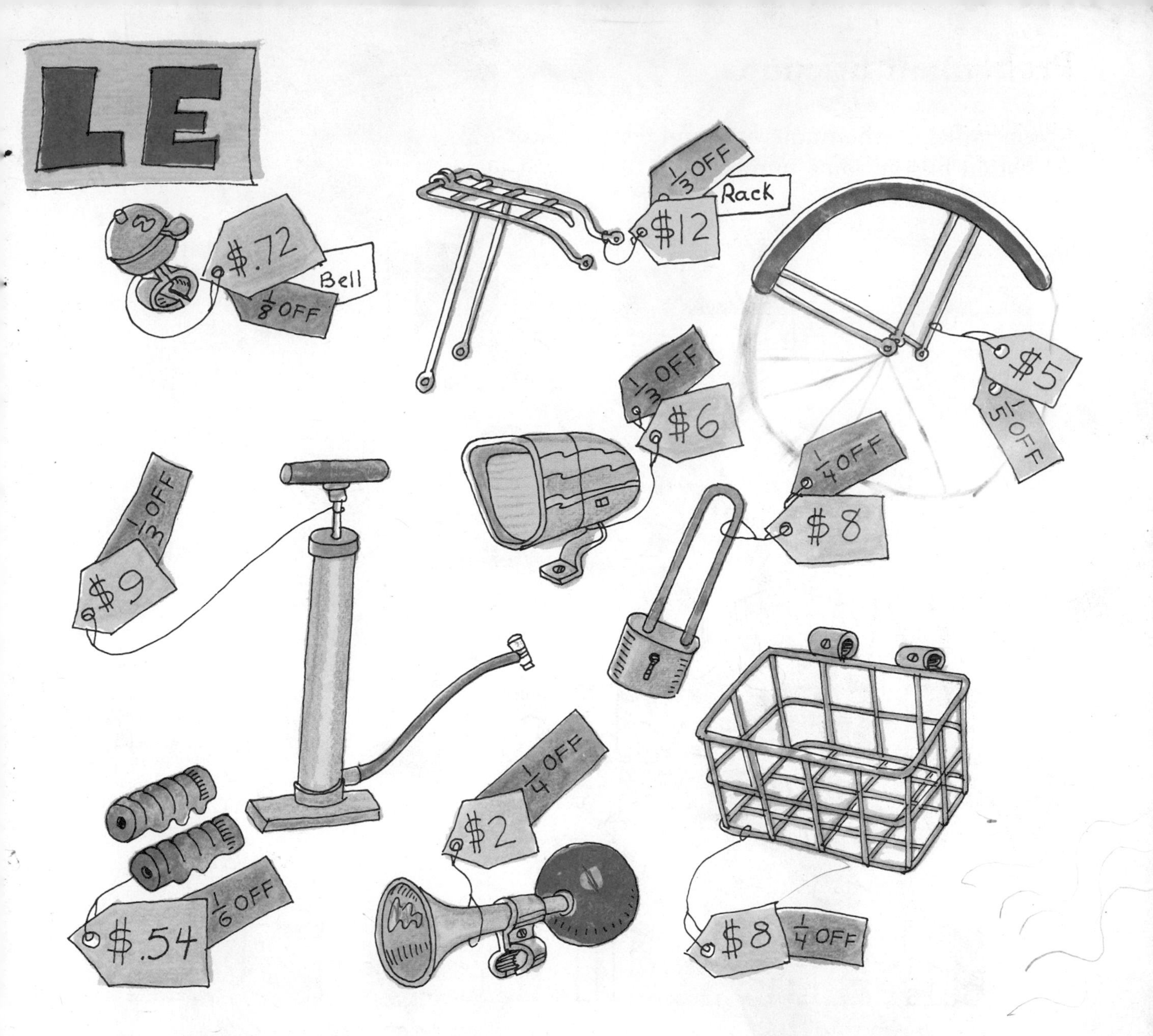

7. What is the sale price of the basket?

8. What is the sale price of the fender?

9. Eric bought a kickstand and a rack. What was the total price?

10. Dave bought a chain and a light. What was the total price?

11. Paul had $5. He bought a horn. How much money did he have left?

★ 12. Mary had $9.40. She bought a bell and a kickstand. How much money did she have left?

Equivalent fractions

If you multiply the numerator and denominator of a fraction by the same number (not 0), you get an equivalent fraction.

Complete.

10. $\frac{1}{4} \overset{\times 2}{\underset{\times 2}{=}} \frac{?}{?}$

11. $\frac{1}{3} \overset{\times 3}{\underset{\times 3}{=}} \frac{?}{?}$

12. $\frac{1}{2} \overset{\times 5}{\underset{\times 5}{=}} \frac{?}{?}$

13. $\frac{2}{3} \overset{\times 4}{\underset{\times 4}{=}} \frac{?}{?}$

14. $\frac{2}{2} \overset{\times 2}{\underset{\times 2}{=}} \frac{?}{?}$

15. $\frac{2}{5} \overset{\times 3}{\underset{\times 3}{=}} \frac{?}{?}$

16. $\frac{1}{6} \overset{\times 3}{\underset{\text{think!}}{=}} \frac{?}{?}$

17. $\frac{1}{8} \overset{\times 2}{\underset{\text{think!}}{=}} \frac{?}{?}$

18. $\frac{3}{8} \overset{\text{think!}}{\underset{\times 2}{=}} \frac{?}{?}$

Copy and complete.

19. a. $\frac{1}{3} = \frac{?}{6}$
b. $\frac{1}{3} = \frac{?}{9}$
c. $\frac{1}{3} = \frac{?}{12}$
d. $\frac{1}{3} = \frac{?}{15}$
e. $\frac{1}{3} = \frac{?}{18}$
f. $\frac{1}{3} = \frac{?}{21}$

20. a. $\frac{1}{4} = \frac{?}{8}$
b. $\frac{1}{4} = \frac{?}{12}$
c. $\frac{1}{4} = \frac{?}{16}$
d. $\frac{1}{4} = \frac{?}{20}$
e. $\frac{1}{4} = \frac{?}{24}$
f. $\frac{1}{4} = \frac{?}{28}$

21. a. $\frac{3}{4} = \frac{?}{8}$
b. $\frac{3}{4} = \frac{?}{12}$
c. $\frac{3}{4} = \frac{?}{16}$
d. $\frac{3}{4} = \frac{?}{20}$
e. $\frac{3}{4} = \frac{?}{24}$
f. $\frac{3}{4} = \frac{?}{28}$

22. a. $\frac{2}{5} = \frac{?}{10}$
b. $\frac{2}{5} = \frac{?}{15}$
c. $\frac{2}{5} = \frac{?}{20}$
d. $\frac{2}{5} = \frac{?}{25}$
e. $\frac{2}{5} = \frac{?}{30}$
f. $\frac{2}{5} = \frac{?}{35}$

Challenge!

Use each digit once to build two equivalent fractions.

23.

24.

25.

26.

Fractions in lowest terms

The **terms** of a fraction are the numerator and the denominator. If you divide the terms of a fraction by the same number (called the **common factor**), you get an equivalent fraction in **lower terms.**

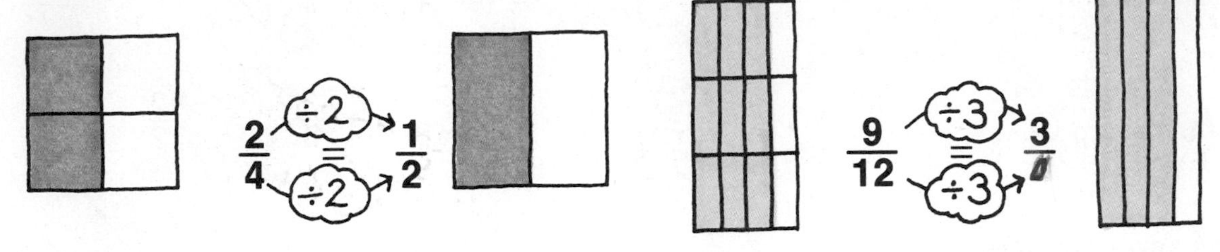

If you divide the terms of a fraction by the **greatest common factor,** you get an equivalent fraction in **lowest terms.**

EXERCISES

Complete.

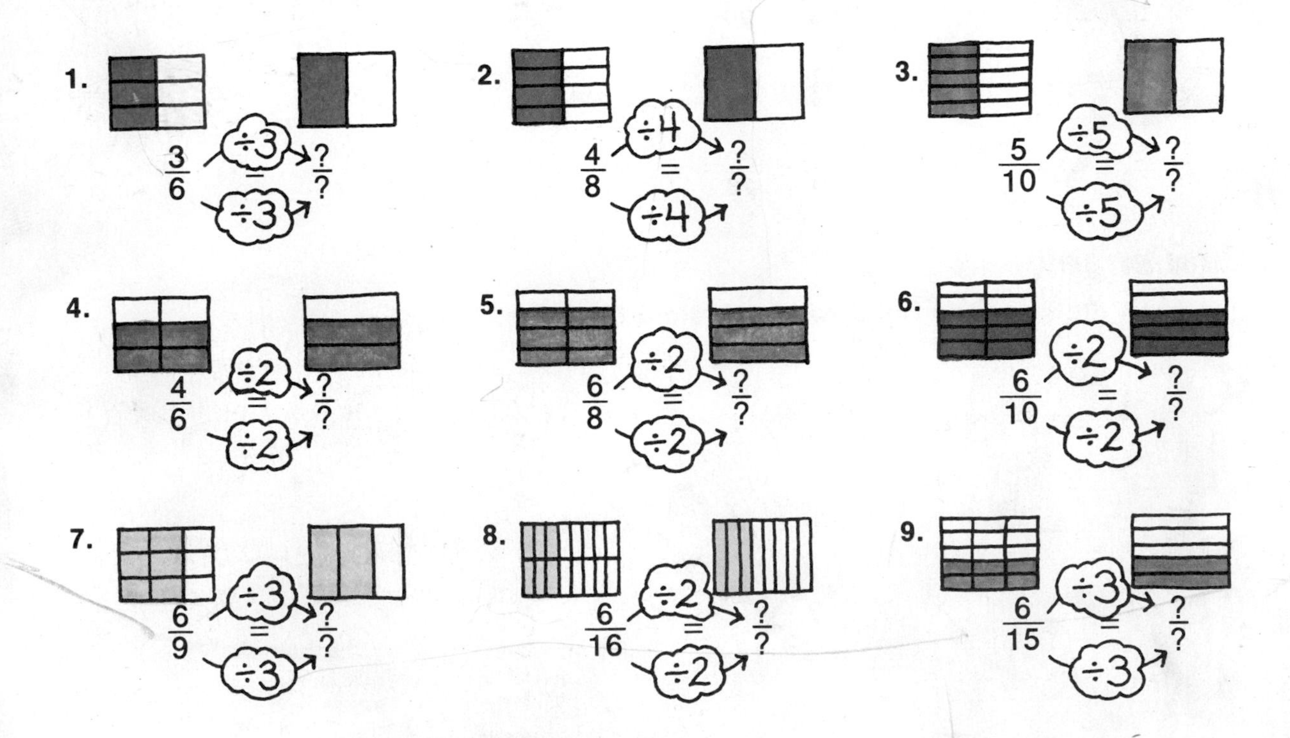

Complete.

10. $\frac{5}{10} \xrightarrow{\div 5} \frac{?}{?}$

11. $\frac{3}{9} \xrightarrow{\div 3} \frac{?}{?}$

12. $\frac{6}{12} \xrightarrow{\div 6} \frac{?}{?}$

13. $\frac{3}{12} \xrightarrow{\div 3} \frac{?}{?}$

14. $\frac{6}{10} \xrightarrow{\div 2} \frac{?}{?}$

15. $\frac{4}{12} \xrightarrow{\div 4} \frac{?}{?}$

16. $\frac{10}{16}$ (÷2 numerator, think denominator) $\frac{?}{?}$

17. $\frac{12}{16}$ (think numerator, ÷4 denominator) $\frac{?}{?}$

18. $\frac{8}{8}$ (÷8 numerator, think denominator) $\frac{?}{?}$

Write each fraction in lowest terms. *Hint:* Divide the numerator and the denominator by the greatest common factor.

19. $\frac{2}{4}$	20. $\frac{2}{10}$	21. $\frac{3}{9}$	22. $\frac{4}{6}$	23. $\frac{5}{10}$	24. $\frac{2}{12}$
25. $\frac{2}{6}$	26. $\frac{6}{8}$	27. $\frac{3}{6}$	28. $\frac{9}{12}$	29. $\frac{2}{16}$	30. $\frac{10}{15}$
31. $\frac{8}{12}$	32. $\frac{6}{6}$	33. $\frac{4}{12}$	34. $\frac{6}{9}$	35. $\frac{4}{8}$	36. $\frac{15}{20}$

Challenge!

Guess my number.

37. My fraction is equivalent to $\frac{2}{3}$. The denominator is 6.

38. The numerator of my fraction is 12. It is equivalent to $\frac{3}{4}$.

39. The denominator of my fraction is 6. The fraction is equivalent to $\frac{12}{18}$.

40. My fraction is equivalent to $\frac{16}{24}$. It is in lowest terms.

More on fractions in lowest terms

Remember, if you divide the numerator and denominator by the greatest common factor, you get the fraction in lowest terms.

You can also divide by a smaller number, then another smaller number, and so on, until the fraction is in lowest terms.

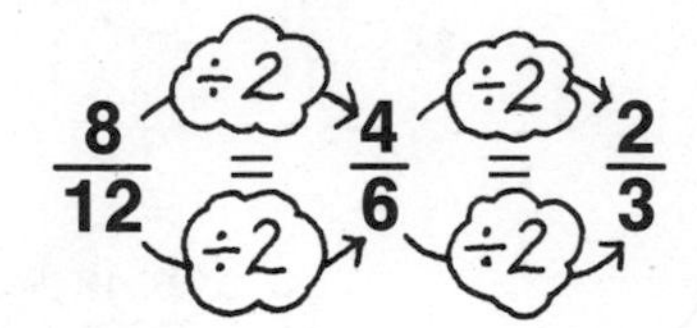

EXERCISES

Write each fraction in lowest terms.

1. $\frac{5}{10}$ 2. $\frac{9}{12}$ 3. $\frac{15}{20}$ 4. $\frac{24}{36}$ 5. $\frac{15}{25}$ 6. $\frac{15}{30}$

7. $\frac{16}{24}$ 8. $\frac{16}{18}$ 9. $\frac{12}{18}$ 10. $\frac{12}{16}$ 11. $\frac{9}{24}$ 12. $\frac{6}{24}$

13. $\frac{9}{36}$ 14. $\frac{8}{24}$ 15. $\frac{18}{27}$ 16. $\frac{18}{24}$ 17. $\frac{42}{56}$ 18. $\frac{35}{40}$

19. $\frac{27}{36}$ 20. $\frac{24}{48}$ 21. $\frac{72}{81}$ 22. $\frac{56}{72}$ 23. $\frac{48}{56}$ 24. $\frac{40}{56}$

SKILL GAME

FRACTION BINGO!

1. Make these fraction cards:

$\frac{6}{9}$	$\frac{8}{10}$	$\frac{2}{12}$	$\frac{4}{14}$
$\frac{3}{15}$	$\frac{6}{16}$	$\frac{15}{18}$	$\frac{15}{20}$
$\frac{3}{21}$	$\frac{3}{24}$	$\frac{9}{27}$	$\frac{7}{28}$

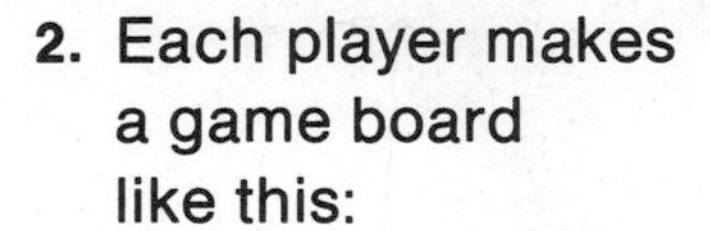

2. Each player makes a game board like this:

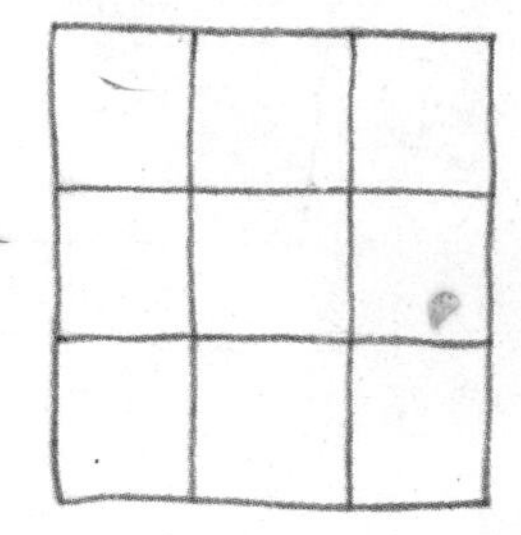

3. Each player fills in the game board with nine of these fractions:

$\frac{1}{3}, \frac{2}{3}, \frac{1}{4}, \frac{3}{4}, \frac{1}{5}, \frac{4}{5}, \frac{1}{6}, \frac{5}{6}, \frac{1}{7}, \frac{2}{7}, \frac{1}{8}, \frac{3}{8}$

4. Put the fraction cards facedown on a table.

5. Turn a fraction card over. If a player's board has an equivalent fraction, the player marks an X on it.

6. Continue turning cards over until someone gets three X's in a row. That person is the winner.

For the game at the top of the page, the next four cards were 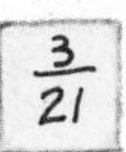 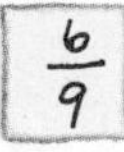 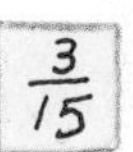 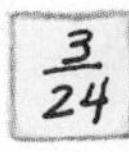 Who won the game?

More about equivalent fractions

Some of the points on the number lines below have been labeled with fractions. The fractions that "line up" are equivalent fractions. They all name the same number.

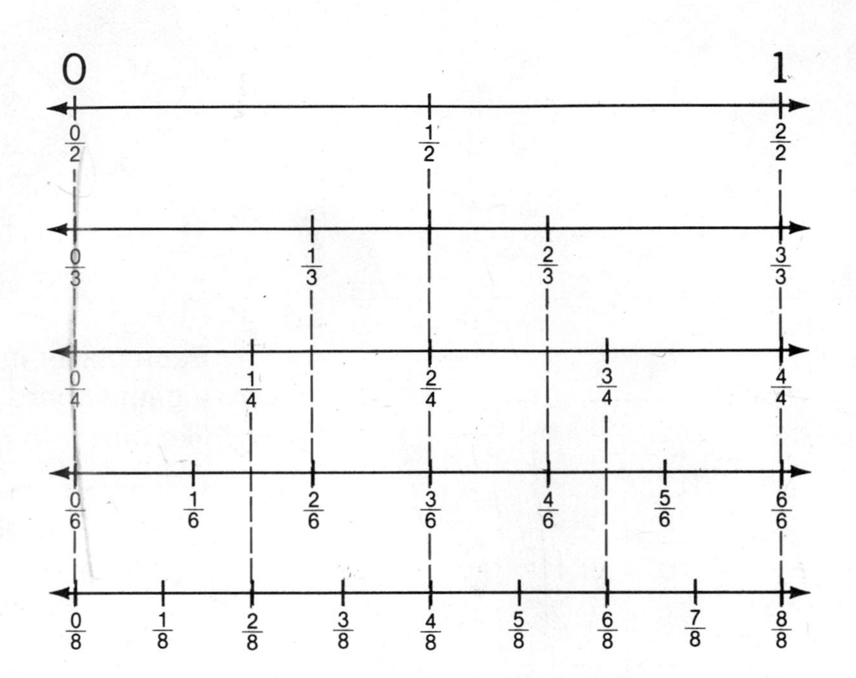

EXERCISES

Copy and complete. Check your work by using the number lines above.

1. $\frac{1}{2} = \frac{?}{8}$	**2.** $\frac{6}{6} = \frac{?}{3}$	**3.** $\frac{1}{2} = \frac{?}{6}$	**4.** $\frac{1}{2} = \frac{?}{4}$
5. $\frac{1}{3} = \frac{?}{6}$	**6.** $\frac{4}{8} = \frac{?}{4}$	**7.** $\frac{2}{8} = \frac{?}{4}$	**8.** $\frac{1}{4} = \frac{?}{8}$
9. $\frac{0}{8} = \frac{?}{2}$	**10.** $\frac{6}{8} = \frac{?}{4}$	**11.** $\frac{2}{3} = \frac{?}{6}$	**12.** $\frac{6}{8} = \frac{?}{4}$
13. $\frac{6}{6} = \frac{?}{2}$	**14.** $\frac{3}{4} = \frac{?}{8}$	**15.** $\frac{4}{8} = \frac{?}{2}$	**16.** $\frac{4}{6} = \frac{?}{3}$

Write each fraction in lowest terms.

17. $\frac{2}{4}$ 18. $\frac{6}{8}$ 19. $\frac{3}{9}$ 20. $\frac{5}{10}$

21. $\frac{4}{6}$ 22. $\frac{9}{15}$ 23. $\frac{10}{14}$ 24. $\frac{14}{16}$

25. $\frac{12}{15}$ 26. $\frac{3}{12}$ 27. $\frac{12}{16}$ 28. $\frac{6}{12}$

29. $\frac{5}{10}$ 30. $\frac{6}{14}$ 31. $\frac{8}{10}$ 32. $\frac{8}{12}$

33. $\frac{9}{12}$ 34. $\frac{4}{8}$ 35. $\frac{3}{6}$ 36. $\frac{2}{12}$

37. $\frac{6}{10}$ 38. $\frac{4}{12}$ 39. $\frac{10}{15}$ 40. $\frac{10}{12}$

41. $\frac{6}{9}$ 42. $\frac{6}{15}$ 43. $\frac{8}{16}$ 44. $\frac{5}{15}$

45. a. Write five fractions equivalent to $\frac{1}{2}$.
 b. Write five fractions equivalent to $\frac{1}{3}$.
 c. Pick a fraction from the first list and a fraction from the second list that have the same denominator.

46. a. Write five fractions equivalent to $\frac{2}{3}$.
 b. Write five fractions equivalent to $\frac{1}{4}$.
 c. Pick a fraction from the first list and a fraction from the second list that have the same denominator.

47. a. Write five fractions equivalent to $\frac{3}{4}$.
 b. Write five fractions equivalent to $\frac{1}{5}$.
 c. Pick a fraction from the first list and a fraction from the second list that have the same denominator.

KEEPING SKILLS SHARP

1. $2\overline{)18}$
2. $4\overline{)32}$
3. $3\overline{)18}$
4. $6\overline{)42}$
5. $8\overline{)64}$
6. $5\overline{)55}$
7. $3\overline{)63}$
8. $4\overline{)48}$
9. $5\overline{)50}$
10. $3\overline{)96}$
11. $4\overline{)52}$
12. $6\overline{)84}$
13. $5\overline{)75}$
14. $2\overline{)96}$
15. $4\overline{)72}$
16. $5\overline{)78}$
17. $7\overline{)93}$
18. $2\overline{)65}$
19. $6\overline{)89}$
20. $3\overline{)85}$

Comparing fractions

Pictures can help you to compare fractions.

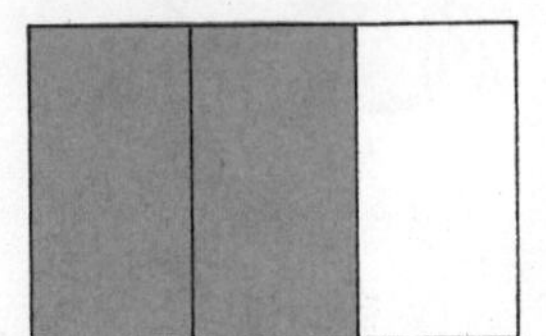 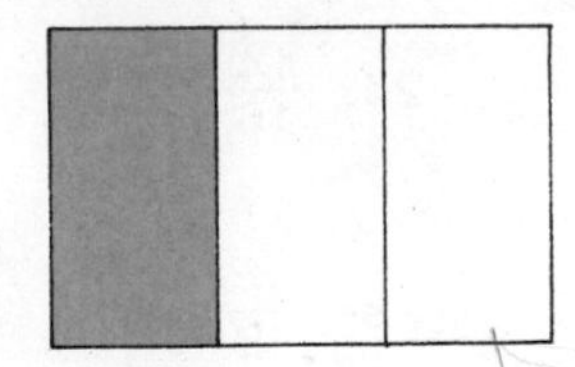 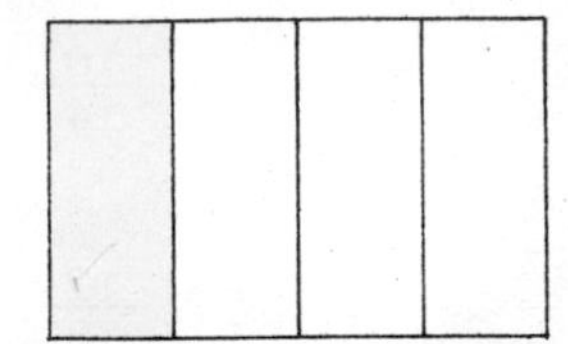 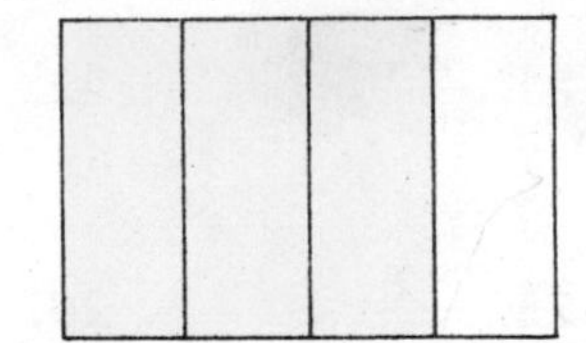

$\frac{2}{3} > \frac{1}{3}$

(is greater than)

$\frac{1}{4} < \frac{3}{4}$

(is less than)

Numbers get greater.

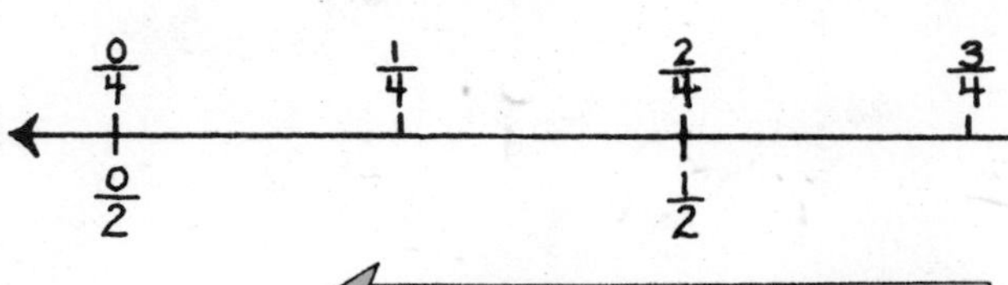

Numbers get less.

$\frac{1}{4} < \frac{1}{2}$

$\frac{3}{4} > \frac{1}{2}$

EXERCISES

< or >?

1.  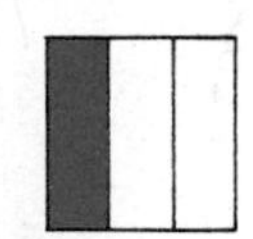

$\frac{1}{2}$ ● $\frac{1}{3}$

2. 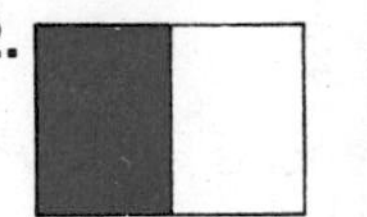 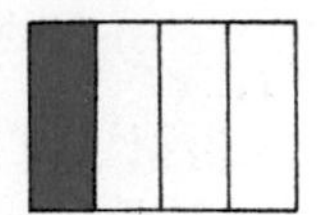

 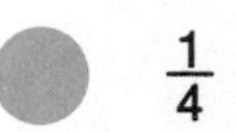

$\frac{1}{2}$ ● $\frac{1}{4}$

3. 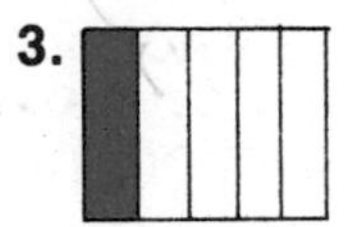 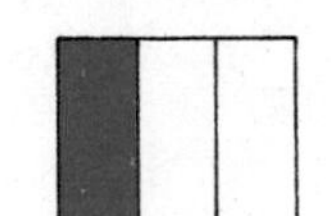

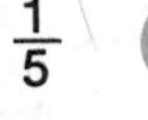 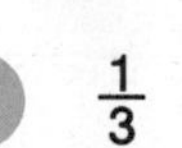

$\frac{1}{5}$ ● $\frac{1}{3}$

4. 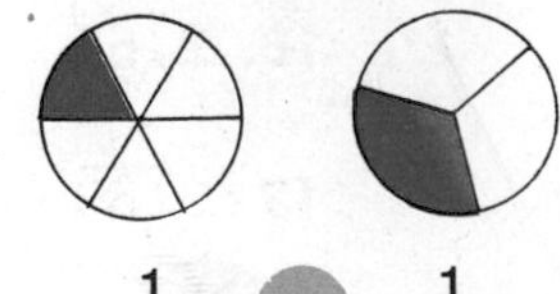

$\frac{1}{6}$ ● $\frac{1}{3}$

5. 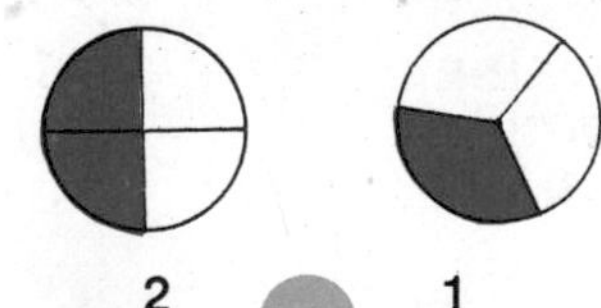

$\frac{2}{4}$ ● $\frac{1}{3}$

6. 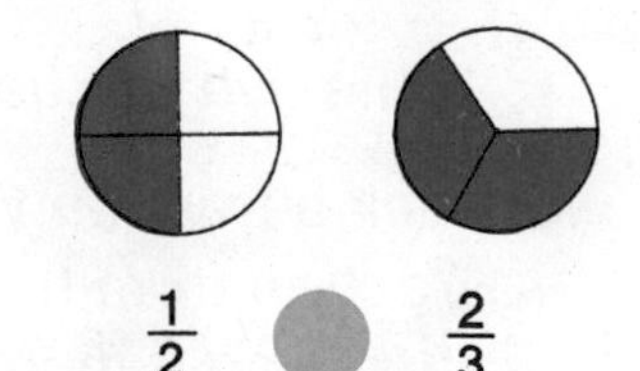

$\frac{1}{2}$ ● $\frac{2}{3}$

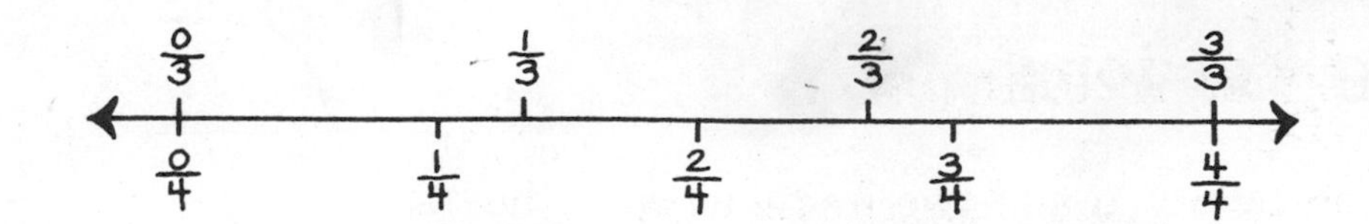

<, >, or =? Use the number line.

7. $\frac{1}{3}$ ● $\frac{2}{3}$	**8.** $\frac{2}{3}$ ● $\frac{3}{3}$	**9.** $\frac{2}{3}$ ● $\frac{1}{3}$	**10.** $\frac{2}{4}$ ● $\frac{0}{4}$
11. $\frac{1}{4}$ ● $\frac{3}{4}$	**12.** $\frac{0}{3}$ ● $\frac{0}{4}$	**13.** $\frac{3}{4}$ ● $\frac{1}{4}$	**14.** $\frac{3}{4}$ ● $\frac{4}{4}$
15. $\frac{1}{3}$ ● $\frac{1}{4}$	**16.** $\frac{3}{4}$ ● $\frac{2}{3}$	**17.** $\frac{2}{4}$ ● $\frac{2}{3}$	**18.** $\frac{3}{3}$ ● $\frac{4}{4}$

<, >, or =?

19. $\frac{1}{6}$ ● $\frac{3}{6}$	**20.** $\frac{5}{9}$ ● $\frac{3}{9}$	**21.** $\frac{3}{8}$ ● $\frac{5}{8}$	**22.** $\frac{4}{9}$ ● $\frac{7}{9}$
23. $\frac{1}{2}$ ● $\frac{1}{4}$	**24.** $\frac{1}{6}$ ● $\frac{1}{3}$	**25.** $\frac{5}{6}$ ● $\frac{1}{2}$	**26.** $\frac{3}{4}$ ● $\frac{5}{8}$

Which would you rather have,

27. $\frac{1}{3}$ of a dozen cookies or $\frac{2}{3}$ of a dozen cookies?

28. $\frac{1}{2}$ of a dollar or $\frac{1}{4}$ of a dollar?

Which is shorter,

29. $\frac{1}{2}$ of an hour or $\frac{3}{4}$ of an hour?

30. $\frac{2}{3}$ of a day or $\frac{3}{4}$ of a day?

Which is more,

31. $\frac{1}{3}$ of a dozen or $\frac{1}{4}$ of a dozen?

32. $\frac{2}{3}$ of 24 children or $\frac{3}{4}$ of 24 children?

Adding fractions

Pictures can help you add two fractions with the same denominator.

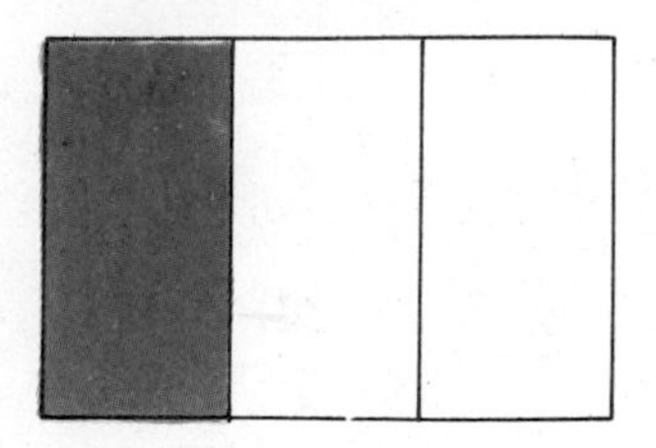

$\frac{1}{3}$

$\frac{1}{3} + \frac{1}{3} = \frac{2}{3}$

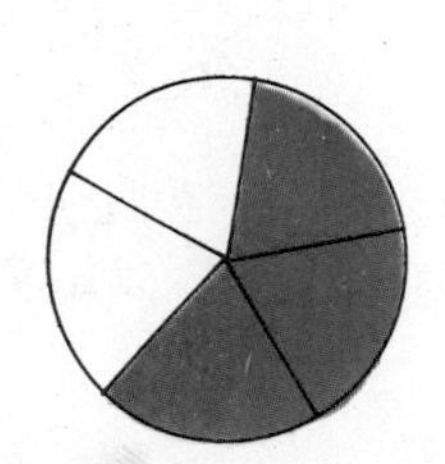

$\frac{3}{5}$

$\frac{3}{5} + \frac{1}{5} = \frac{4}{5}$

When adding two fractions that have the same denominator:

Step 1. Add the numerators to get the numerator of the sum.

Step 2. Use the same denominator for the denominator of the sum.

EXERCISES

Give each sum.

1.  $\frac{2}{4} + \frac{1}{4}$

2.  $\frac{2}{5} + \frac{2}{5}$

3. 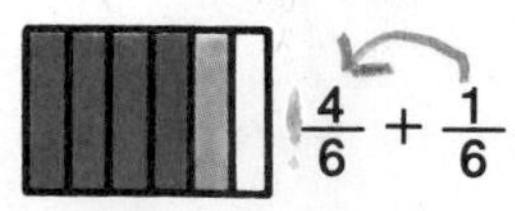 $\frac{4}{6} + \frac{1}{6}$

4. 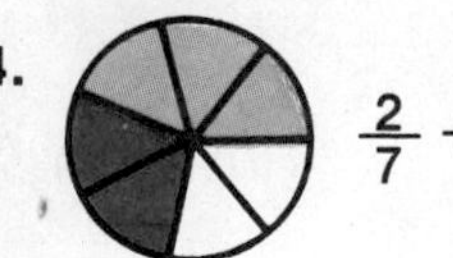 $\frac{2}{7} + \frac{3}{7}$

5. 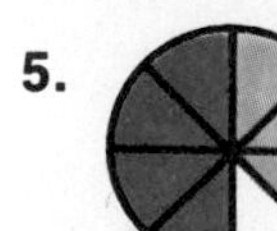 $\frac{4}{8} + \frac{3}{8}$

6. 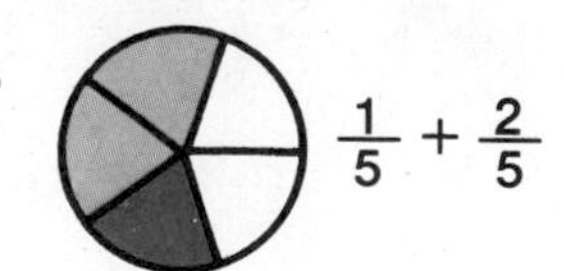 $\frac{1}{5} + \frac{2}{5}$

Give each sum.

7. $\frac{1}{3} + \frac{1}{3}$ 8. $\frac{2}{5} + \frac{2}{5}$ 9. $\frac{0}{2} + \frac{1}{2}$ 10. $\frac{3}{9} + \frac{2}{9}$

11. $\frac{1}{8} + \frac{4}{8}$ 12. $\frac{3}{8} + \frac{2}{8}$ 13. $\frac{1}{10} + \frac{6}{10}$ 14. $\frac{2}{5} + \frac{1}{5}$

Add.

15. $\frac{0}{4} + \frac{3}{4}$ 16. $\frac{4}{6} + \frac{1}{6}$ 17. $\frac{4}{7} + \frac{2}{7}$ 18. $\frac{2}{5} + \frac{1}{5}$ 19. $\frac{1}{3} + \frac{1}{3}$ 20. $\frac{2}{9} + \frac{5}{9}$

21. $\frac{1}{7} + \frac{5}{7}$ 22. $\frac{0}{12} + \frac{5}{12}$ 23. $\frac{4}{12} + \frac{3}{12}$ 24. $\frac{2}{8} + \frac{3}{8}$ 25. $\frac{3}{9} + \frac{5}{9}$ 26. $\frac{1}{6} + \frac{4}{6}$

Add. Write each sum in lowest terms.

27. $\frac{1}{4} + \frac{1}{4}$ 28. $\frac{1}{8} + \frac{1}{8}$ 29. $\frac{1}{6} + \frac{1}{6}$ 30. $\frac{1}{10} + \frac{1}{10}$ 31. $\frac{5}{8} + \frac{1}{8}$ 32. $\frac{2}{9} + \frac{4}{9}$

33. $\frac{5}{12} + \frac{1}{12}$ 34. $\frac{3}{10} + \frac{5}{10}$ 35. $\frac{3}{8} + \frac{3}{8}$ 36. $\frac{5}{10} + \frac{1}{10}$ 37. $\frac{1}{6} + \frac{3}{6}$ 38. $\frac{1}{10} + \frac{3}{10}$

KEEPING SKILLS SHARP

Give the least common multiple.

1. 2, 3   2. 3, 4   3. 2, 8   4. 2, 4   5. 3, 5

6. 3, 6   7. 6, 4   8. 8, 4   9. 2, 6   10. 3, 9

11. 3, 7   12. 8, 6   13. 5, 7   14. 6, 9   15. 5, 6

More about adding

The fractions on each strip are equivalent.

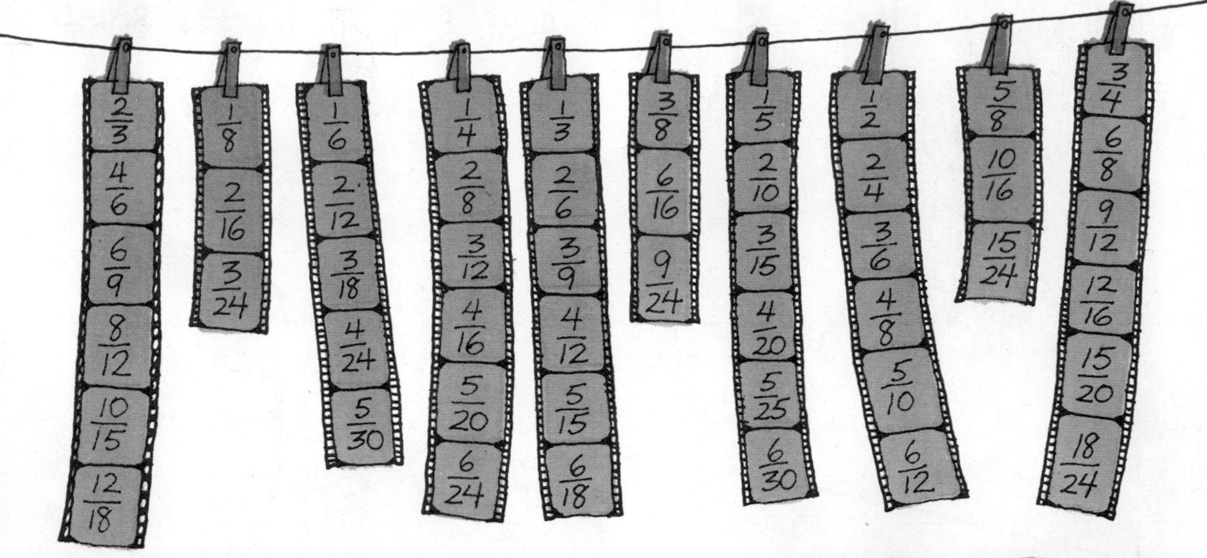

You can use the equivalent-fraction strips to add fractions that have different denominators.

EXAMPLE 1.

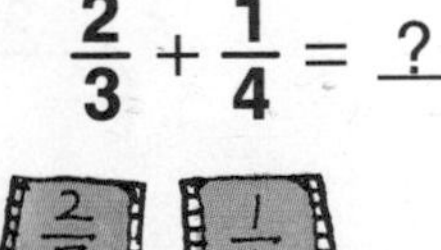

$\frac{2}{3} + \frac{1}{4} = \underline{?}$

$\frac{2}{3} + \frac{1}{4} = \frac{8}{12} + \frac{3}{12}$

$= \frac{11}{12}$

EXAMPLE 2.

$\frac{1}{2} + \frac{3}{8} = \underline{?}$

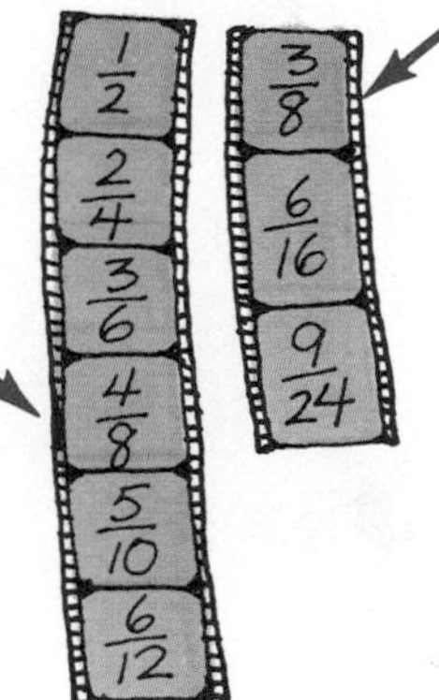

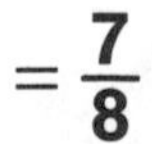

$\frac{1}{2} + \frac{3}{8} = \frac{4}{8} + \frac{3}{8}$

$= \frac{7}{8}$

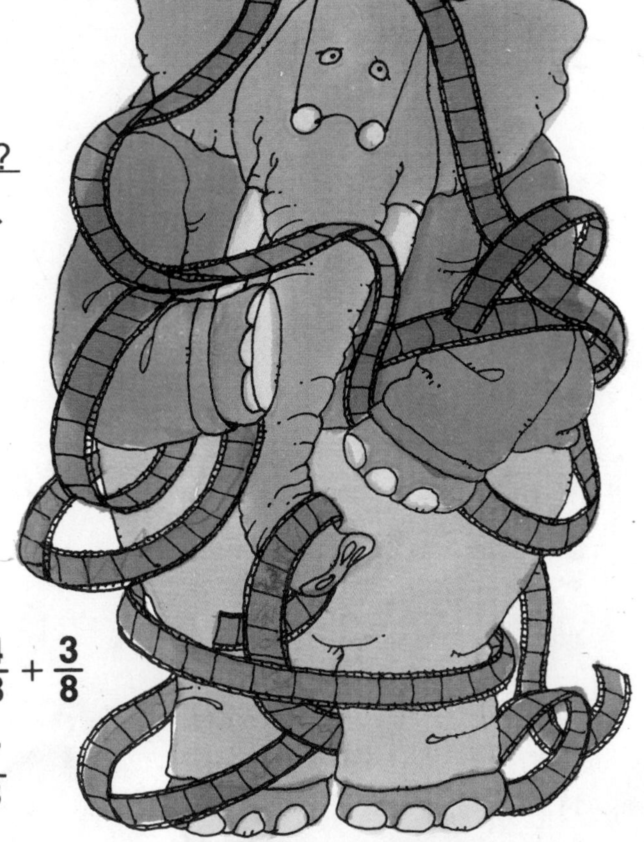

EXERCISES

Give each sum.

1. $\frac{1}{3} + \frac{1}{2}$

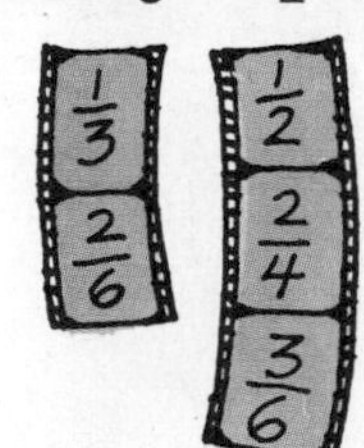

2. $\frac{1}{8} + \frac{1}{2}$

3. $\frac{1}{2} + \frac{3}{8}$

4. $\frac{1}{6} + \frac{1}{4}$

5. $\frac{1}{4} + \frac{5}{8}$

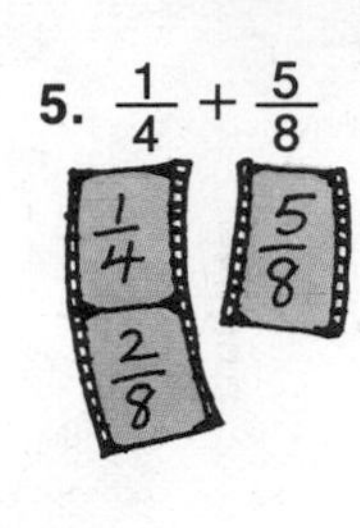

6. $\frac{1}{3} + \frac{1}{4}$

Give each sum. If you need to, use the fraction strips.

7. $\frac{1}{4} + \frac{1}{8}$
8. $\frac{1}{4} + \frac{3}{8}$
9. $\frac{2}{3} + \frac{1}{6}$
10. $\frac{1}{5} + \frac{2}{3}$
11. $\frac{1}{2} + \frac{1}{5}$
12. $\frac{3}{4} + \frac{1}{6}$
13. $\frac{1}{6} + \frac{1}{5}$
14. $\frac{1}{5} + \frac{3}{4}$
15. $\frac{1}{6} + \frac{3}{8}$
16. $\frac{1}{8} + \frac{1}{6}$
17. $\frac{1}{5} + \frac{1}{4}$
18. $\frac{1}{6} + \frac{1}{4}$
19. $\frac{1}{2} + \frac{1}{4}$
20. $\frac{1}{5} + \frac{1}{3}$
21. $\frac{5}{8} + \frac{1}{6}$

Solve.

22. Robert mowed $\frac{1}{2}$ of his lawn before dinner and $\frac{1}{4}$ of it after dinner. What fraction of the lawn did he mow?

23. Jan bought $\frac{1}{4}$ pound of Swiss cheese and $\frac{5}{8}$ pound of American cheese. How many pounds of cheese did she buy?

Subtracting fractions

You can draw pictures to help you subtract fractions.

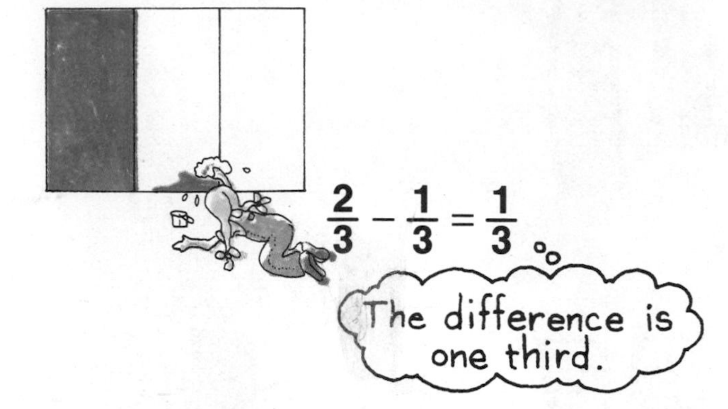

$\frac{4}{5} - \frac{1}{5} = \frac{3}{5}$

The difference is three fifths.

When subtracting two fractions that have the same denominator:

Step 1. Subtract the numerators to get the numerator of the difference.

Step 2. Use the same denominator for the denominator of the difference.

EXERCISES

Give each difference.

1. $\frac{3}{4} - \frac{2}{4}$

2. 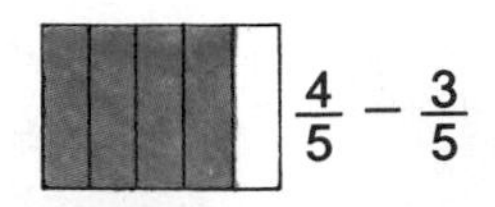 $\frac{4}{5} - \frac{3}{5}$

3. 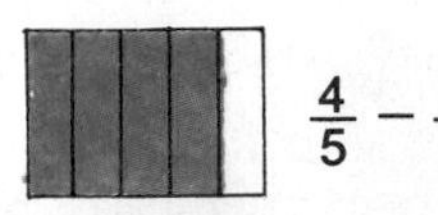 $\frac{4}{5} - \frac{2}{5}$

4. 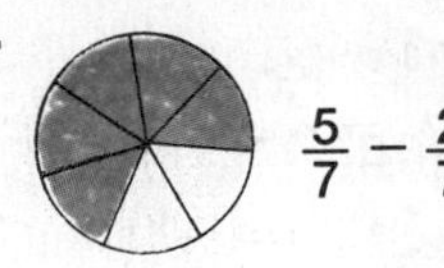 $\frac{5}{7} - \frac{2}{7}$

5. 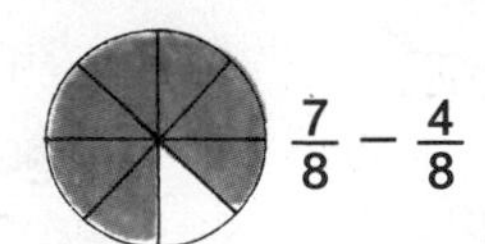 $\frac{7}{8} - \frac{4}{8}$

6. 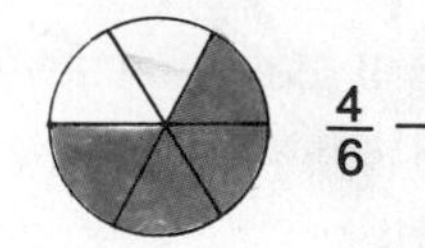 $\frac{4}{6} - \frac{3}{6}$

7. $\frac{3}{5} - \frac{2}{5}$

8. $\frac{5}{5} - \frac{1}{5}$

9. $\frac{3}{6} - \frac{2}{6}$

10. $\frac{2}{3} - \frac{1}{3}$

11. $\frac{6}{7} - \frac{4}{7}$

12. $\frac{3}{3} - \frac{2}{3}$

13. $\frac{5}{8} - \frac{2}{8}$

14. $\frac{3}{5} - \frac{3}{5}$

Subtract.

15. $\frac{3}{9} - \frac{1}{9}$

16. $\frac{5}{8} - \frac{4}{8}$

17. $\frac{2}{10} - \frac{1}{10}$

18. $\frac{4}{5} - \frac{3}{5}$

19. $\frac{5}{8} - \frac{0}{8}$

20. $\frac{3}{6} - \frac{2}{6}$

21. $\frac{8}{9} - \frac{3}{9}$

22. $\frac{5}{6} - \frac{4}{6}$

23. $\frac{9}{12} - \frac{2}{12}$

24. $\frac{7}{12} - \frac{2}{12}$

25. $\frac{8}{10} - \frac{5}{10}$

26. $\frac{7}{8} - \frac{7}{8}$

Subtract. Then write the answer in lowest terms.

27. $\frac{5}{8} - \frac{1}{8}$

$\frac{4}{8} = \frac{1}{2}$

28. $\frac{7}{10} - \frac{5}{10}$

29. $\frac{5}{6} - \frac{3}{6}$

30. $\frac{7}{8} - \frac{1}{8}$

31. $\frac{8}{9} - \frac{2}{9}$

32. $\frac{10}{12} - \frac{2}{12}$

Solve.

33. Mio had $\frac{2}{3}$ of a pound of flour. She used $\frac{1}{3}$ of a pound. What fraction of a pound did she have left?

34. Jack bought $\frac{2}{4}$ of a pound of cheese. He ate $\frac{1}{4}$ of a pound. What fraction of a pound did he have left?

35. Laurel needs $\frac{2}{3}$ of a pound of white flour and $\frac{1}{4}$ of a pound of rye flour for some rolls. What fraction of a pound does she need?

36. Roberto needs $\frac{6}{8}$ of a pound of ham to make sandwiches. He has $\frac{2}{8}$ of a pound. What fraction of a pound does he still need?

More about subtracting

You can use these equivalent-fraction strips to subtract fractions with different denominators.

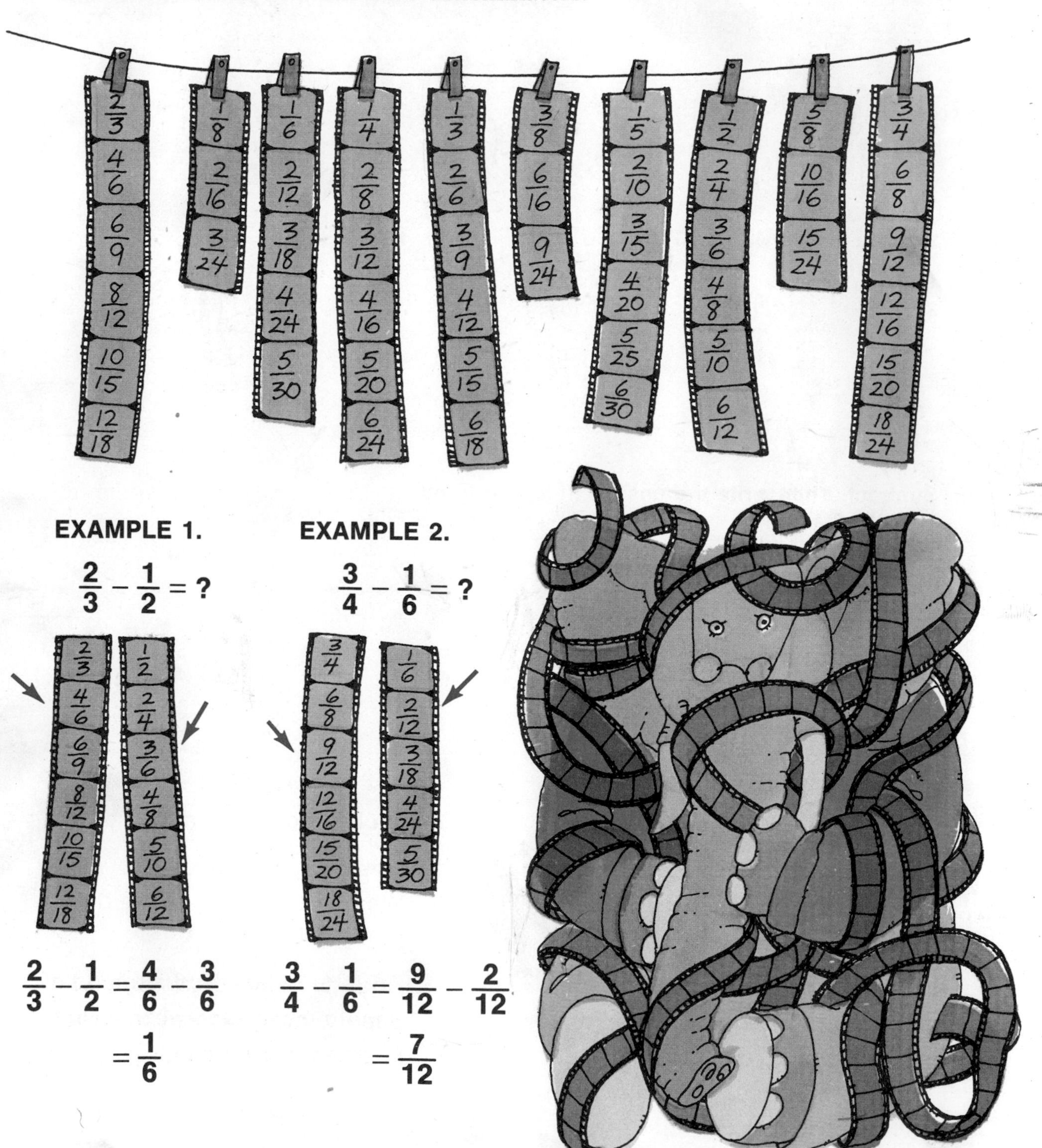

EXAMPLE 1.

$\frac{2}{3} - \frac{1}{2} = ?$

$$\frac{2}{3} - \frac{1}{2} = \frac{4}{6} - \frac{3}{6}$$
$$= \frac{1}{6}$$

EXAMPLE 2.

$\frac{3}{4} - \frac{1}{6} = ?$

$$\frac{3}{4} - \frac{1}{6} = \frac{9}{12} - \frac{2}{12}$$
$$= \frac{7}{12}$$

EXERCISES

Give each difference.

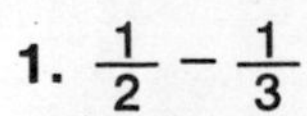

1. $\frac{1}{2} - \frac{1}{3}$

2. $\frac{5}{8} - \frac{1}{2}$

3. $\frac{2}{3} - \frac{1}{4}$

4. $\frac{3}{8} - \frac{1}{6}$

5. $\frac{3}{4} - \frac{5}{8}$

6. $\frac{3}{4} - \frac{1}{8}$

Give each difference. If you need to, use the fraction strips.

7. $\frac{1}{3} - \frac{1}{6}$
8. $\frac{1}{4} - \frac{1}{8}$
9. $\frac{3}{8} - \frac{1}{4}$
10. $\frac{1}{5} - \frac{1}{6}$
11. $\frac{2}{3} - \frac{1}{5}$
12. $\frac{1}{4} - \frac{1}{6}$
13. $\frac{3}{4} - \frac{1}{5}$
14. $\frac{1}{2} - \frac{1}{5}$
15. $\frac{5}{8} - \frac{1}{6}$
16. $\frac{2}{3} - \frac{1}{2}$
17. $\frac{1}{4} - \frac{1}{5}$
18. $\frac{1}{6} - \frac{1}{8}$
19. $\frac{1}{3} - \frac{1}{5}$
20. $\frac{1}{2} - \frac{3}{8}$
21. $\frac{3}{4} - \frac{1}{6}$

Solve.

22. Kristen bought $\frac{3}{8}$ yard of blue ribbon and $\frac{1}{2}$ yard of red ribbon. How many yards of ribbon did she buy?

23. Andrew had $\frac{2}{3}$ dozen eggs. He used $\frac{1}{6}$ dozen in a cake. What fraction of a dozen did he have left?

Adding and subtracting fractions

You can add and subtract fractions without making long lists of equivalent fractions.

EXAMPLE 1. $\frac{1}{2} + \frac{2}{5}$

Step 1. Find the least common multiple of the denominators.

$$\frac{1}{2} + \frac{2}{5}$$

Step 2. Use the least common multiple for the common denominator.

$$\frac{1}{2} + \frac{2}{5} = \frac{}{10} + \frac{}{10}$$

Step 3. Complete the equivalent fractions.

$$\frac{1}{2} + \frac{2}{5} = \frac{5}{10} + \frac{4}{10}$$

Step 4. Add.

$$\frac{1}{2} + \frac{2}{5} = \frac{5}{10} + \frac{4}{10} = \frac{9}{10}$$

EXAMPLE 2. $\frac{5}{8} - \frac{1}{4}$

Step 1. Find the least common multiple of the denominators.

$$\frac{5}{8} - \frac{1}{4}$$

Step 2. Use the least common multiple for the common denominator.

$$\frac{5}{8} - \frac{1}{4} = \frac{}{8} - \frac{}{8}$$

Step 3. Complete the equivalent fractions.

$$\frac{5}{8} - \frac{1}{4} = \frac{5}{8} - \frac{2}{8}$$

Step 4. Subtract.

$$\frac{5}{8} - \frac{1}{4} = \frac{5}{8} - \frac{2}{8} = \frac{3}{8}$$

EXERCISES

Add or subtract. Write answers in lowest terms. ***Hint:*** **First find the least common multiple of the denominators.**

1. $\frac{1}{2} + \frac{2}{5}$	**2.** $\frac{1}{2} - \frac{2}{5}$	**3.** $\frac{2}{3} + \frac{1}{9}$	**4.** $\frac{2}{3} - \frac{1}{9}$
5. $\frac{1}{6} + \frac{1}{9}$	**6.** $\frac{1}{6} - \frac{1}{9}$	**7.** $\frac{2}{3} + \frac{1}{7}$	**8.** $\frac{2}{3} - \frac{1}{7}$
9. $\frac{1}{2} + \frac{1}{7}$	**10.** $\frac{1}{2} - \frac{1}{7}$	**11.** $\frac{5}{8} + \frac{1}{4}$	**12.** $\frac{5}{8} - \frac{1}{4}$
13. $\frac{1}{2} + \frac{1}{6}$	**14.** $\frac{1}{2} - \frac{1}{6}$	**15.** $\frac{5}{6} + \frac{1}{8}$	**16.** $\frac{5}{6} - \frac{1}{8}$
17. $\frac{3}{5} + \frac{1}{4}$	**18.** $\frac{3}{5} - \frac{1}{4}$	**19.** $\frac{2}{3} + \frac{1}{5}$	**20.** $\frac{2}{3} - \frac{1}{5}$
21. $\frac{1}{3} + \frac{2}{9}$	**22.** $\frac{1}{3} - \frac{2}{9}$	**23.** $\frac{1}{2} + \frac{1}{8}$	**24.** $\frac{1}{2} - \frac{1}{8}$
25. $\frac{5}{6} + \frac{1}{9}$	**26.** $\frac{5}{6} - \frac{1}{9}$	**27.** $\frac{1}{2} + \frac{1}{9}$	**28.** $\frac{1}{2} - \frac{1}{9}$
29. $\frac{1}{4} + \frac{1}{5}$	**30.** $\frac{1}{4} - \frac{1}{5}$	**31.** $\frac{1}{3} + \frac{1}{6}$	**32.** $\frac{1}{3} - \frac{1}{6}$

Challenge!

This circle graph shows how Lois spent her allowance. What fraction of her allowance did she spend for

33. a record and movies?

34. clothes and movies?

35. food and movies?

36. all except clothes?

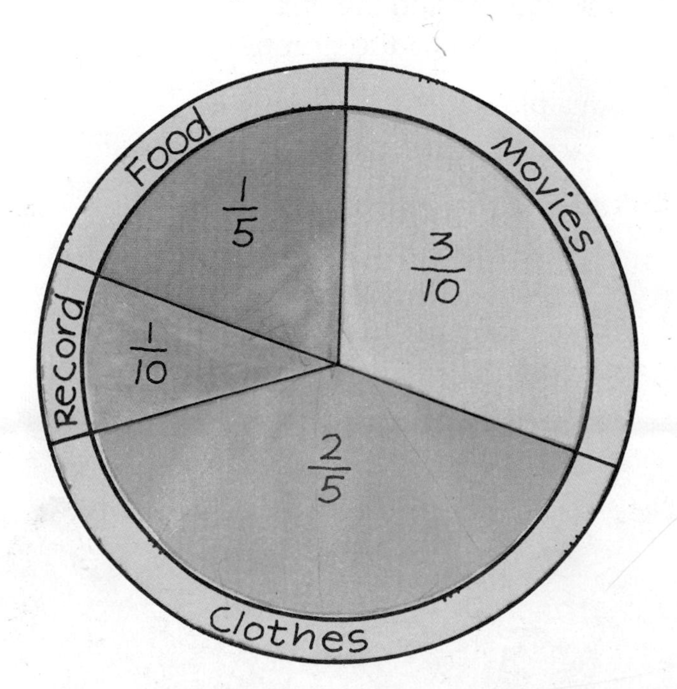

Problem solving

The River City Band is made up of students from all the schools in River City. The graph shows the number of students and the instruments that they play.

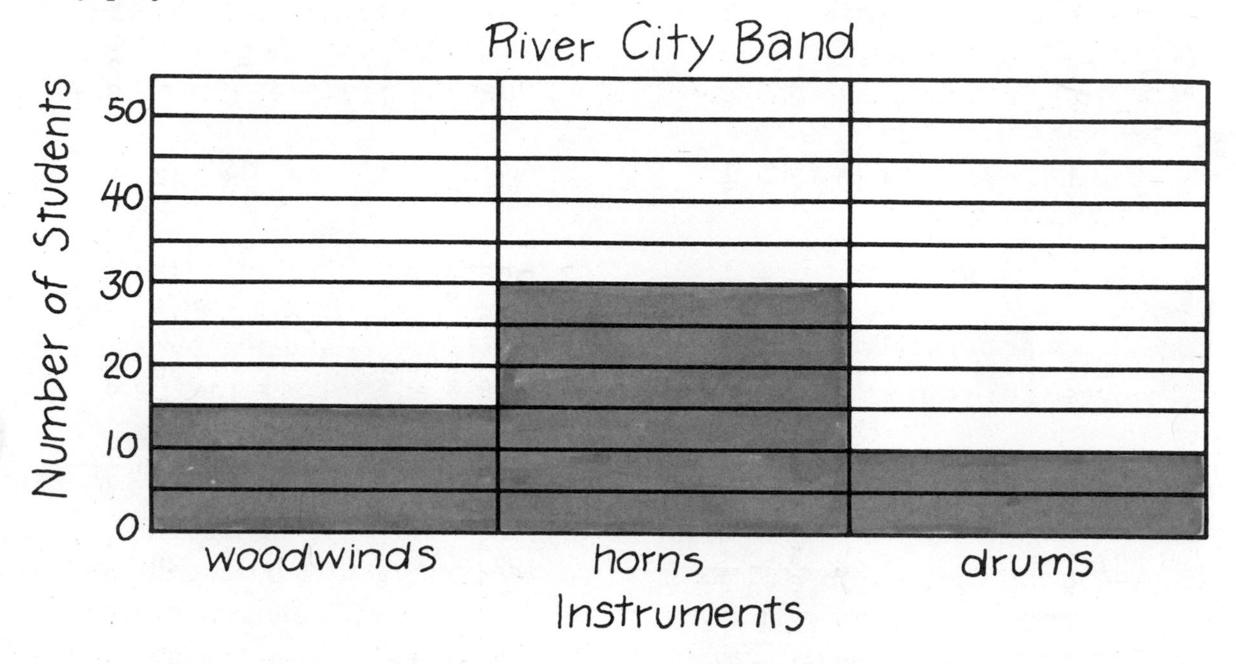

1. How many band members are represented by the graph?

2. What instruments are played by 30 students?

3. Boys play $\frac{1}{3}$ of the woodwinds. How many boys play woodwinds?

4. Girls play $\frac{3}{5}$ of the drums. How many girls play drums?

5. The first bus carries the band members that play drums or woodwinds. The second bus carries the band members that play horns. Which bus has more passengers?

6. $\frac{3}{5}$ of the horn players own their horns. How many horn players own their horns?

7. Mike rents his flute for $3.50 per week. Will $30 be enough money to rent the flute for 9 weeks?

8. $\frac{1}{3}$ of the band members attend Central High School. $\frac{1}{6}$ attend Fairview Junior High School. What fraction of the band members attend the two schools?

9. The Memorial Day parade route was 5 kilometers long. A band concert was held at the halfway point and lasted 25 minutes. The concert started at 1:40. When did it finish?

10. The band practices 35 minutes each day. How many minutes does the band practice in 5 days?

11. 48 of the band members marched in a parade. They marched in 8 equal rows. How many band members were in each row?

12. Kristi practiced at home for 84 minutes last week. She practiced for 4 nights and she practiced the same amount of time each night. How many minutes did she practice each night?

13. The Spring Concert was attended by 480 people. Tickets cost $3.00 per person. The concert started at 7:30 and ended at 9:15. How many minutes long was the concert?

★14. Sharon wants to buy a trumpet that costs $279.00. Her mother agreed to pay $\frac{1}{3}$ of the cost. Sharon has $190 saved. Does she have enough money?

Changing mixed numbers and whole numbers to fractions

$2\frac{1}{2}$ sandwiches

2 sandwiches

$$2\frac{1}{2} = \frac{5}{2}$$

2 and 1 half equals 5 halves.

$2\frac{1}{2}$ is a **mixed number.**

$$2 = \frac{8}{4}$$

2 equals 8 fourths.

EXERCISES

Write a mixed number and a fraction.

1.

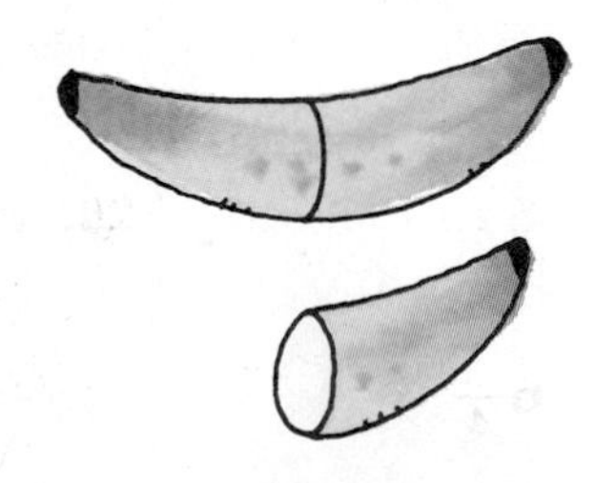

2.

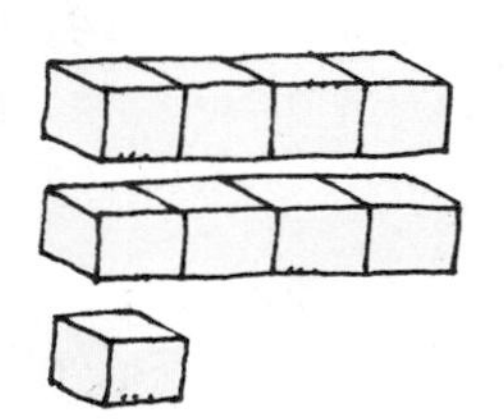

3.

4.

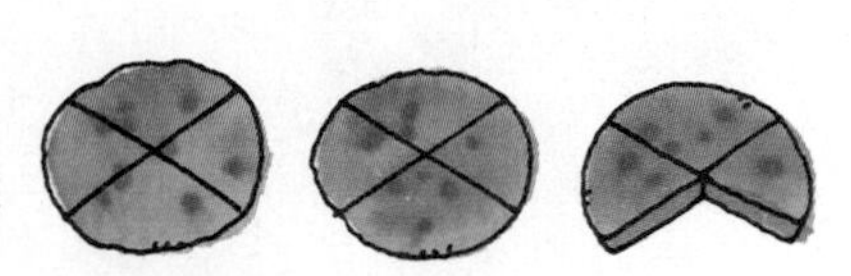

5.

6.

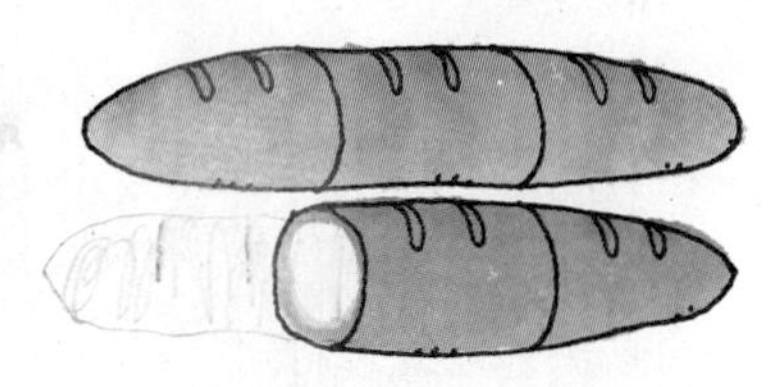

Write a whole number and a fraction.

7.

8.

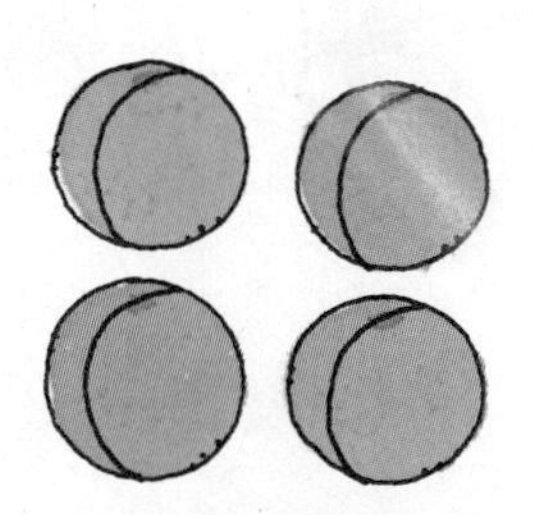

9.

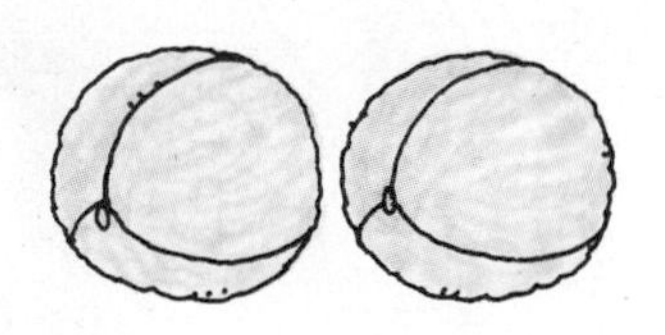

10. Draw a picture to show that $1\frac{3}{4} = \frac{7}{4}$.

11. Draw a picture to show that $3 = \frac{9}{3}$.

Draw a picture. Then complete with a fraction.

12. $1\frac{1}{4} = \underline{?}$ 13. $3\frac{1}{2} = \underline{?}$ 14. $1\frac{2}{3} = \underline{?}$ 15. $3\frac{1}{4} = \underline{?}$

Complete.

16. $1\frac{1}{3} = \frac{?}{3}$ 17. $1\frac{1}{4} = \frac{?}{4}$ 18. $1 = \frac{?}{3}$ 19. $2\frac{1}{2} = \frac{?}{2}$

20. $2 = \frac{?}{4}$ 21. $2\frac{3}{4} = \frac{?}{4}$ 22. $3 = \frac{?}{2}$ 23. $3\frac{1}{3} = \frac{?}{3}$

Change to fractions.

24. $3\frac{2}{3}$ 25. $3\frac{1}{4}$ 26. $3\frac{3}{4}$ 27. $4\frac{1}{2}$ 28. $5\frac{1}{2}$

29. $4\frac{1}{3}$ 30. $5\frac{1}{4}$ 31. $2\frac{5}{6}$ 32. $6\frac{1}{4}$ 33. $4\frac{2}{3}$

34. $5\frac{3}{4}$ 35. $8\frac{1}{10}$ 36. $4\frac{1}{4}$ 37. $3\frac{3}{10}$ 38. $6\frac{1}{3}$

KEEPING SKILLS SHARP

Add or subtract. Watch the signs.

1. $\begin{array}{r} 503 \\ -256 \\ \hline \end{array}$ 2. $\begin{array}{r} 783 \\ +159 \\ \hline \end{array}$ 3. $\begin{array}{r} 864 \\ +118 \\ \hline \end{array}$ 4. $\begin{array}{r} 286 \\ +99 \\ \hline \end{array}$ 5. $\begin{array}{r} 566 \\ -277 \\ \hline \end{array}$

6. $\begin{array}{r} 421 \\ +288 \\ \hline \end{array}$ 7. $\begin{array}{r} 277 \\ +193 \\ \hline \end{array}$ 8. $\begin{array}{r} 468 \\ +232 \\ \hline \end{array}$ 9. $\begin{array}{r} 674 \\ -177 \\ \hline \end{array}$ 10. $\begin{array}{r} 300 \\ -145 \\ \hline \end{array}$

Changing fractions to mixed numbers or whole numbers

fraction → $\frac{3}{2}$ = $1\frac{1}{2}$ ← mixed number

A mixed number has a whole-number part and a fraction part.

EXERCISES

Give the mixed number.

1. 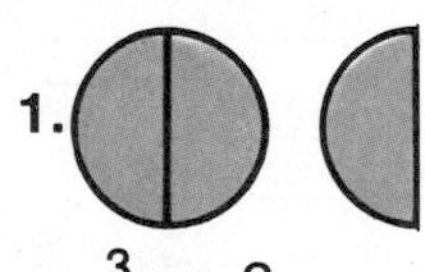

$\frac{3}{2}$ = ?

2. $\frac{5}{4}$ = ?

3. 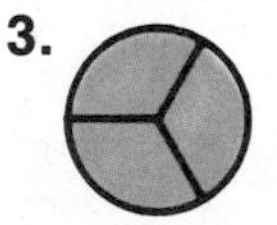

$\frac{5}{3}$ = ?

4. $\frac{8}{4}$ = ?

5. $\frac{4}{3}$ = ?
6. $\frac{5}{2}$ = ?
7. $\frac{6}{5}$ = ?
8. $\frac{7}{3}$ = ?
9. $\frac{5}{4}$ = ?
10. $\frac{7}{5}$ = ?
11. $\frac{7}{4}$ = ?
12. $\frac{8}{5}$ = ?

Give the whole number.

13. $\frac{8}{2}$ = ?
14. $\frac{10}{2}$ = ?
15. $\frac{15}{3}$ = ?
16. $\frac{16}{4}$ = ?

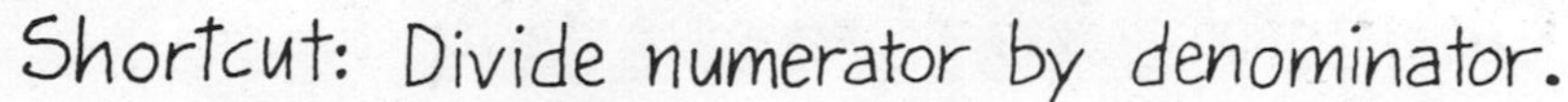

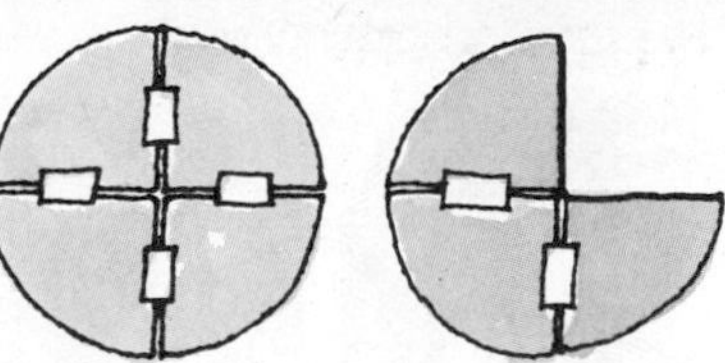

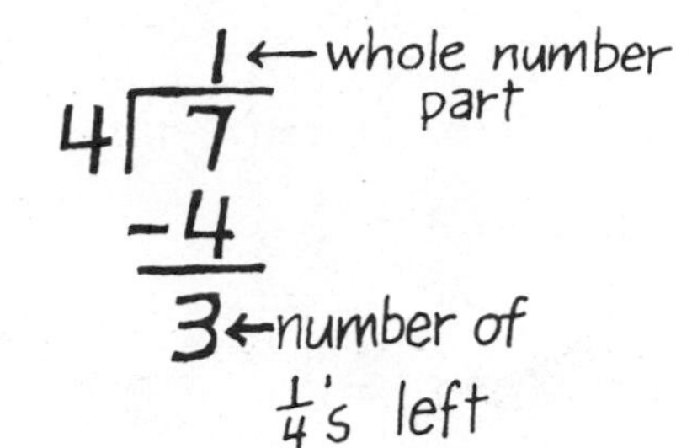

$$\frac{7}{4} = 1\frac{3}{4}$$

Change to mixed numbers or whole numbers.

17. $\frac{9}{4}$
18. $\frac{8}{5}$
19. $\frac{3}{2}$
20. $\frac{5}{4}$
21. $\frac{10}{5}$
22. $\frac{8}{3}$
23. $\frac{10}{3}$
24. $\frac{12}{3}$
25. $\frac{11}{5}$
26. $\frac{7}{2}$
27. $\frac{20}{5}$
28. $\frac{14}{3}$
29. $\frac{12}{4}$
30. $\frac{17}{5}$
31. $\frac{20}{3}$

Add. Give your answers as mixed numbers or whole numbers.

32. $\frac{5}{4} + \frac{2}{4}$
33. $\frac{2}{3} + \frac{2}{3}$
34. $\frac{3}{4} + \frac{1}{4}$
35. $\frac{5}{3} + \frac{2}{3}$
36. $\frac{3}{4} + \frac{6}{4}$
37. $\frac{3}{2} + \frac{4}{2}$
38. $\frac{5}{6} + \frac{2}{6}$
39. $\frac{7}{5} + \frac{3}{5}$
40. $\frac{4}{5} + \frac{4}{5}$
41. $\frac{4}{5} + \frac{3}{5}$
42. $\frac{7}{3} + \frac{2}{3}$
43. $\frac{2}{5} + \frac{7}{5}$

CHAPTER CHECKUP

What fraction is red? [pages 172–177]

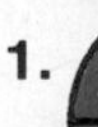

1. 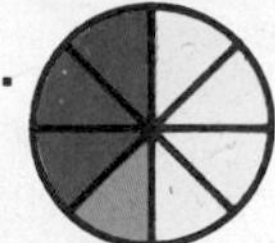

2. 

3. 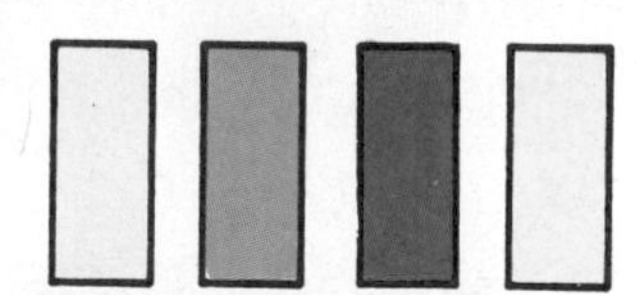

Complete. [pages 178–181]

4. $\frac{1}{2}$ of 8 = ?　　5. $\frac{1}{4}$ of 20 = ?　　6. $\frac{1}{3}$ of 18 = ?

7. $\frac{2}{3}$ of 18 = ?　　8. $\frac{3}{4}$ of 12 = ?　　9. $\frac{5}{6}$ of 18 = ?

Find the sale price. [pages 182–183]

10. 

11. 

Complete the equivalent fractions. [pages 184–191]

12. $\frac{1}{2} = \frac{?}{4}$　　13. $\frac{1}{4} = \frac{?}{12}$　　14. $\frac{2}{3} = \frac{?}{6}$　　15. $\frac{6}{8} = \frac{?}{4}$

$<$ or $>$? [pages 192–193]

16. $\frac{1}{2}$  $\frac{1}{3}$　　17. $\frac{1}{8}$  $\frac{1}{4}$　　18. $\frac{1}{2}$ 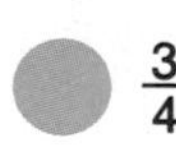 $\frac{3}{4}$　　19. $\frac{3}{8}$  $\frac{1}{4}$

Add. [pages 194–197, 202–203]

20. $\frac{2}{5} + \frac{1}{5}$　　21. $\frac{2}{8} + \frac{3}{8}$　　22. $\frac{3}{8} + \frac{1}{2}$　　23. $\frac{2}{3} + \frac{1}{4}$

Subtract. [pages 198–203]

24. $\frac{3}{8} - \frac{2}{8}$　　25. $\frac{7}{9} - \frac{4}{9}$　　26. $\frac{3}{4} - \frac{1}{3}$　　27. $\frac{5}{6} - \frac{1}{4}$

Complete. [pages 206–209]

	28.	29.	30.	31.	32.	33.
Mixed number	$2\frac{1}{3}$	$3\frac{3}{4}$	?	?	$2\frac{1}{2}$	?
Fraction	?	?	$\frac{5}{3}$	$\frac{7}{2}$	?	$\frac{9}{4}$

CHAPTER PROJECT

Probability experiment

1. You will need three coins for this experiment.
2. Toss all three coins 60 times and record the outcome of each toss as shown above.
3. What fraction (in lowest terms) of the outcomes were

   a. 3 heads? **b.** 2 heads and 1 tail?
   c. 1 head and 2 tails? **d.** 3 tails?

4. From your experiment, do you think your chances are better for getting 3 heads or 2 heads and 1 tail?
5. You may wish to do the experiment with four coins.

CHAPTER REVIEW

What fraction is colored?

1. 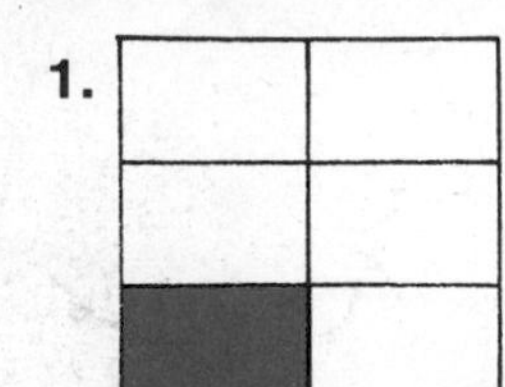

   ● parts colored
   ■ parts in all

2. ● parts colored
   ■ parts in all

What fraction of the crayons are blue?

3. 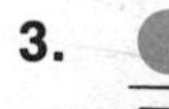 blue crayons
    crayons in all

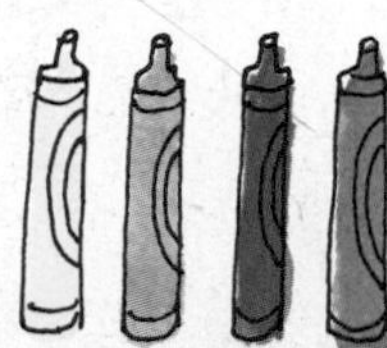

4. ● blue crayons
   ■ crayons in all

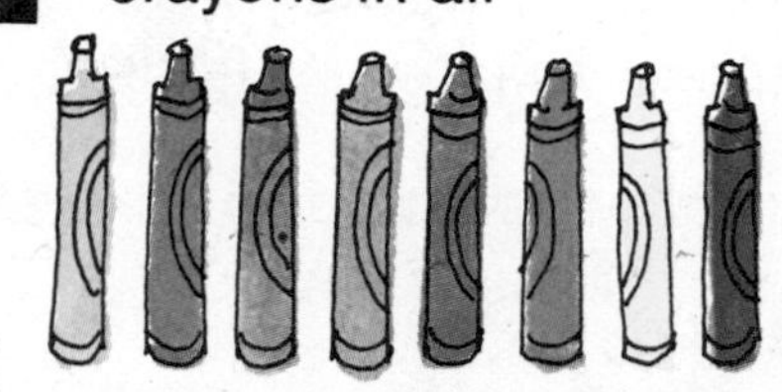

Complete.

5. $\frac{1}{4}$ of 8 = ?

6. $\frac{2}{3}$ of 12 = ?

Give the equivalent fraction.

7. $\frac{1}{2} = \frac{?}{?}$ (×3, ×3)

8. $\frac{1}{3} = \frac{?}{?}$ (×2, ×2)

9. $\frac{6}{8} = \frac{?}{?}$ (÷2, ÷2)

$\frac{0}{8}$ $\frac{1}{8}$ $\frac{2}{8}$ $\frac{3}{8}$ $\frac{4}{8}$ $\frac{5}{8}$ $\frac{6}{8}$ $\frac{7}{8}$

Use the number line to find these sums.

10. $\frac{1}{8} + \frac{2}{8}$

11. $\frac{3}{8} + \frac{2}{8}$

12. $\frac{4}{8} + \frac{3}{8}$

Use the number line to find these differences.

13. $\frac{5}{8} - \frac{2}{8}$

14. $\frac{6}{8} - \frac{5}{8}$

15. $\frac{7}{8} - \frac{2}{8}$

CHAPTER CHALLENGE

Think about putting these marbles in a bag, mixing them up, and then picking a marble out of the bag.

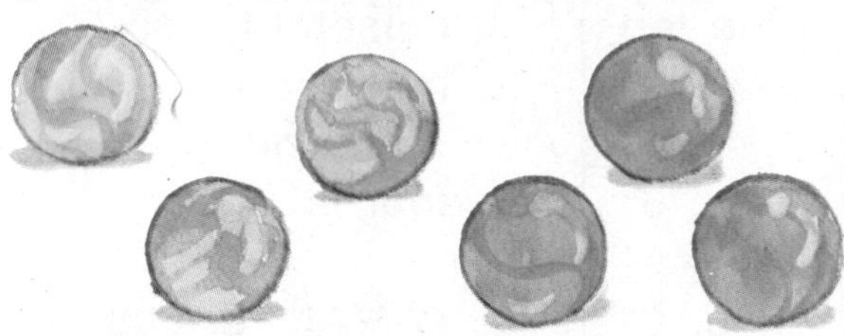

1. Which color would you have the greatest chance of picking?
2. Which color would you have the least chance of picking?

The probability (or chance) of picking a yellow marble is

$\frac{1}{6}$ ← number of yellow marbles / number of marbles

3. What is the probability (in lowest terms) of picking
   a. a green marble?
   b. a blue marble?

Think about putting these marbles in a bag, mixing them up, and then picking a marble out of the bag.

4. Give the probability (in lowest terms) of picking
   a. a red marble.
   b. a blue marble.
   c. a green marble.
   d. a yellow marble.
   e. a marble that is not red.
   f. a marble that is not yellow.

MAJOR CHECKUP
STANDARDIZED FORMAT

Choose the correct letter.

1. 938 rounded to the nearest ten is

a. 930
b. 900
c. 940
d. none of these

2. Which digit in 58,267 is in the ten thousands place?

a. 6
b. 8
c. 2
d. 5

3. The standard numeral for six hundred eight thousand, twenty-four is

a. 608,240 b. 608,024
c. 680,024 d. 608,204

4. Add.

$$\begin{array}{r} 3675 \\ +2859 \\ \hline \end{array}$$

a. 6534
b. 6524
c. 5424
d. none of these

5. Add.

$$\begin{array}{r} 35 \\ 26 \\ 58 \\ +29 \\ \hline \end{array}$$

a. 147
b. 138
c. 148
d. none of these

6. Subtract.

$$\begin{array}{r} 524 \\ -278 \\ \hline \end{array}$$

a. 246
b. 256
c. 346
d. none of these

7. How many minutes from 8:30 P.M. to 9:20 P.M.?

a. 10
b. 40
c. 50
d. none of these

8. How much money?
1 five-dollar bill
3 one-dollar bills
1 half-dollar
2 dimes
2 nickels

a. \$4.80 b. \$8.80
c. \$8.50 d. none of these

9. Multiply.

$$\begin{array}{r} 78 \\ \times 6 \\ \hline \end{array}$$

a. 468
b. 428
c. 426
d. 482

10. Divide.
$4\overline{)93}$

a. 14 R1
b. 24 R1
c. 23 R1
d. 33 R1

11. There are 65 players on teams. Each team has 5 players. How many teams are there?

a. 15 b. 70
c. 325 d. none of these

12. Each of 27 boys is 4 feet tall. Each of 35 girls is 4 feet tall. How many children are there?

a. 70 b. 62
c. 248 d. none of these

Measurement

8

Centimeter

The **centimeter (cm)** is a unit for measuring length in the metric system.

The length of the nail is between 5 and 6 centimeters. It is nearer to 6 centimeters. The length of the nail measured to the **nearest centimeter** is 6 centimeters.

EXERCISES

Measure to the nearest centimeter.

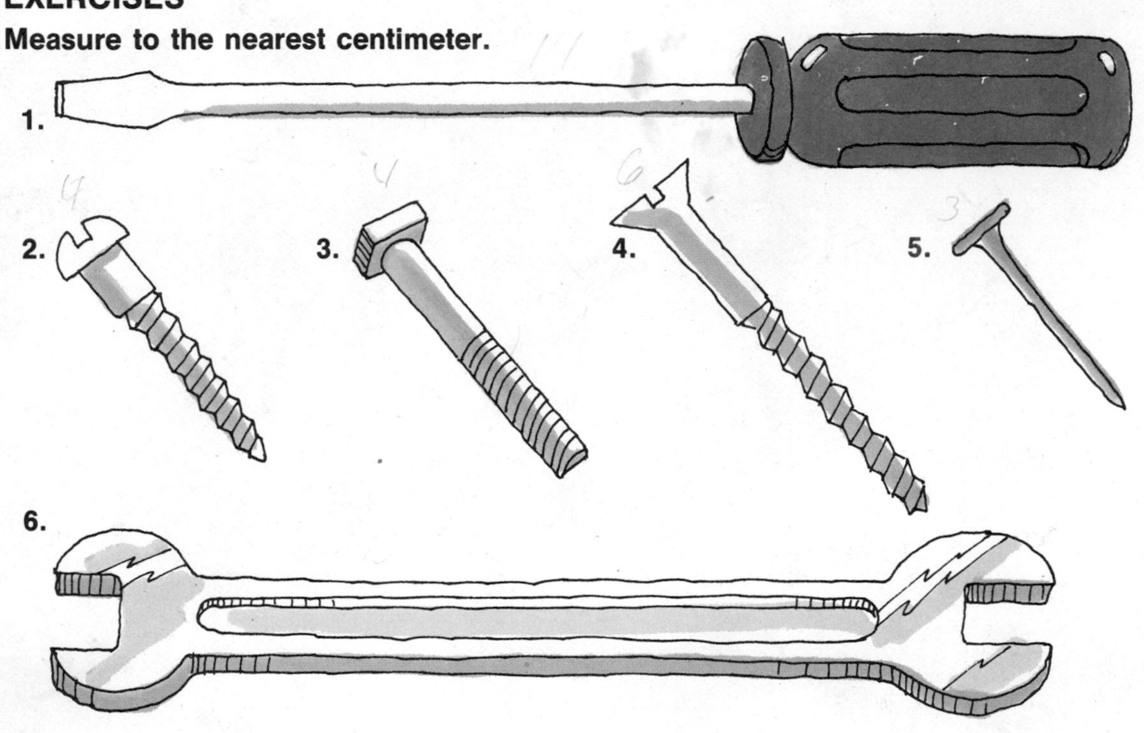

Draw sticks with these lengths.

7. 9 cm
8. 14 cm
9. 18 cm
10. 11 cm

Draw a stick that is between

11. 8 cm and 9 cm, but nearer 8 cm.
12. 6 cm and 7 cm, but nearer 7 cm.

The turtle decided to go for a swim.

Measure each part of its path to the water in centimeters.

16. From START to

17. From to

18. From to

19. From to

20. From to

21. From to

22. From to END

23. How far did it walk in all to get to the water?

Measure to the nearest centimeter

24. the length of your little finger.

25. the width of your hand.

26. the length of your foot.

27. around your wrist. [If you don't have a centimeter tape, use a string and a ruler.]

28. around your waist.

29. your height.

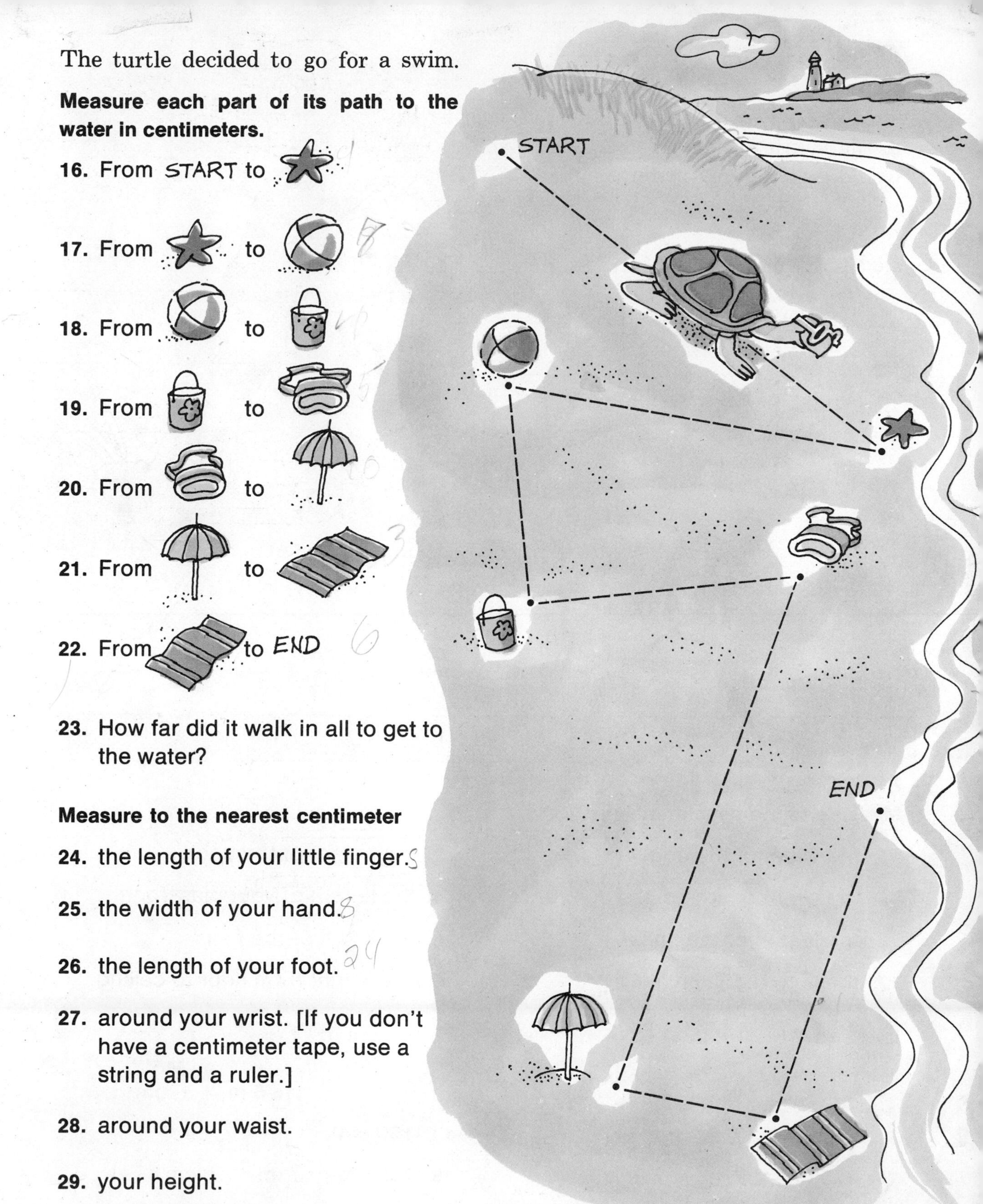

Meters and kilometers

The **meter (m)** is used to measure longer lengths in the metric system. If you took a "giant step," it would be close to 1 meter long.

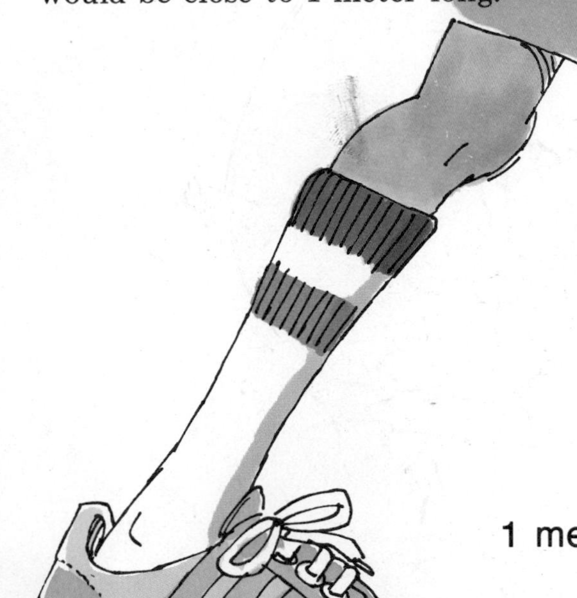

1 meter (m) = 100 cm

0 Centimeters 50 100

EXERCISES

Measure to the nearest meter

1. the width of the door.
2. the height of the door.
3. the width of your classroom.
4. the length of your classroom.
5. the length of the hall.
6. the width of the hall.
7. the length of a chalktray.
8. the height from floor to ceiling.

Complete.

9. 1 m = ? cm
10. 3 m = ? cm
11. 5 m = ? cm
12. 200 cm = ? m
13. 150 cm = 1 m + ? cm
14. 135 cm = ? m + ? cm
15. 215 cm = ? m + ? cm

The **kilometer (km)** is used to measure long distances in the metric system. If you took 1000 giant steps, you would have walked about 1 kilometer.

1 kilometer (km) = 1000 m

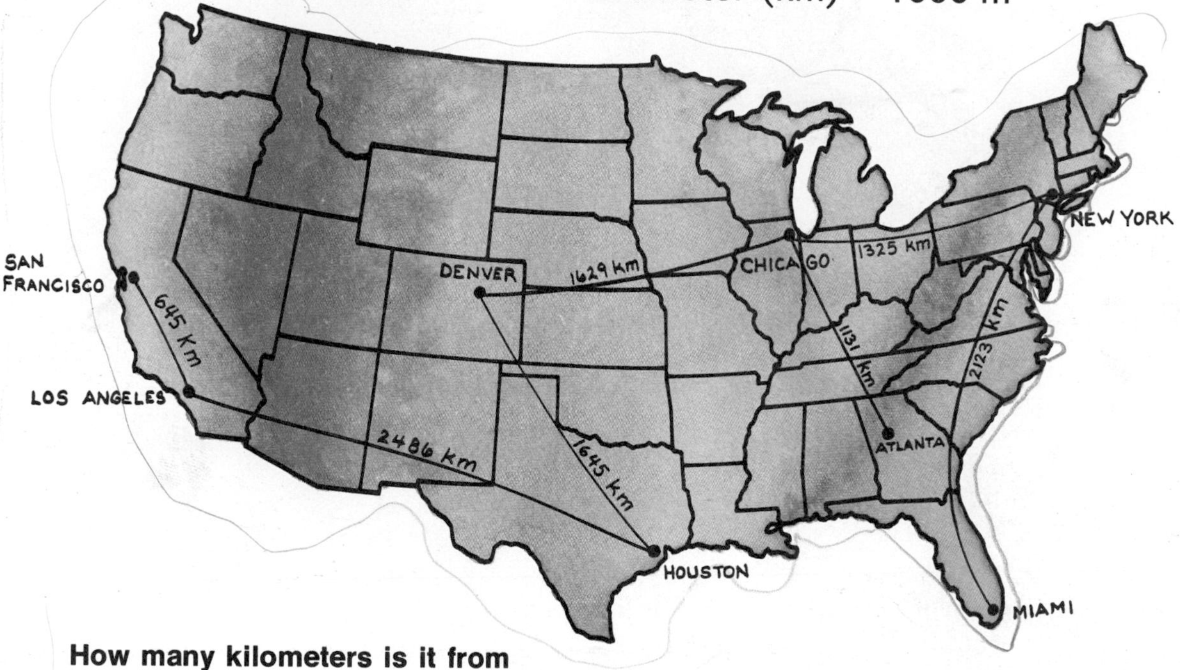

How many kilometers is it from

16. New York to Miami?

17. Chicago to Atlanta?

18. Denver to Houston?

Give the total distance in kilometers for each of these road trips.

19. Denver–Chicago–New York

20. Houston–Los Angeles–San Francisco

Complete.

21. 1 km = ? m

22. 2 km = ? m

23. 4000 m = ? km

★ **24.** 3800 m = ? km + ? m

Lay out a 50-meter course on the school grounds or in the gym. Find out how long it takes you to walk 1 kilometer.

Perimeter

The distance around a figure is called the **perimeter** of the figure.

You can find the perimeter of a figure by adding the lengths of its sides.

EXERCISES

Give the perimeter of each figure.

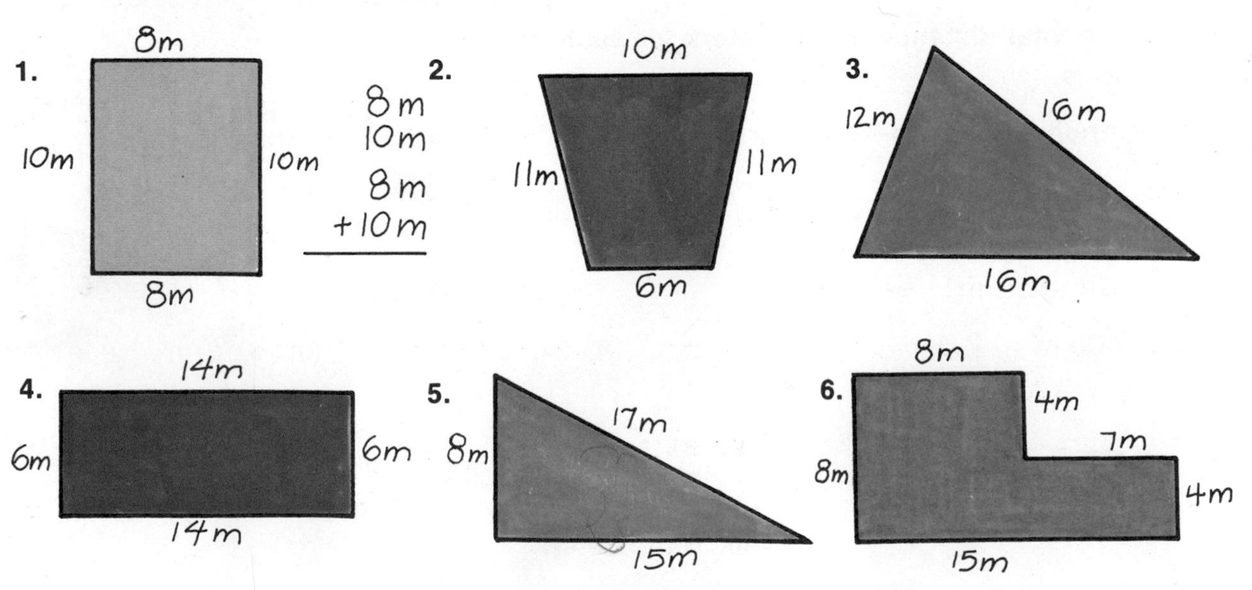

Measure each side in centimeters and give the perimeter.

7.

8.

9.

10.

11.

12.

Solve.

13. A large lawn is shaped like a rectangle. It is 42 meters long and 26 meters wide. How much fence will be needed to go around the lawn?

14. Find the perimeter of
- **a.** your desk top.
- **b.** this page in your math book.
- ★ **c.** the bottom of your shoe.

KEEPING SKILLS SHARP

1. 24×3
2. 35×5
3. 53×2
4. 47×4
5. 58×6
6. 65×7
7. 74×7
8. 93×6
9. 80×9
10. 76×2
11. 84×3
12. 95×8

Area

To find the **area** of this region, we pick a unit

1 cm
1 cm
1 square centimeter

and count the units (square-centimeter tiles) that it takes to cover the region.

The area is 24 square centimeters.

To find the area of squares and rectangles, you can multiply.

There are 6 columns of 4 tiles each.
There are 6 × 4 tiles, or 24 tiles.

Area = length × width
Area = 6 cm × 4 cm = 24 square centimeters

EXERCISES

Give each area.

1.

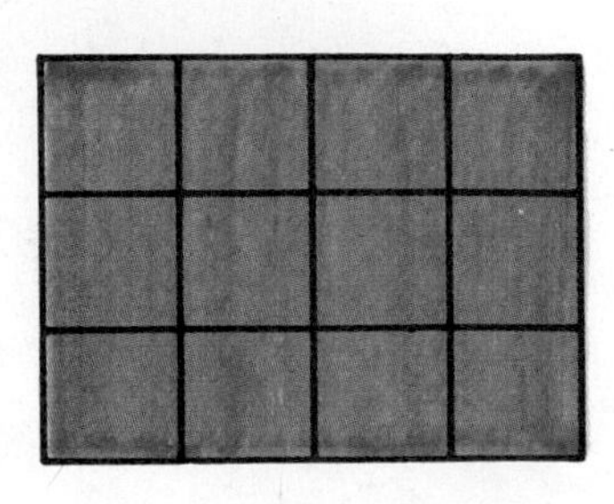

2.

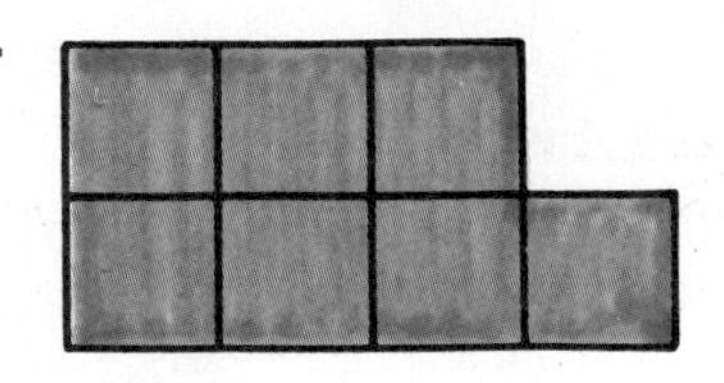

3.

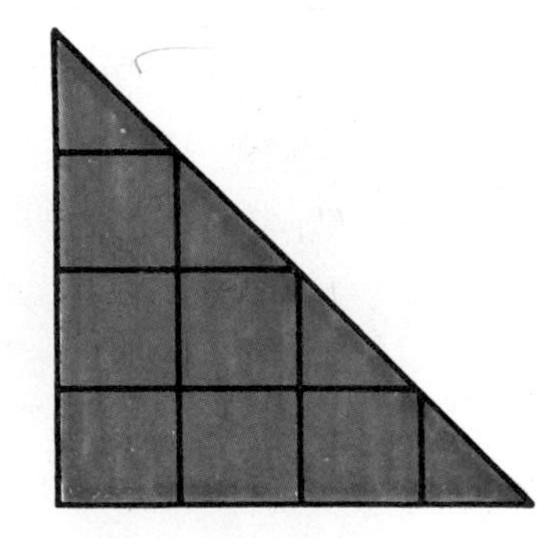

Give each area.

4.

5.

6.

7. Copy and complete this table.

				Figure		
	A	B	C	D	E	F
Length in cm	3					
Width in cm	1					
Area in square cm	3					

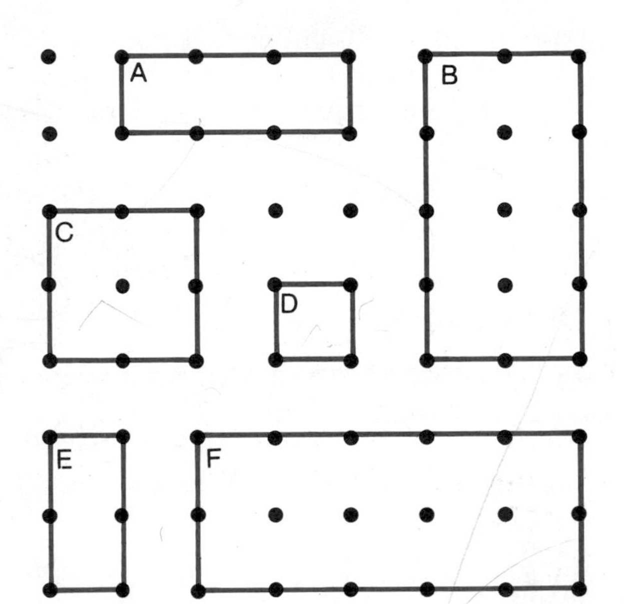

Solve.

8. A rectangular garden plot is 9 meters long and 7 meters wide. What is the area?

9. A square patio is 8 meters on a side. What is its area?

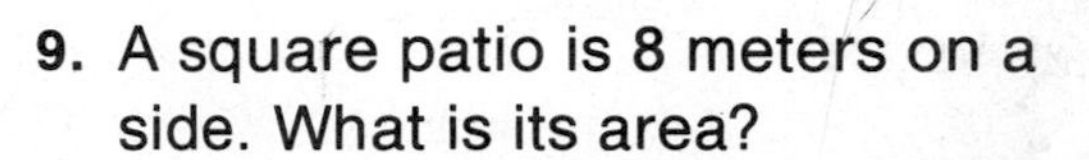

Challenge!

10. a. Get or draw a piece of square-centimeter graph paper.

b. The green figure has an area of 12 square centimeters. Draw some other figures that have an area of 12 square centimeters.

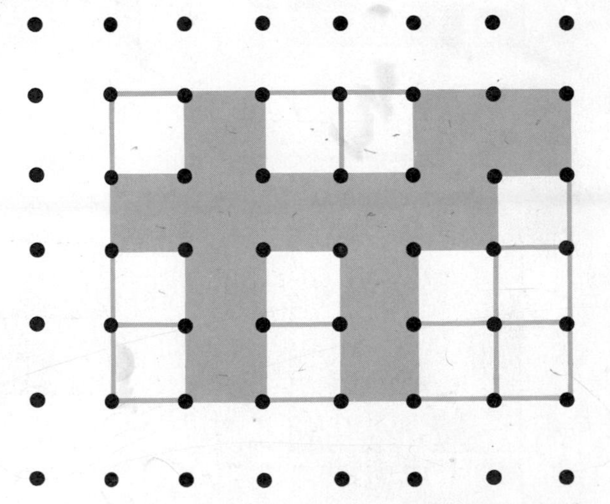

Volume

To find the **volume** of this box, we pick a unit

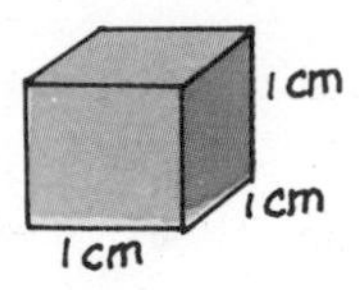

and count the units (cubic-centimeter blocks) that will fill the box.

The volume is 16 cubic centimeters.

To find the volume of a box like the one shown above, you can multiply.
There are 2×4 blocks in a layer.
There are 2 layers.

Volume = $2 \times 4 \times 2$ blocks = 16 blocks

Volume = 2 cm × 4 cm × 2 cm = 16 cubic cm

EXERCISES

Give each volume.

1. 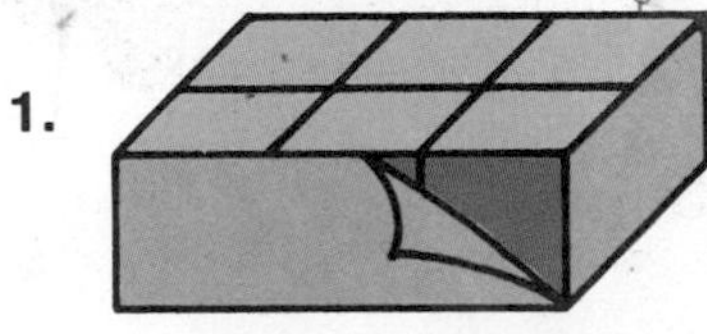

2. 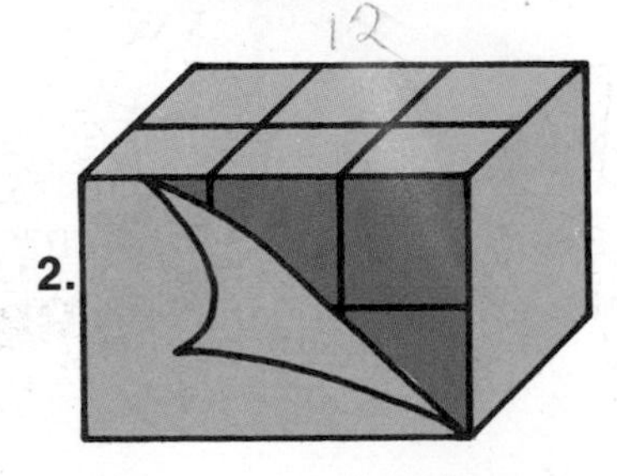

3. 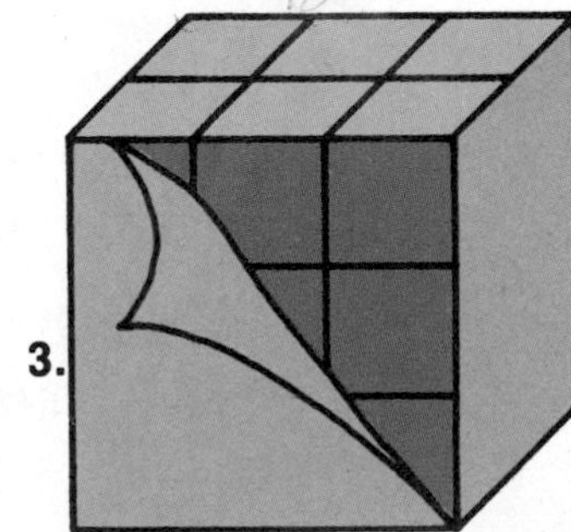

4. 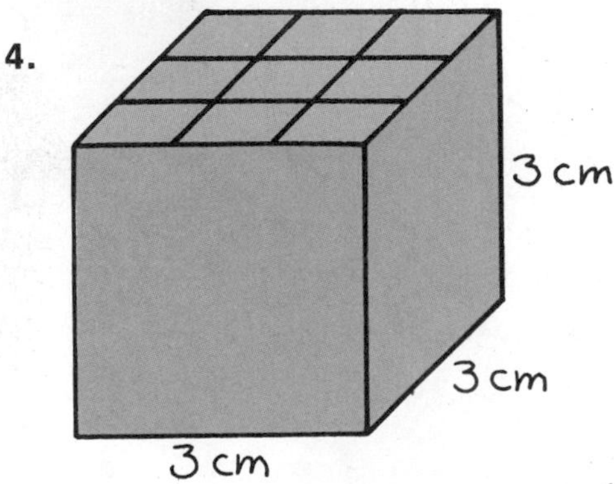

5. 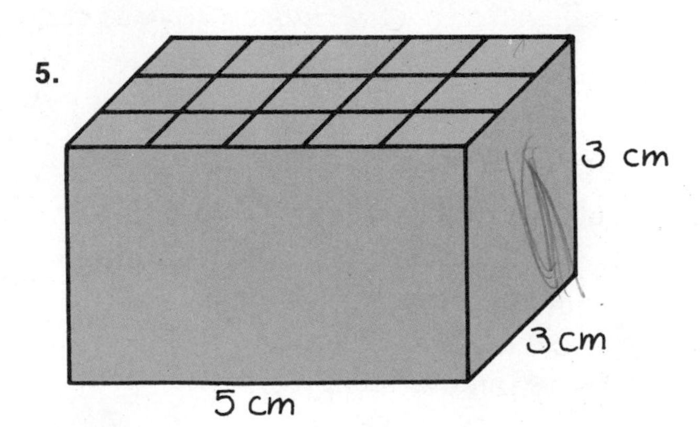

Complete this table.

	Measurements of box in cm			
	Length cm	Width cm	Height cm	Volume cubic cm
6.	2	3	4	?
7.	3	2	5	?
8.	2	2	6	?
9.	5	2	4	?

Is the question about length, area, or volume?

10. How much string do you have?

11. How much sand will the box hold?

12. How much paper is needed to cover the bulletin board?

13. How much water is in the fish tank?

14. How much carpet is on the floor?

15. How far do you live from school?

Solve.

16. A fence is to be built around a square garden. The garden is 19 meters on each side. How many meters of fence are needed?

17. A room is 5 meters wide and 8 meters long. How many square meters of carpet are needed to cover the floor?

18. A form 2 meters long, 3 meters wide, and 2 meters deep is to be filled with concrete. How many cubic meters of concrete should be ordered?

19. A large fish tank is 12 meters long, 4 meters wide, and 3 meters deep. How many cubic meters of water will it hold?

Liquid volume and weight—metric

1 **milliliter** of water will fill this cube:

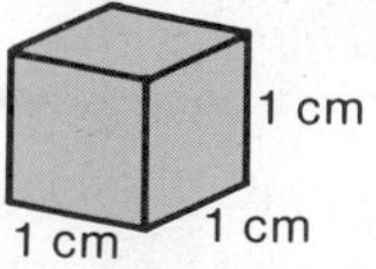

1000 milliliters (mL) = 1 liter (L)

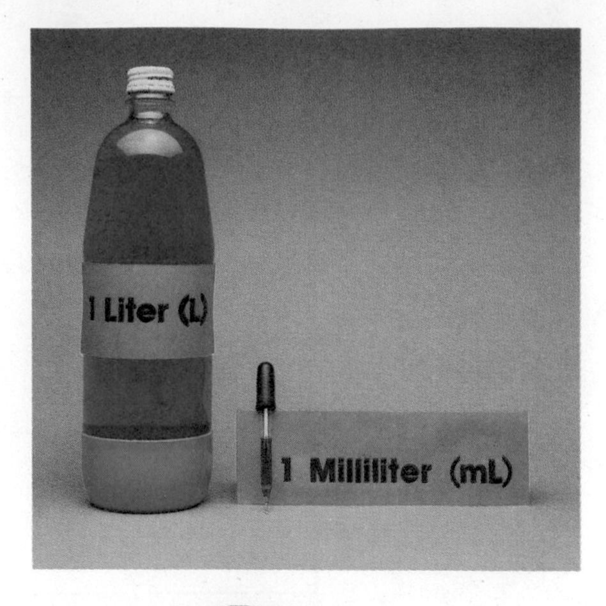

EXERCISES

Choose the answer that seems right.

1. 

The amount of gasoline in a car tank is

a. 60 mL **b.** 60 L

2. 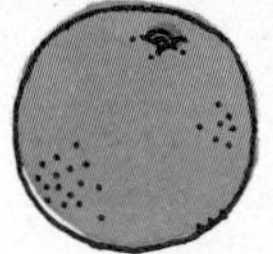

The amount of juice in an orange is

a. 50 mL **b.** 50 L

3. 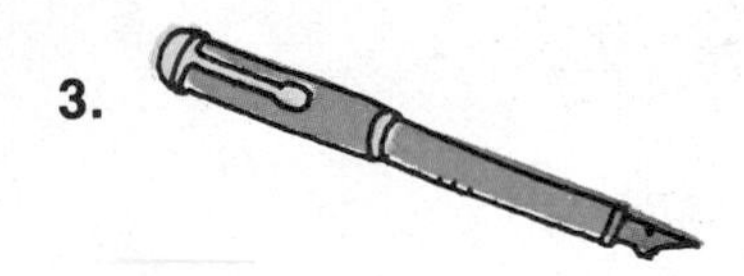

The amount of ink in a pen is

a. 2 mL **b.** 2 L

4. 

The amount of water in a washing machine is

a. 140 mL **b.** 140 L

Complete.

5. There are ? mL in 3 L.

6. There are ? L in 5000 mL.

Collect some containers and estimate how many liters they will hold.

Check your estimates by filling each container.

The water in this cube weighs 1 **gram** (g).

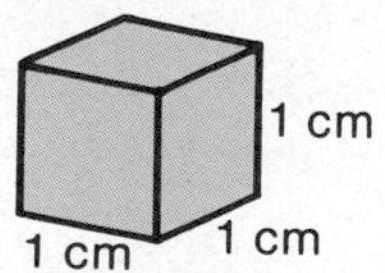

1000 grams (g) = 1 kilogram (kg)

EXERCISES

Choose the answer that seems right.

7.

The bicycle weighs

a. 12 g **b.** 12 kg

8.

The hamburger weighs

a. 170 g **b.** 170 kg

9.

The penny weighs

a. 3 g **b.** 3 kg

10.

The car weighs

a. 900 g **b.** 900 kg

Complete.

11. There are _?_ g in 2 kg.

12. There are _?_ kg in 4000 g.

Weigh yourself in kilograms.

Temperature is sometimes measured on the **Celsius** scale. Get a Celsius thermometer. Keep a record of the temperature at 12:00 noon each day for two weeks. Make a graph.

Problem solving–drawing pictures

1. Study and understand.
2. Plan and do.
3. Answer and check.

Drawing a picture can help you solve a problem.

EXAMPLE.

The zoo is 35 kilometers east of Charlene's house. The amusement park is 47 kilometers east of Charlene's house. How far is it from the zoo to the amusement park?

1. Study and understand.

   Draw a simple picture.

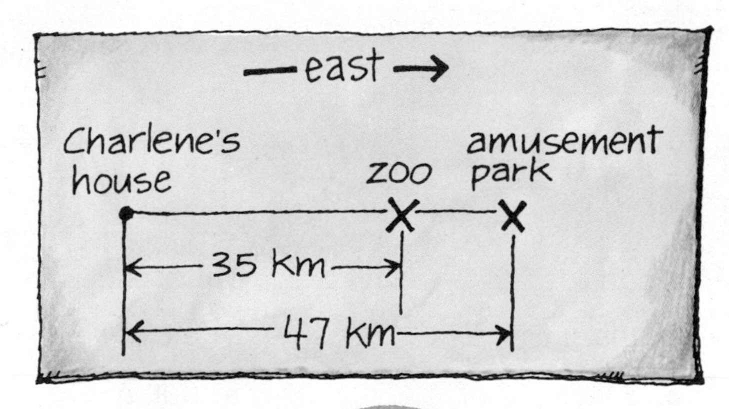

2. Plan and do.

   Use your picture to plan what to do.

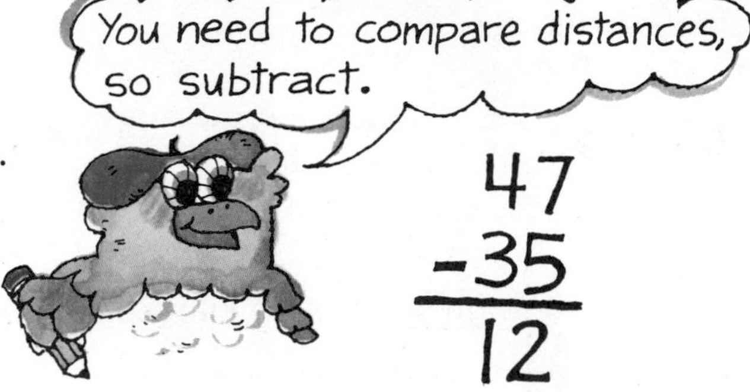

3. Answer and check.

   It is 12 kilometers from the zoo to the amusement park.

Solve.

1. The zoo is 35 kilometers east of Charlene's house. The science museum is 16 kilometers west of Charlene's house. How far is it from the zoo to the science museum?

2. Bob has a rectangular garden beside his house. The 20-meter side of the garden is protected by the house. So, he only needs to build a fence on 3 sides. The garden is 15 meters wide. How much fence does he need?

3. One wall of Sarah's bedroom is 4 meters long and 3 meters high. In the wall there is a window that is 1 meter wide and 2 meters high. The wall, except for the window, is to be painted. How many square meters will be painted?

4. Lois used some mosaic tiles to make a rectangular cover. Each side of a tile was 1 centimeter long. The cover was 10 centimeters wide and 12 centimeters long. How many tiles were in the cover?

★ 5. A swimming pool is to be used for a swimming meet. The pool must be divided into 8 lanes down the 50-meter length of the pool. How much rope will be needed to make the lanes?

★ 6. Simon used some mosaic tiles to make a rectangular cover. Each side of a tile was 2 centimeters long. The cover was 10 centimeters wide and 12 centimeters long. How many tiles were in the cover?

7. Paul's fish tank is 60 centimeters long, 30 centimeters wide, and 40 centimeters high. How much water does he use to fill it to 5 centimeters from the top? (*Hint:* Draw and label a box.)

Here is how to draw a box.

Draw a rectangle.

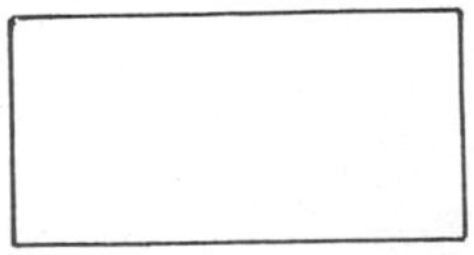

Draw 3 parallel segments all the same length.

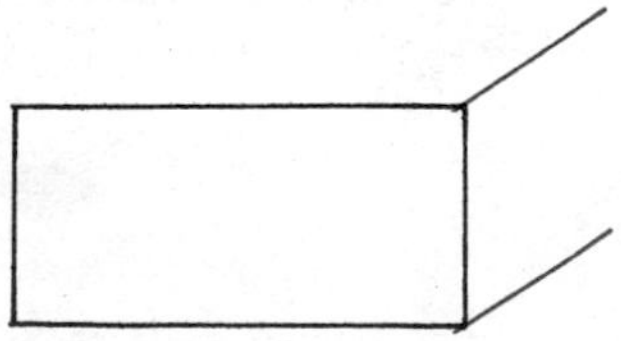

Join the ends of the segments.

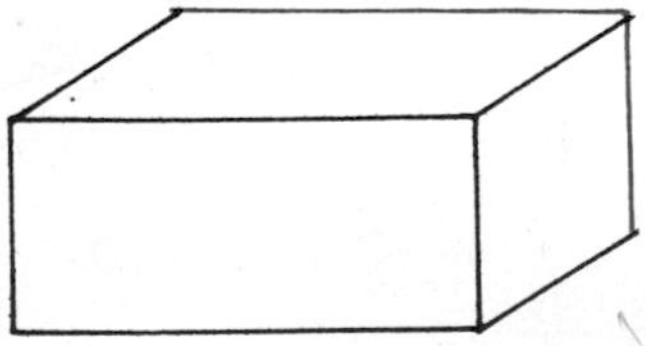

Inch

The customary system is another measurement system. The **inch (in.)** is a unit for measuring length in the customary system.

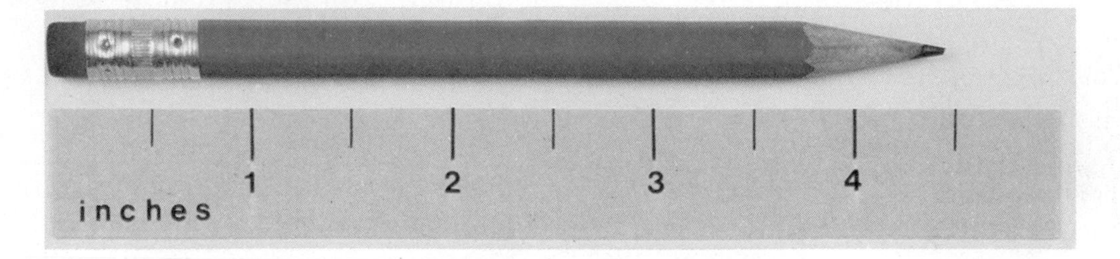

The length of the pencil measured to the nearest inch is 4 inches. The length of the pencil measured to the nearest $\frac{1}{2}$ inch is $4\frac{1}{2}$ in.

The same pencil can be measured with a ruler marked in $\frac{1}{4}$ inches and a ruler marked in $\frac{1}{8}$ inches.

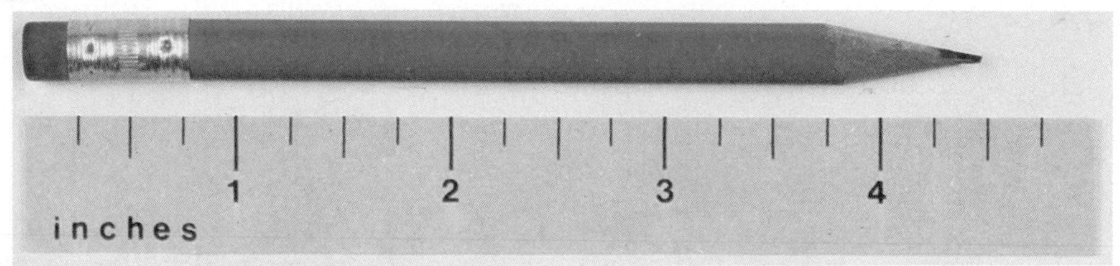

The length of the pencil measured to the nearest $\frac{1}{4}$ inch is $4\frac{1}{2}$ in.

The length of the pencil measured to the nearest $\frac{1}{8}$ inch is $4\frac{3}{8}$ in.

EXERCISES

Draw sticks with these lengths.

1. 3 in.
2. 4 in.
3. $2\frac{1}{2}$ in.
4. $3\frac{1}{2}$ in.
5. $2\frac{1}{4}$ in.
6. $2\frac{3}{4}$ in.
7. $3\frac{3}{8}$ in.
8. $2\frac{7}{8}$ in.

Measure the length of each rope to the nearest inch and $\frac{1}{2}$ inch.

9.

10.

11. 12.

13.

Measure the length of each rope to the nearest $\frac{1}{4}$ inch and $\frac{1}{8}$ inch.

14.

15.

16.

Challenge!

Measure

17. the height of your desk top to the nearest inch.
18. the width of your desk top to the nearest $\frac{1}{2}$ inch.
19. the thickness of your desk top to the nearest $\frac{1}{4}$ inch.
20. the width of your math book to the nearest $\frac{1}{8}$ inch.

Inch, foot, yard, and mile

In the customary system, the **foot, yard,** and **mile** are units used to measure longer lengths.

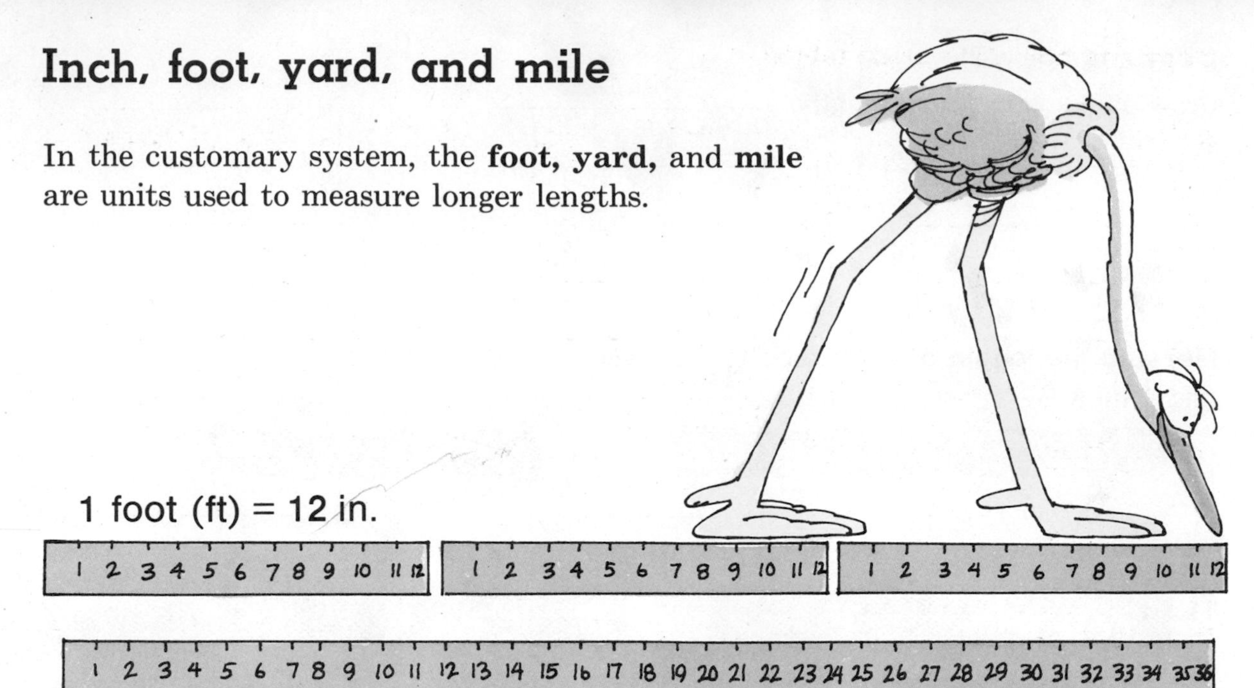

1 foot (ft) = 12 in.

1 yard (yd) = 3 ft

1 yard (yd) = 36 in.

The mile is used to measure long distances.

1 mile (mi) = 5280 ft

1 mile (mi) = 1760 yd

You would have to take about 1800 giant steps to walk a mile.

EXERCISES

1. Measure the length of a chalkboard to the nearest foot.
2. Measure the width of a chalkboard to the nearest foot.
3. What is the perimeter of the chalkboard?
4. Measure the height of the door to the nearest foot.
5. Measure the length of your classroom to the nearest yard.
6. Measure the width of your classroom to the nearest yard.
7. Measure the height of the ceiling to the nearest yard.

Copy and complete these tables.

8.

Feet	Inches
1	12
2	?
3	?
4	?
5	?
6	?

12
×2

9.

Yards	Feet
1	3
2	?
3	?
4	?
5	?
6	?

10.

Yards	Inches
1	36
2	?
3	?
4	?
5	?
6	?

Inch, foot, or mile?
Tell which unit you would use to measure

11. the length of a pencil.
12. the width of a book.
13. the height of a building.
14. the length of a river.
15. the perimeter of your classroom.
16. the distance between two cities.

Solve.

17. By automobile, it is 707 miles from Boston to Detroit and 269 miles from Detroit to Chicago. How many miles is a Boston–Detroit–Chicago trip?

18. It is 4931 miles from New York to Los Angeles by water. By air, it is 2451 miles. How many miles less is it by air?

KEEPING SKILLS SHARP

1. $2\overline{)84}$ 2. $3\overline{)96}$ 3. $4\overline{)48}$ 4. $6\overline{)66}$ 5. $5\overline{)50}$

6. $3\overline{)72}$ 7. $4\overline{)52}$ 8. $6\overline{)96}$ 9. $8\overline{)96}$ 10. $5\overline{)65}$

11. $3\overline{)58}$ 12. $4\overline{)73}$ 13. $5\overline{)74}$ 14. $6\overline{)87}$ 15. $8\overline{)94}$

Perimeter, area, and volume

You can add to find the perimeter.

$$\begin{array}{r} 7\text{ ft} \\ 4\text{ ft} \\ 7\text{ ft} \\ +4\text{ ft} \\ \hline 22\text{ ft} \end{array}$$

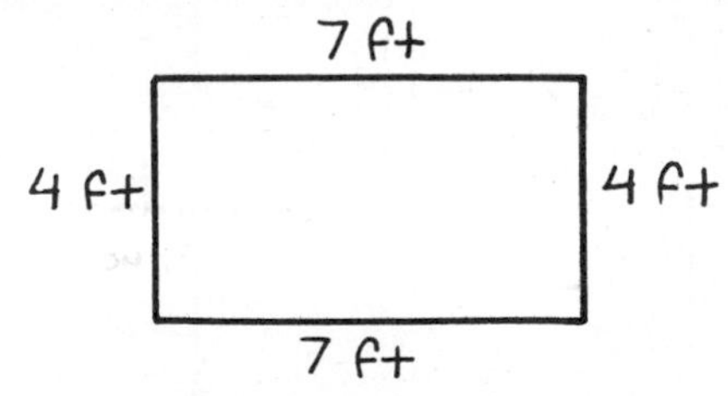

You can multiply to find the area.

Area = length × width
= 7 ft × 4 ft
= 28 square ft

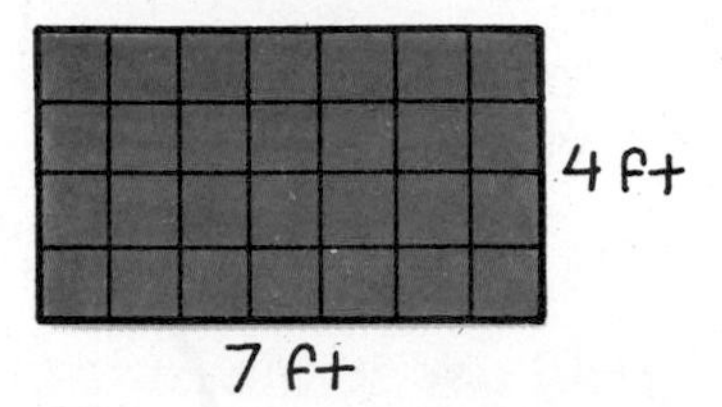

You can multiply to find the volume.

Volume = length × width × height
= 7 ft × 4 ft × 3 ft
= 84 cubic ft

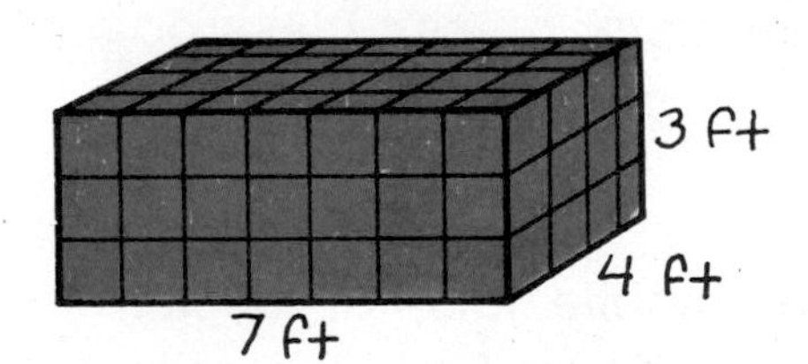

EXERCISES

Give each perimeter.

1.

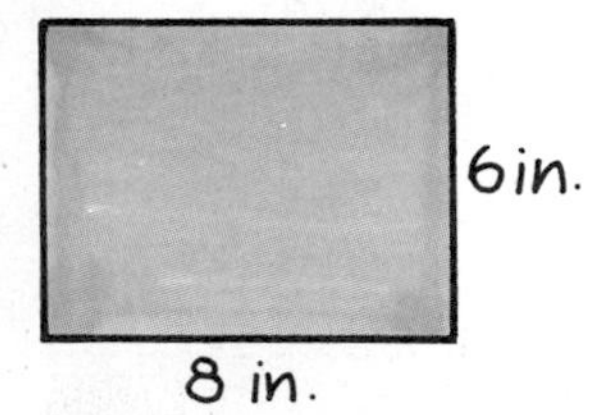

2.

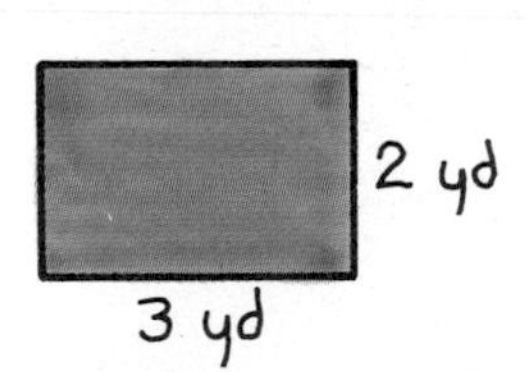

3.

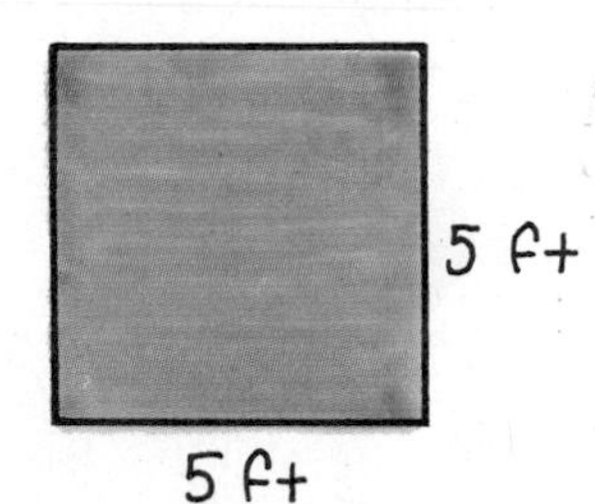

4.

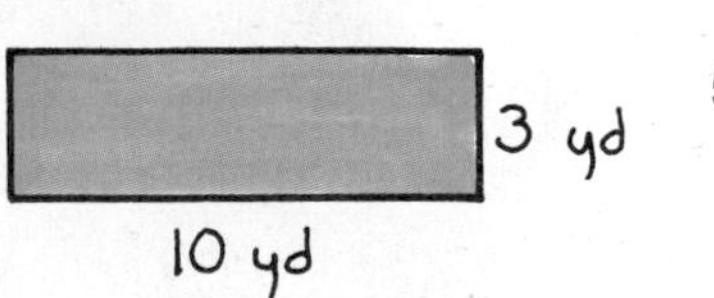

5.

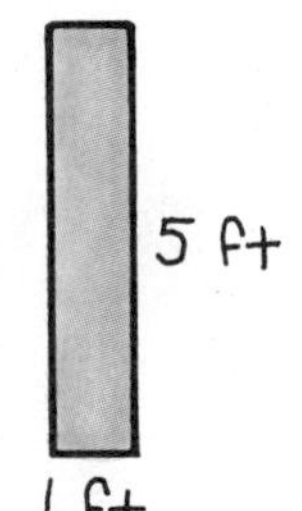

6.

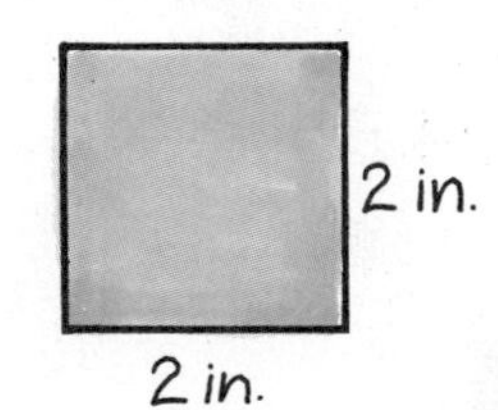

Give each area.

7.
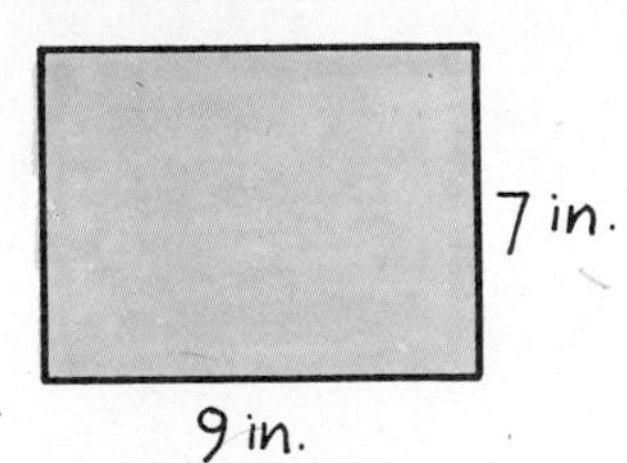

8.
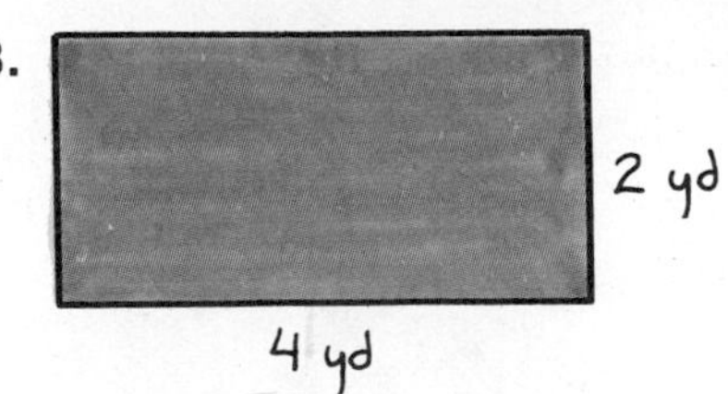

9.
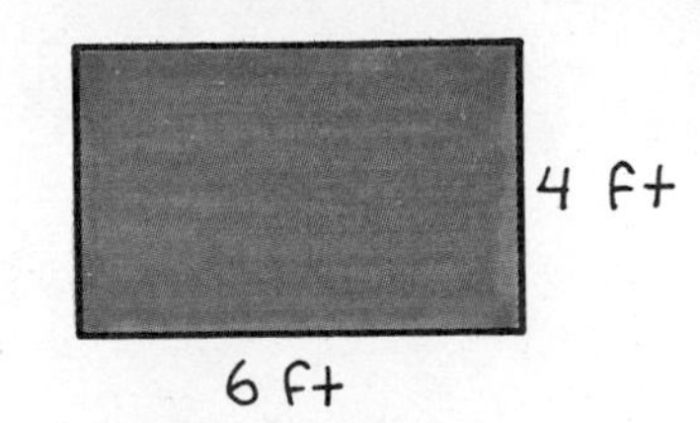

10.
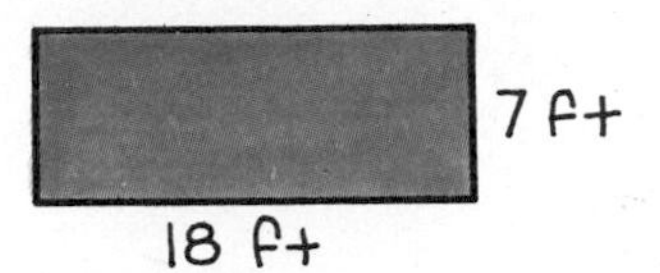

11.
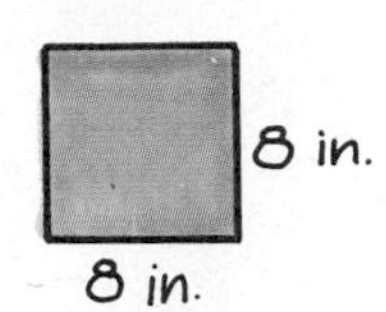

12.
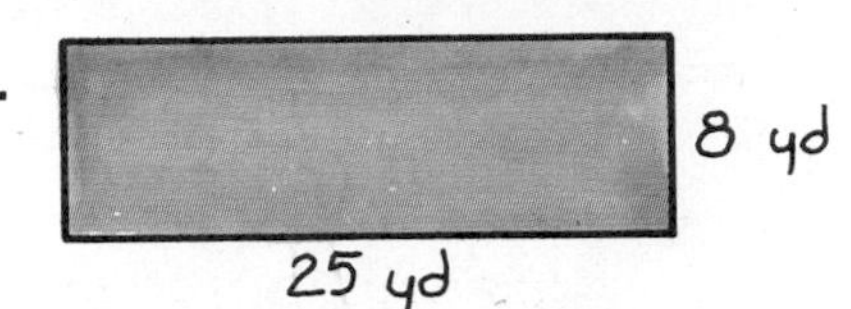

Give each volume.

13.
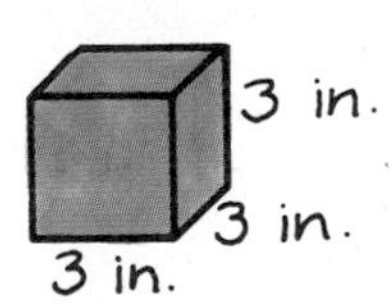

14.
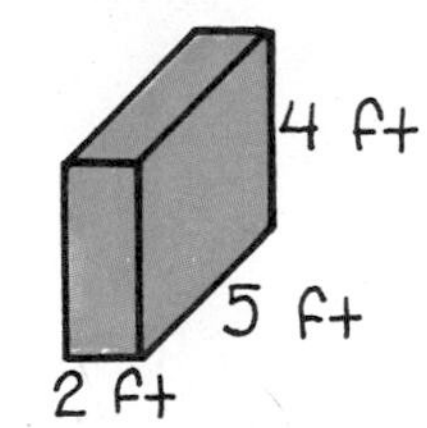

15.
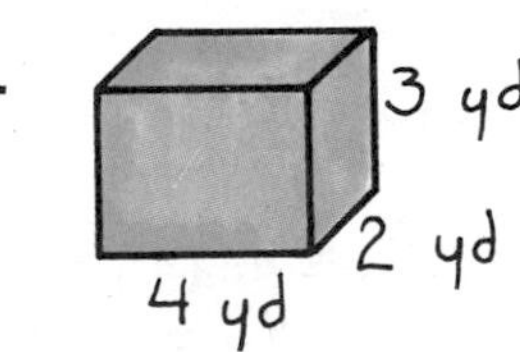

16.
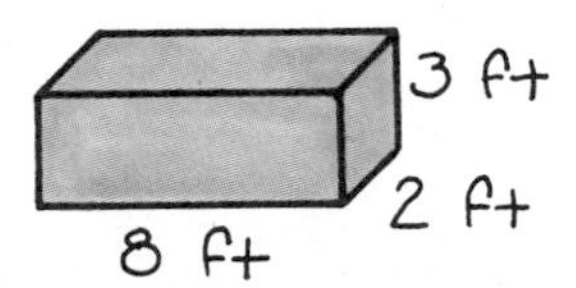

17.
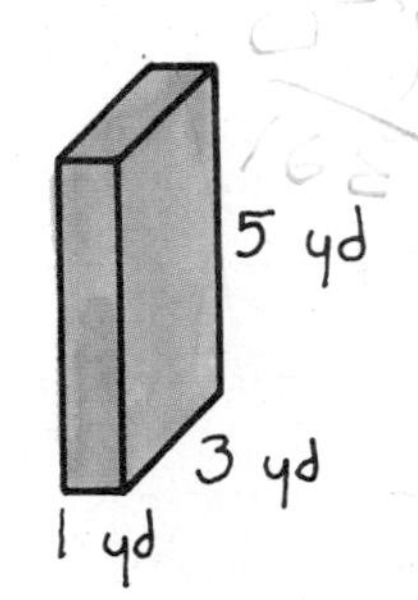

18.
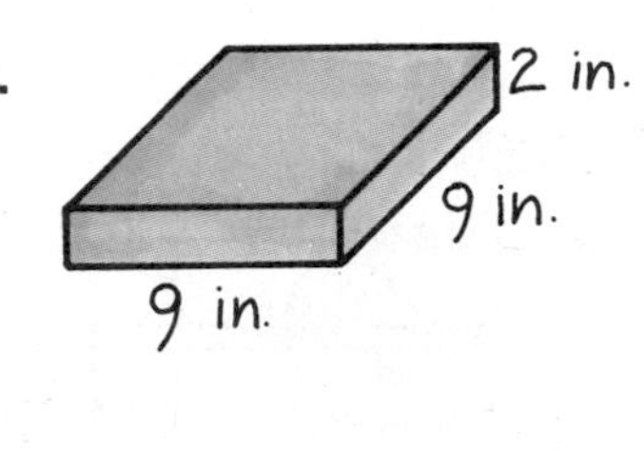

Solve.

19. Look at the picture of the fish tank.
   a. How high is it?
   b. How much water will it hold?
   c. How much tin is needed to make a cover for the top?

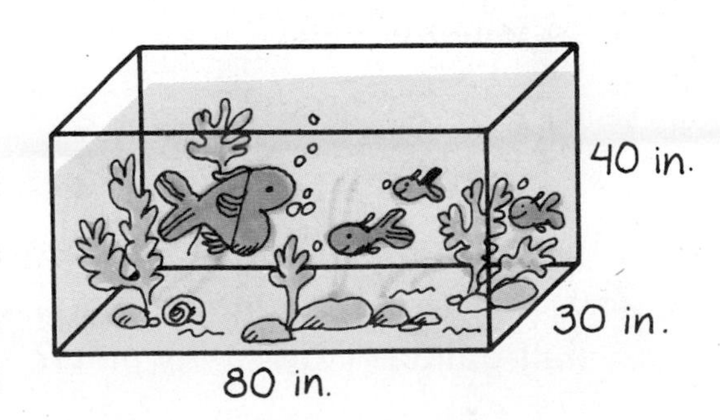

Liquid volume and weight—customary

These units are used to measure liquid volumes in the customary system.

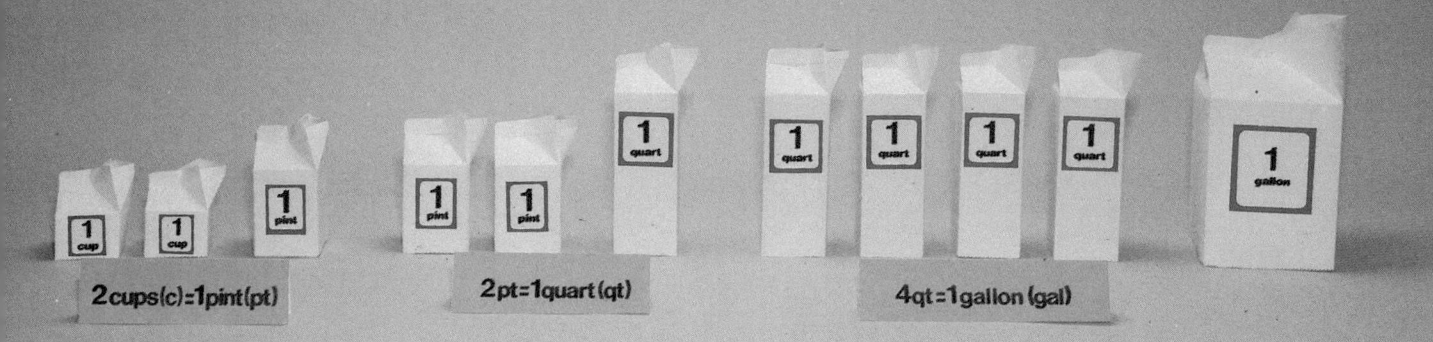

EXERCISES

Complete.

1. 2 pt = _?_ c
2. 2 pt = _?_ qt
3. 6 c = _?_ pt
4. 8 qt = _?_ gal
5. 3 qt = _?_ pt
6. 1 gal = _?_ pt

Complete.

7.

Quarts	1	2	3	4	5	6
Pints	?	?	?	?	?	?
Cups	?	?	?	?	?	?

Which is more,

8. 3 pints or 1 quart?
9. 2 pints or 5 cups?
10. 5 quarts or 11 pints?
11. 1 gallon or 3 quarts?
12. 1 gallon or 9 cups?
13. 30 cups or 7 quarts?

★ 14. Some things, such as milk, come in half-gallon containers. How many quarts are there in one half-gallon? How many pints?

The ounce (oz) and the pound (lb) are units used for measuring weight in the customary system.

16 ounces (oz) = 1 pound (lb)

EXERCISES

Choose the answer that seems right.

15. a. 8 oz
b. 8 lb

16. a. 2 oz
b. 2 lb

17. a. 10 oz
b. 10 lb

18. a. 4 oz
b. 4 lb

Complete.

19.	16 oz	? lb
20.	18 oz	1 lb ? oz
21.	24 oz	? lb ? oz
22.	30 oz	? lb ? oz
23.	32 oz	? lb
24.	36 oz	? lb ? oz

Complete.

25. $\frac{1}{2}$ lb = ? oz **26.** $\frac{1}{4}$ lb = ? oz

27. $\frac{3}{4}$ lb = ? oz **28.** $\frac{5}{8}$ lb = ? oz

Weigh yourself in pounds.

Temperature is sometimes measured on the Fahrenheit scale. Study a Fahrenheit thermometer.
Graph the daily high temperature for two weeks in degrees Fahrenheit. (You can find this information in a newspaper.)

CHAPTER CHECKUP

Measure to the nearest centimeter. [pages 216–217]

1.

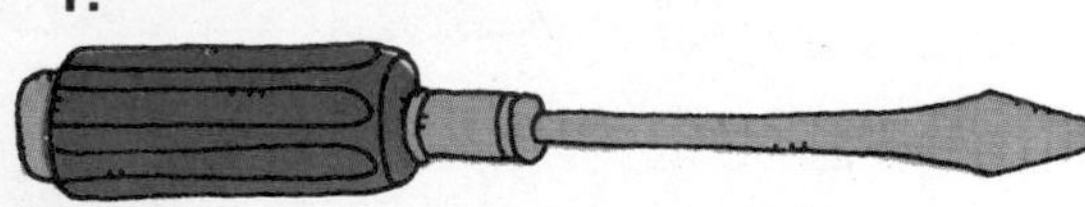

2.

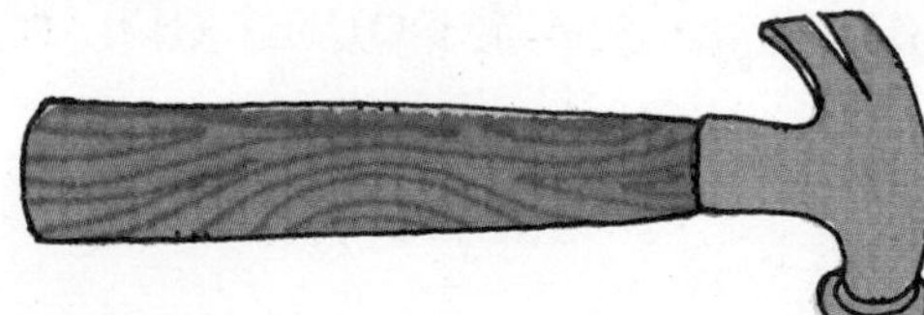

Complete. [pages 218–219]

3. 1 meter = ? cm

4. 1 kilometer = ? meters

5. Give the perimeter of the rectangle. [pages 220–221]

6. Give the area of the rectangle. [pages 222–223]

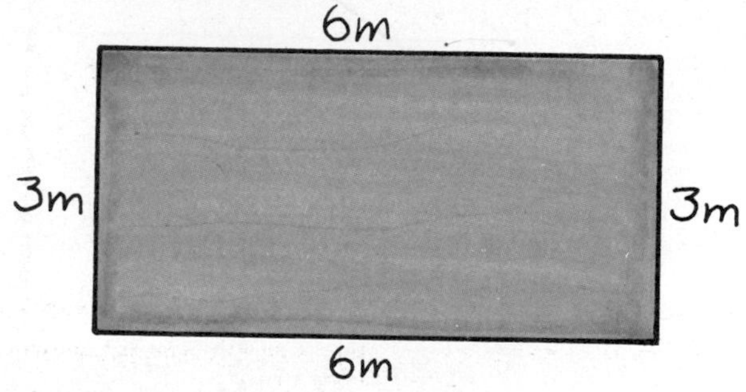

7. Give the volume. [pages 224–225]

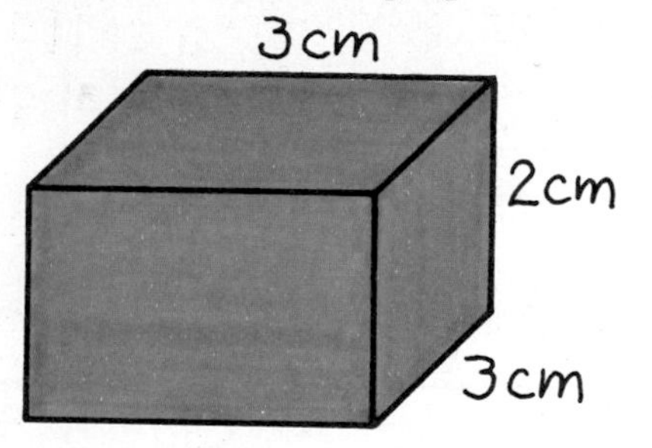

Complete. [pages 226–227]

8. 1 liter = ? milliliters

9. 1 kilogram = ? grams

Measure to the nearest $\frac{1}{4}$ inch. [pages 230–231]

10.

11.

Complete. [pages 232–237]

12. 1 foot = ? inches

13. 1 yard = ? feet

14. 1 pint = ? cups

15. 1 quart = ? pints

16. 1 gallon = ? quarts

17. 1 pound = ? ounces

CHAPTER PROJECT

1. Measure your arm span and your height to the nearest centimeter.
2. Compare your measurements.

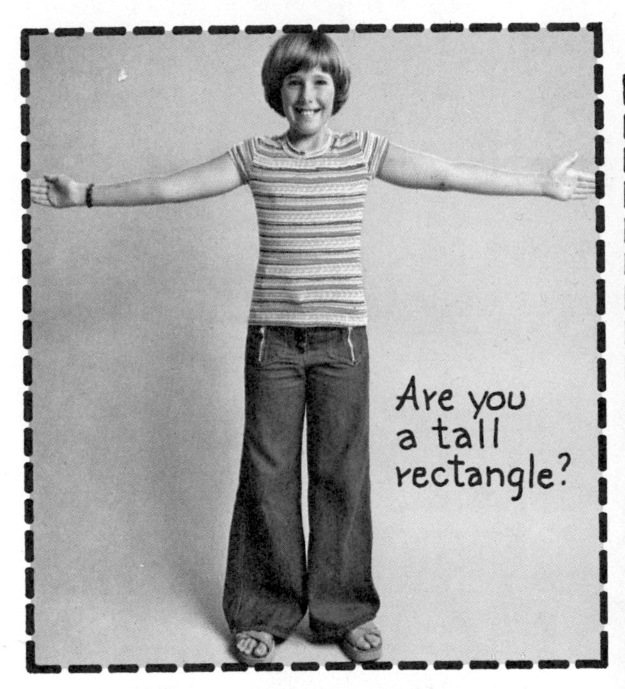

3. Find the number of children in your class who are squares, tall rectangles, and short rectangles.
4. Make a bar graph of what you found.
5. Tell some things that your graph shows.

CHAPTER REVIEW

1. Measure to the nearest centimeter.

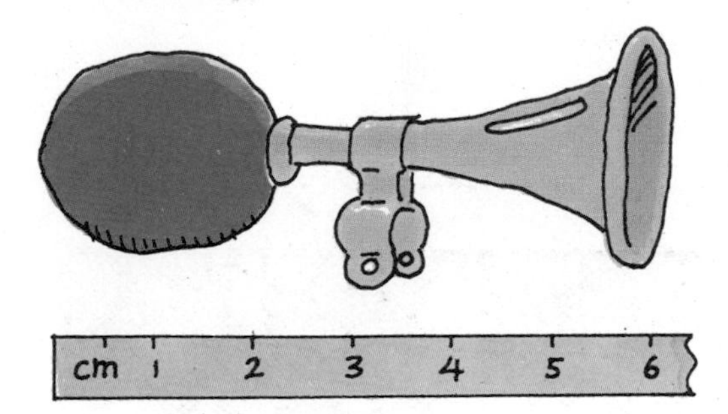

2. Give the perimeter.

3. Give the area.

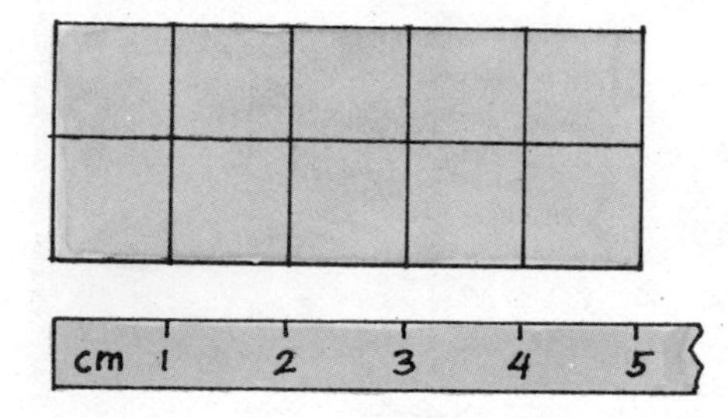

4. Give the volume.

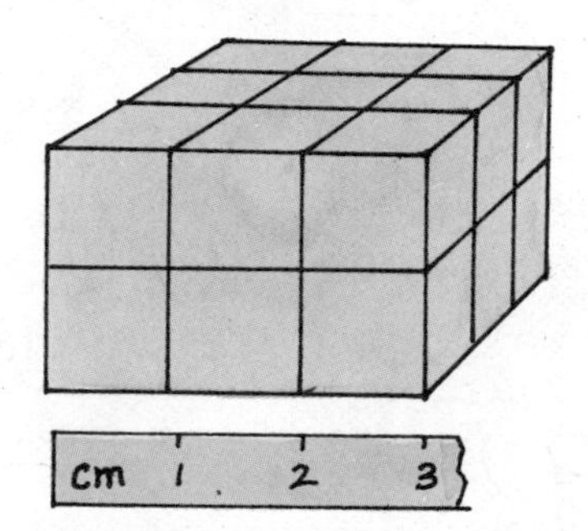

5. Measure to the nearest $\frac{1}{2}$ inch.

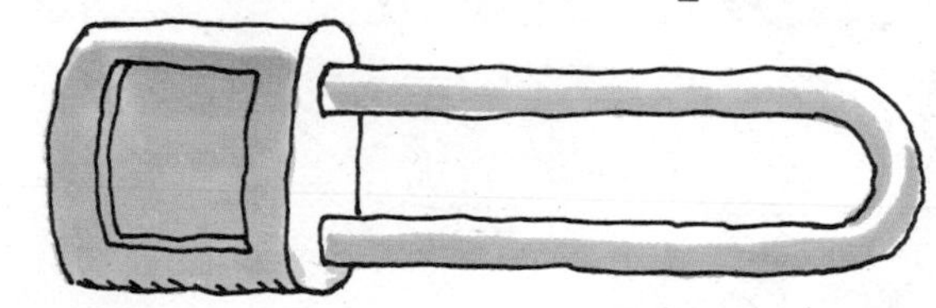

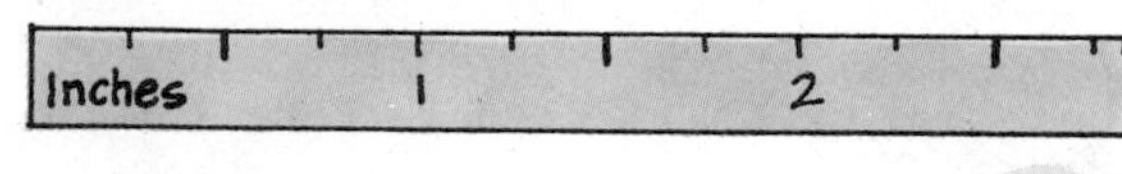

6. Measure to the nearest $\frac{1}{4}$ inch.

Complete.

7. 12 in. = _?_ ft

8. 1 yd = _?_ ft

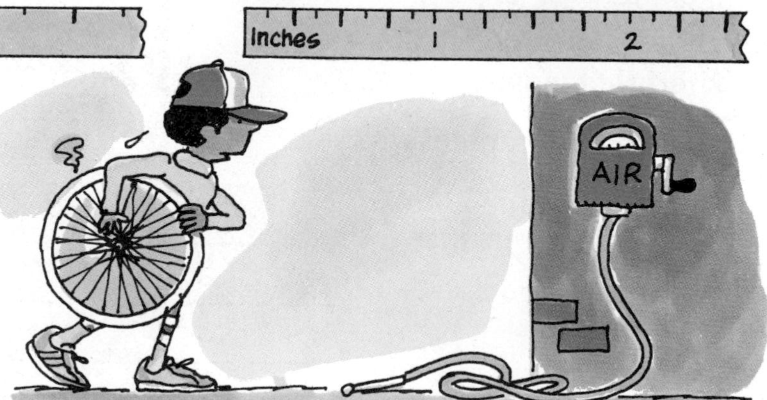

CHAPTER CHALLENGE

This solid has been covered with centimeter graph paper.

If we add the areas of all six sides, we get the **surface area** of the solid.

Top	18 square cm
Bottom	18 square cm
Front	12 square cm
Back	12 square cm
Right side	6 square cm
Left side	6 square cm
	72 square cm

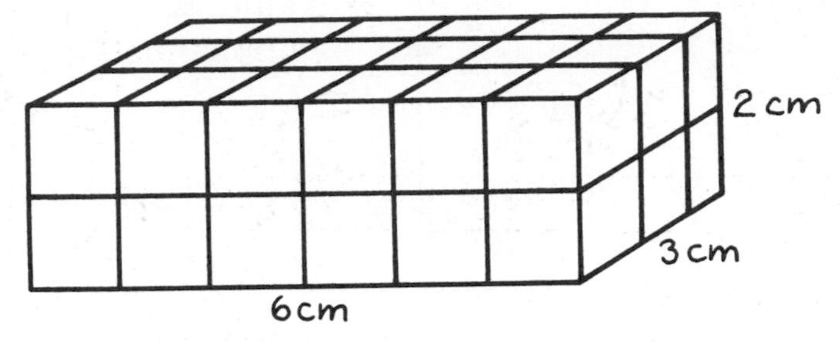

Give the surface area of each solid.

1.

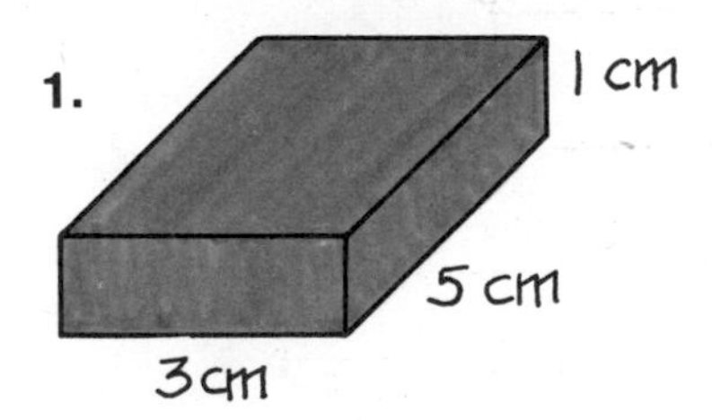

2.

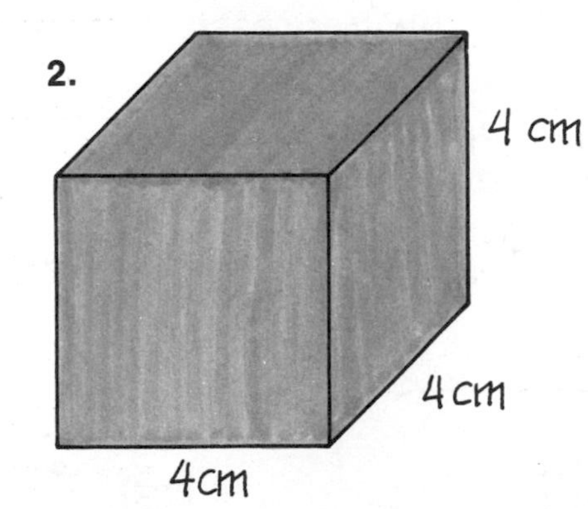

3.

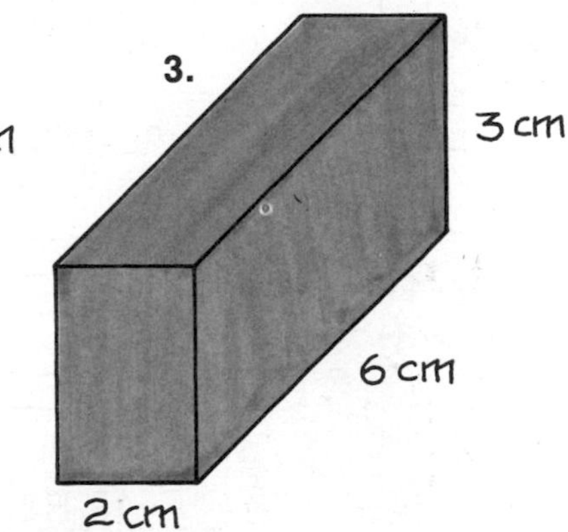

4.

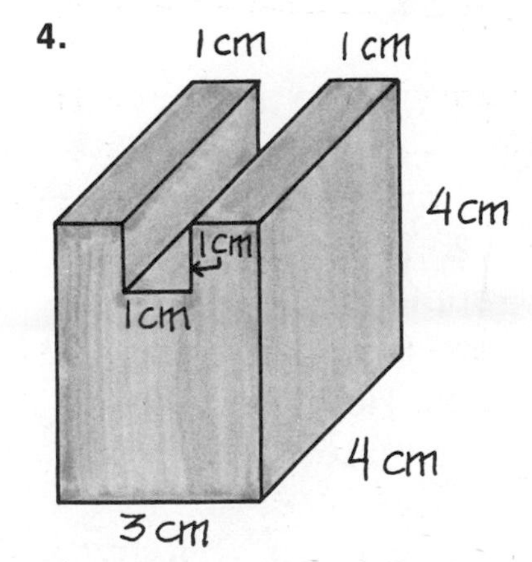

5. Copy and complete this table for cubes (blocks).

Length of each side in cm	Volume in cubic cm	Surface area in square cm
1	1	6
2	?	?
3	?	?
4	?	?

Choose the correct letter.

1. 3568 rounded to the nearest hundred is

a. 3500
b. 4000
c. 3570
d. 3600

2. Which digit in 375,289 is in the hundred thousands place?

a. 2
b. 5
c. 7
d. 3

3. Add.

$$\begin{array}{r} 293 \\ 148 \\ +256 \\ \hline \end{array}$$

a. 687
b. 597
c. 697
d. none of these

4. Subtract.

$$\begin{array}{r} 602 \\ -258 \\ \hline \end{array}$$

a. 344
b. 444
c. 354
d. none of these

5. What time is shown?

a. quarter past twelve
b. quarter to one
c. quarter to twelve
d. quarter past one

6. Which number is a multiple of 6?

a. 12
b. 9
c. 3
d. none of these

7. Divide.

$4\overline{)78}$

a. 19 R2
b. 29 R2
c. 12 R2
d. none of these

8. Complete.

$\frac{2}{3}$ of 18 = ?

a. 9
b. 6
c. 8
d. none of these

9. Add.

$\frac{1}{4} + \frac{3}{8} = ?$

a. $\frac{4}{12}$
b. $\frac{5}{8}$
c. $\frac{4}{8}$
d. $\frac{7}{8}$

10. Subtract.

$\frac{5}{6} - \frac{1}{3} = ?$

a. $\frac{1}{3}$
b. $\frac{4}{6}$
c. $\frac{4}{9}$
d. $\frac{1}{2}$

11. There were 8 buses. There were 40 people on each bus. How many people were there in all?

a. 320 **b.** 5
c. 48 **d.** none of these

12. There are 75 apples in boxes. There are 5 apples in each box. How many boxes are there?

a. 70 **b.** 375
c. 15 **d.** none of these

Multiplication by 1- and 2-Digit Factors

9

READY OR NOT?

Multiply.

1.	2 ×6	**2.**	8 ×3
3.	2 ×4	**4.**	3 ×3
5.	6 ×7	**6.**	7 ×7
7.	7 ×8	**8.**	6 ×3
9.	6 ×5	**10.**	9 ×6
11.	3 ×6	**12.**	9 ×4
13.	12 ×4	**14.**	23 ×2
15.	20 ×4	**16.**	41 ×2
17.	52 ×4	**18.**	31 ×7
19.	54 ×2	**20.**	61 ×4
21.	28 ×2	**22.**	36 ×3
23.	43 ×5	**24.**	52 ×9

Multiplying a 3-digit number

Corn is planted in long straight rows. If 175 seeds are planted in each row, how many seeds would be planted in 3 rows?

EXAMPLE.

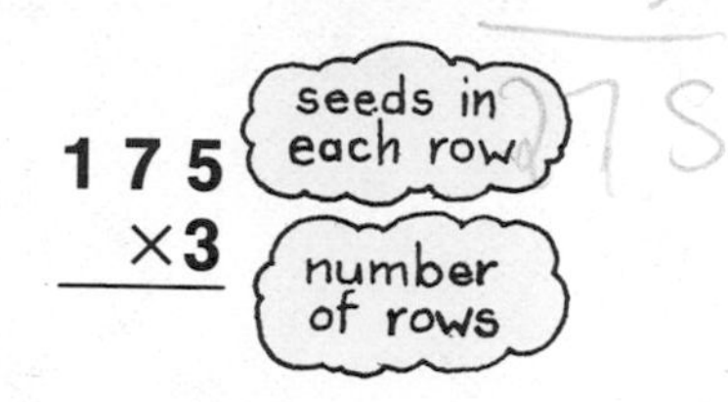

Step 1. Multiply ones and regroup.

$$\begin{array}{r} {}^{1}\\ 175 \\ \times 3 \\ \hline 5 \end{array}$$

Step 2. Multiply tens, add, and regroup.

$$\begin{array}{r} {}^{2\ 1}\\ 175 \\ \times 3 \\ \hline 25 \end{array}$$

(3×7) + 1 = 22

Step 3. Multiply hundreds and add.

$$\begin{array}{r} {}^{2\ 1}\\ 175 \\ \times 3 \\ \hline 525 \end{array}$$

(3×1) + 2 = 5

525 seeds would be planted in 3 rows.

EXERCISES

Multiply.

1. 214 ×2	**2.** 303 ×3	**3.** 323 ×2
4. 324 ×3	**5.** 216 ×4	**6.** 103 ×5
7. 152 ×4	**8.** 261 ×3	**9.** 150 ×4

10. 118 ×4	**11.** 217 ×3	**12.** 308 ×4	**13.** 121 ×4	**14.** 102 ×4
15. 181 ×4	**16.** 271 ×3	**17.** 380 ×4	**18.** 222 ×4	**19.** 111 ×6
20. 283 ×3	**21.** 394 ×2	**22.** 282 ×4	**23.** 141 ×5	**24.** 231 ×4
25. 126 ×4	**26.** 196 ×5	**27.** 148 ×6	**28.** 109 ×8	**29.** 138 ×4
30. 168 ×5	**31.** 167 ×5	**32.** 126 ×3	**33.** 284 ×3	**34.** 296 ×3
35. 157 ×5	**36.** 287 ×3	**37.** 175 ×4	**38.** 359 ×2	**39.** 248 ×3
40. \$3.75 ×2	**41.** \$1.46 ×6	**42.** \$1.29 ×3	**43.** \$1.24 ×8	**44.** \$1.79 ×5

Solve.

45. There are 75 tomato plants in each of 3 rows. How many tomato plants are there?

46. There are 85 pepper plants in all. There are the same number in each of 5 rows. How many pepper plants are in each row?

More about multiplying 3-digit numbers

EXAMPLE.

$$\begin{array}{r} 538 \\ \times 6 \\ \hline \end{array}$$

Step 1.

Multiply ones and regroup.

$$\begin{array}{r} {}^{4} \\ 538 \\ \times 6 \\ \hline 8 \end{array}$$

Step 2.

Multiply tens, add, and regroup.

$$\begin{array}{r} {}^{2\,4} \\ 538 \\ \times 6 \\ \hline 28 \end{array}$$

Step 3.

Multiply hundreds and add.

$$\begin{array}{r} {}^{2\,4} \\ 538 \\ \times 6 \\ \hline 3228 \end{array}$$

32 hundreds is 3 thousands and 2 hundreds.

EXERCISES

Multiply.

1. 358×5
2. 643×4
3. 482×6
4. 952×3
5. 554×2
6. 683×7
7. 397×8
8. 749×5
9. 376×7
10. 874×9
11. 755×8
12. 576×5
13. 892×3
14. 675×6
15. 453×8
16. 556×5
17. 653×6
18. 759×3
19. 914×6
20. 371×4
21. 853×4
22. 493×9
23. 942×3
24. 784×7
25. 256×8

26. 300 $\times 5$

27. 800 $\times 9$

28. 900 $\times 6$

29. 700 $\times 7$

30. 800 $\times 4$

31. 629 $\times 5$

32. 158 $\times 4$

33. 346 $\times 6$

34. 792 $\times 3$

35. 502 $\times 7$

36. \$6.28 $\times 2$ = \$12.56

37. \$3.07 $\times 5$

38. \$9.34 $\times 8$

39. \$7.61 $\times 6$

40. \$4.75 $\times 4$

Give each product.

41. 264×3

42. 526×5

43. 387×4

44. 195×2

45. 468×6

46. 706×7

47. 257×9

48. 676×6

Solve.

49. An elephant weighs 4900 kilograms. A hippo weighs 2270 kilograms less. What is the weight of the hippo?

50. Each day an elephant eats 68 kilograms of grass. How much does it eat in a week?

51. Each month the zoo buys 585 kilograms of bananas. How much does it buy in 6 months?

52. The zoo keeper evenly divides 68 packages of sunflower seeds among 4 bird cages. How many does she put in each cage?

KEEPING SKILLS SHARP

Add. Give each sum in lowest terms.

1. $\frac{1}{3} + \frac{1}{3}$

2. $\frac{1}{4} + \frac{1}{2}$

3. $\frac{1}{4} + \frac{1}{3}$

4. $\frac{2}{3} + \frac{1}{4}$

5. $\frac{1}{2} + \frac{3}{8}$

6. $\frac{3}{10} + \frac{1}{5}$

7. $\frac{2}{3} + \frac{1}{6}$

8. $\frac{1}{5} + \frac{2}{5}$

9. $\frac{7}{10} + \frac{1}{5}$

10. $\frac{3}{8} + \frac{1}{4}$

Multiplying a 4-digit number

Larger numbers are multiplied in the same way as smaller numbers. Study this example.

$$\begin{array}{r} 4178 \\ \times 7 \\ \hline 29{,}246 \end{array}$$

Remember that we can use rounding to estimate a product. An estimate can tell us whether our answer makes sense.

Since 29,246 (our answer) is close to 28,000 (our estimate), our answer makes sense.

EXERCISES

First multiply. Then estimate the product to see if your answer makes sense.

1. 78×9
2. 94×8
3. 766×5
4. 911×7
5. 835×6
6. 2130×4
7. 3896×7
8. 5235×6
9. 7869×5
10. 8350×8

Multiply.

11. 4328 ×2
12. 5210 ×4
13. 6348 ×3
14. 2065 ×6
15. 3154 ×5

16. $52.81 ×7 $369.67
17. $73.06 ×8
18. $58.29 ×5
19. $74.65 ×6
20. $38.42 ×9

21. 396 × 5
22. 784 × 6
23. 923 × 8

24. 7124 × 2
25. 5638 × 4
26. 6341 × 5

Solve.

27. How much for 4 kg?

28. How much for 3 tickets?

29. A dozen doughnuts cost $1.79. A dozen cookies cost $.99. How much do 6 dozen doughnuts cost?

30. A car traveled 94 kilometers in 2 hours. How many kilometers did it average per hour?

31. A rectangular lot is 9 meters wide and 27 meters long. What is its area?

32. A certain style of jeans costs $18.79. How much do 3 pairs cost?

Challenge!

Multiply.

33. 12,345,679 ×9

34. Can you guess this product?

12,345,679 ×18

Hint: Compare exercises 33 and 34.

Addition, subtraction, multiplication, and division

Remember to work inside the grouping symbols first.

$$(160 \times 3) + 320 = 800$$

$$\begin{array}{r} 160 \\ \times 3 \\ \hline 480 \\ +320 \\ \hline 800 \end{array}$$

EXERCISES

Compute.

1. (58 × 3) + 4
2. 58 × (3 + 4)
3. (167 × 5) − 2
4. 167 × (5 − 2)
5. (532 − 4) × 2
6. 532 − (4 × 2)
7. (3821 × 5) + 253
8. (4816 × 2) − 4816
9. (5621 − 387) × 6
10. (7135 + 2816) × 8

Tell what you would do to solve each problem. (*Hint:* Each problem has two steps.)

11. Ms. Chaisson bought 2 hammers for $■ each and a saw for $■. What was the total cost?
12. Al bought a sander for $■ and a router for $■. He gave the clerk $■. How much change did he get?
13. Steve could buy 3 packages of sandpaper for $■ and a paintbrush for $■. How much would 1 package of sandpaper and a paintbrush cost?
14. Susan bought 3 cans of paint for $■ each. She gave the clerk $■. How much change did she get?

Build the Greatest Product

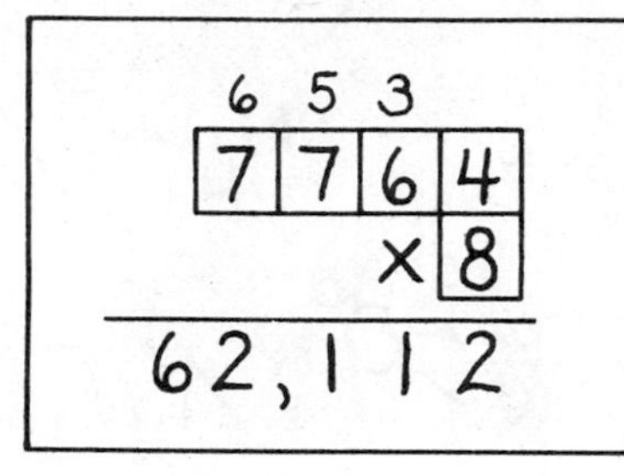

3 5 5
6 4 7 8
× 7
45,3 4 6

Which product is greater?

1. 3504 × 8 4830 × 5

2. 5768 × 9 7985 × 6

3. 7385 × 4 4538 × 7

4. 9473 × 6 6347 × 9

Play the game.

1. Choose a leader.
2. Make two cards for each digit.

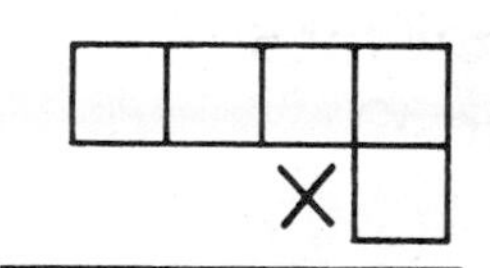

3. Each player draws a table.
4. As the leader picks a digit, write it in your table.
5. Repeat step 4 until your table is filled in.
6. Multiply. The player who builds the greatest product wins the game.

Problem solving

Solve.

1. Jeanette rode the roller coaster 3 times. How much did she spend for roller-coaster tickets?

2. How much do 5 merry-go-round tickets cost?

3. What is the total cost of 2 rides on the whip and 2 rides on the bumper cars?

4. What is the total cost of 3 Ferris wheel rides and 2 flying-saucer rides?

TICKETS	
Merry-Go-Round	45¢
Roller Coaster	85¢
Ferris Wheel	55¢
Whip	60¢
Flying Saucer	40¢
Bumper Cars	75¢
Monster House	35¢
Speedboat	45¢

5. Which costs more, 4 monster-house tickets or 3 speedboat tickets?

6. Ruth had $3.40. She bought 2 bumper-car tickets. How much money did she have left?

7. Chris bought 3 roller-coaster tickets. He gave the ticket seller $3.00. How much change did he get back?

8. Maria had $5.00. She bought 2 speedboat tickets and a roller-coaster ticket. How much money did she have left?

★ 9. Alan got $5 for his birthday. He wanted to go on each ride once. Did he have enough money?

★ 10. Juan had $2.05. He rode the merry-go-round twice and the bumper cars once. What other rides could he take?

KEEPING SKILLS SHARP

1. $\frac{1}{2}$ of 16 = ?
2. $\frac{1}{3}$ of 24 = ?
3. $\frac{1}{5}$ of 30 = ?
4. $\frac{1}{4}$ of 16 = ?
5. $\frac{1}{8}$ of 48 = ?
6. $\frac{1}{3}$ of 30 = ?
7. $\frac{2}{3}$ of 27 = ?
8. $\frac{3}{4}$ of 32 = ?
9. $\frac{3}{8}$ of 48 = ?
10. $\frac{3}{5}$ of 50 = ?
11. $\frac{7}{8}$ of 56 = ?
12. $\frac{2}{3}$ of 18 = ?
13. $\frac{5}{6}$ of 30 = ?
14. $\frac{3}{4}$ of 24 = ?
15. $\frac{5}{8}$ of 40 = ?
16. $\frac{2}{3}$ of 30 = ?
17. $\frac{3}{8}$ of 56 = ?
18. $\frac{4}{7}$ of 70 = ?

Multiplying by 10

To find the total number of marbles, you can multiply.

$$\begin{array}{r} 10 \\ \times 6 \\ \hline 60 \end{array} \qquad \begin{array}{r} 6 \\ \times 10 \\ \hline 60 \end{array}$$

(6 tens)

$$\begin{array}{r} 9 \\ \times 10 \\ \hline 90 \end{array} \quad \begin{array}{r} 12 \\ \times 10 \\ \hline 120 \end{array} \quad \begin{array}{r} 18 \\ \times 10 \\ \hline 180 \end{array} \quad \begin{array}{r} 25 \\ \times 10 \\ \hline 250 \end{array}$$

Do you see an easy way to multiply a number by 10?

EXERCISES

Multiply.

1. 13 × 10	**2.** 15 × 10	**3.** 19 × 10	**4.** 22 × 10	**5.** 26 × 10
6. 34 × 10	**7.** 53 × 10	**8.** 50 × 10	**9.** 62 × 10	**10.** 92 × 10
11. 124 × 10	**12.** 156 × 10	**13.** 132 × 10	**14.** 175 × 10	**15.** 190 × 10
16. 210 × 10	**17.** 234 × 10	**18.** 353 × 10	**19.** 400 × 10	**20.** 526 × 10
21. 58 × 10	**22.** 672 × 10	**23.** 70 × 10	**24.** 803 × 10	**25.** 49 × 10
26. 864 × 10	**27.** 230 × 10	**28.** 78 × 10	**29.** 600 × 10	**30.** 347 × 10

Complete.

31. 5 dimes are worth ? pennies.

32. 16 dimes are worth ? pennies.

33. 3 dollars are worth ? dimes.

34. 36 dollars are worth ? dimes.

35. 20 dimes are worth ? pennies.

36. 50 dollars are worth ? dimes.

Solve.

37.

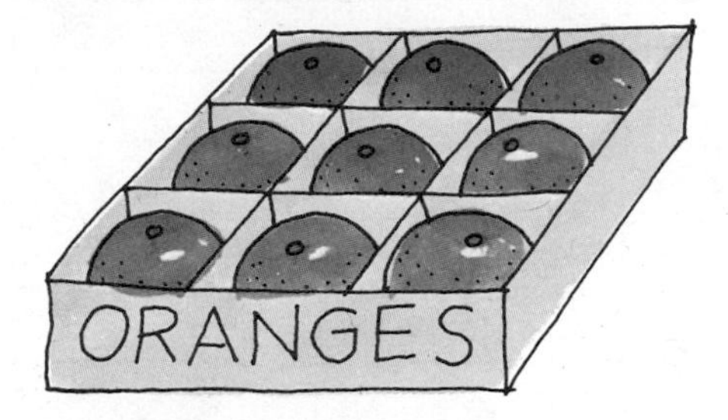

How many oranges in 10 packages?

38.

How many apples in 10 packages?

39. There are 10 players on a team. There are 6 players on another team. How many players are there in all?

40. There are 10 teams. There are 6 players on each team. How many players are there in all?

41. There are 10 players on team A. There are 6 fewer players on team B. How many players are there on team B?

42. There are 10 players on team A. There are 6 more players on team B. How many players are there on team B?

43. Jack puts 10 pictures on each page of his picture album. How many pictures can he put on 36 pages?

44. June has 43 pieces of track for her electric train. Each piece is 10 cm long. How long is the longest track she can build?

Challenge!

Find the missing input or output.

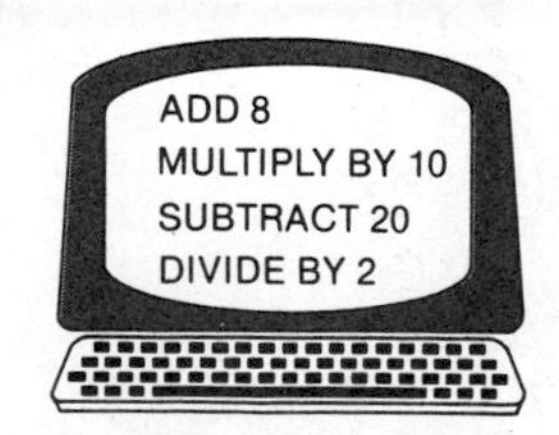

	Input	Output
	52	290
45.	14	?
46.	29	?
47.	?	140

Multiplying by tens

How would you find the number of eggs in all?

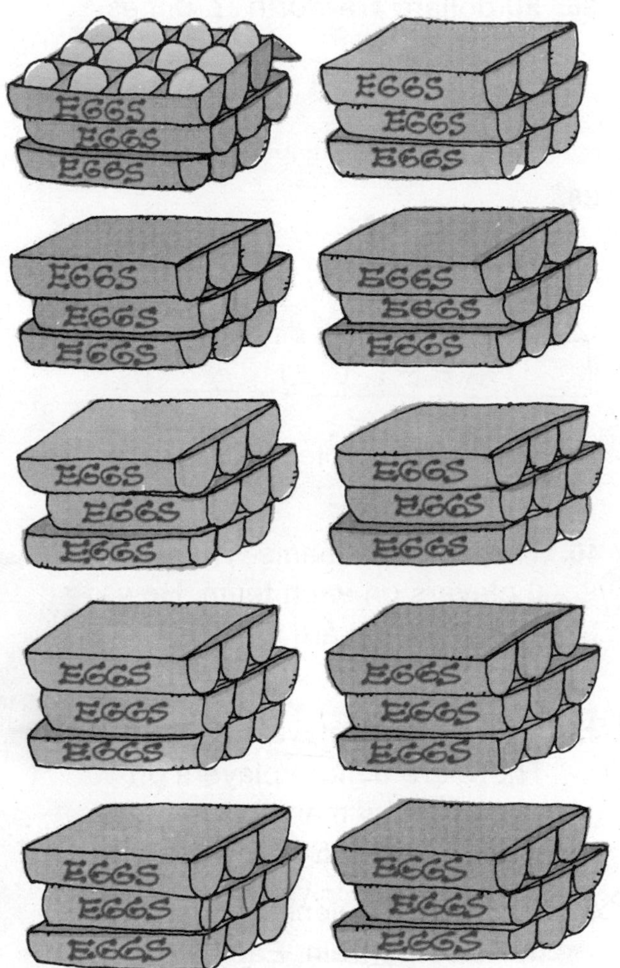

Here is how Sharon found the number of eggs.

Sharon
12 in each carton
x3 cartons
36 in each stack

36 in each stack
x 10 stacks
360 eggs

To find this product:

$$\begin{array}{r l} 12 & \text{in each carton} \\ \times 30 & \text{cartons} \\ \hline \end{array}$$

she first multiplied 12 by 3. Then she multiplied that answer by 10.

She could have shortened her work by doing this:

EXERCISES

Multiply.

1. 42 ×20	Multiply by 2 and then by 10.	
2. 31 ×30	Multiply by 3 and then by 10.	
3. 43 ×20	Multiply by 2 and then by 10.	
4. 16 ×40	Multiply by 4 and then by 10.	

You can write the 0 first and then multiply by 4.

$$\begin{array}{r} \overset{1}{5}3 \\ \times 40 \\ \hline 2120 \end{array}$$

Shortcut

Multiply. Write the 0 first.

5. $\begin{array}{r} 34 \\ \times 20 \\ \hline \end{array}$ **6.** $\begin{array}{r} 21 \\ \times 30 \\ \hline \end{array}$ **7.** $\begin{array}{r} 23 \\ \times 30 \\ \hline \end{array}$

8. $\begin{array}{r} 36 \\ \times 20 \\ \hline \end{array}$ **9.** $\begin{array}{r} 30 \\ \times 30 \\ \hline \end{array}$ **10.** $\begin{array}{r} 48 \\ \times 50 \\ \hline \end{array}$ **11.** $\begin{array}{r} 46 \\ \times 40 \\ \hline \end{array}$

12. $\begin{array}{r} 68 \\ \times 50 \\ \hline \end{array}$ **13.** $\begin{array}{r} 93 \\ \times 60 \\ \hline \end{array}$ **14.** $\begin{array}{r} 82 \\ \times 70 \\ \hline \end{array}$ **15.** $\begin{array}{r} 84 \\ \times 50 \\ \hline \end{array}$

16. $\begin{array}{r} 284 \\ \times 30 \\ \hline \end{array}$ **17.** $\begin{array}{r} 358 \\ \times 40 \\ \hline \end{array}$ **18.** $\begin{array}{r} 472 \\ \times 30 \\ \hline \end{array}$ **19.** $\begin{array}{r} 700 \\ \times 50 \\ \hline \end{array}$

20. $\begin{array}{r} \$6.03 \\ \times 60 \\ \hline \$361.80 \end{array}$ **21.** $\begin{array}{r} \$3.81 \\ \times 80 \\ \hline \end{array}$ **22.** $\begin{array}{r} \$7.49 \\ \times 70 \\ \hline \end{array}$ **23.** $\begin{array}{r} \$8.05 \\ \times 60 \\ \hline \end{array}$ **24.** $\begin{array}{r} \$7.45 \\ \times 50 \\ \hline \end{array}$

Solve.

25.

How much will 20 stamps cost?

26.

36 COOKIES

How many cookies in 30 boxes?

27. Mark earns $13 each week. It costs $575 to go to summer camp. Does he earn enough in 40 weeks to be able to go to summer camp?

28. Joanna earns $1.25 an hour for baby-sitting. One month she baby-sat 5 hours each Saturday. There were 4 Saturdays in the month. How much did she earn?

Multiplying by a 2-digit number

There are 24 hours in a day. To find how many hours in 23 days, multiply:

$$\begin{array}{r} 24 \\ \times 23 \\ \hline \end{array}$$

Here is how to find the product:

Step 1. Multiply by 3.

$$\begin{array}{r} {}^{1} \\ 24 \\ \times 23 \\ \hline 72 \end{array}$$ hours in 3 days

Step 2. Multiply by 20.

$$\begin{array}{r} {}^{1} \\ 24 \\ \times 23 \\ \hline 72 \\ 480 \\ \hline \end{array}$$ hours in 20 days

Step 3. Add.

$$\begin{array}{r} {}^{1} \\ 24 \\ \times 23 \\ \hline 72 \\ 480 \\ \hline 552 \end{array}$$ hours in 23 days

EXERCISES

Multiply.

1.
$$\begin{array}{r} 42 \\ \times 23 \\ \hline 126 \\ 840 \\ \hline 966 \end{array}$$
(126 = 3 × 42; 840 = 20 × 42; 966 = 23 × 42)

2.
$$\begin{array}{r} 43 \\ \times 32 \\ \hline 86 \\ 1290 \\ \hline \end{array}$$
(86 = 2 × 43; 1290 = 30 × 43)

3.
$$\begin{array}{r} 32 \\ \times 26 \\ \hline 192 \end{array}$$
(192 = 6 × 32; 20 × 32)

4. 54×43 **5.** 78×53 **6.** 65×48 **7.** 53×49 **8.** 49×53

9. 54×32 **10.** 58×42 **11.** 79×19 **12.** 65×25 **13.** 80×52

14. 74×29 **15.** 93×50 **16.** 76×43 **17.** 85×74 **18.** 96×44

19. 54×63 **20.** 63×63 **21.** 69×41 **22.** 77×55 **23.** 86×48

24. 52×37 **25.** 78×58 **26.** 65×49 **27.** 79×27 **28.** 95×63

Solve.

29. How many hours are there in 56 days?

30. How many weeks are there in 42 days?

31. How many minutes are there in 85 hours?

32. How many seconds are there in 72 minutes?

33. How many inches are there in 36 feet?

34. How many yards are there in 51 feet?

35. How many days are there in 63 weeks?

36. How many eggs are there in 54 dozen eggs?

Multiplying by a 2-digit number

Sylvia had 28 boxes of baseball cards.
There were 116 cards in each box.
How many cards did she have?

Step 1. Multiply 116 by 8.

$$\begin{array}{r} {}^{1}{}^{4} \\ 116 \\ \times 28 \\ \hline 928 \end{array}$$

Step 2. Multiply 116 by 20.

$$\begin{array}{r} {}^{1} \\ {}^{1}{}^{4} \\ 116 \\ \times 28 \\ \hline 928 \\ 2320 \end{array}$$

Step 3. Add.

$$\begin{array}{r} {}^{1} \\ {}^{1}{}^{4} \\ 116 \\ \times 28 \\ \hline 928 \\ 2320 \\ \hline 3248 \end{array}$$

Estimate the product to see if this answer makes sense.

116	Round to nearest 100	**100**
×28	Round to nearest 10	**×30**
		3000

Sylvia had 3248 baseball cards.

EXERCISES

Estimate each product.

1. $89 \times 52 \rightarrow 90 \times 50 = ?$
2. $73 \times 41 \rightarrow 70 \times 40 = ?$
3. $68 \times 56 \rightarrow 70 \times 60 = ?$
4. $112 \times 46 \rightarrow 100 \times 50 = ?$
5. $283 \times 34 \rightarrow 300 \times 30 = ?$
6. $419 \times 85 \rightarrow 400 \times 90 = ?$

First multiply. Then estimate to see if the answer makes sense.

7. 72×40
8. 82×36
9. 59×54
10. 76×18
11. 93×27
12. 48×86
13. 92×52
14. 78×23
15. 65×49
16. 49×65
17. 115×36
18. 782×40
19. 592×36
20. 435×48
21. 673×25

Challenge!

Guess my number.

22. I am the product of 427 and 83.
23. I am 54 greater than the product of 226 and 36.
24. I am 137 less than the product of 483 and 29.

KEEPING SKILLS SHARP

Subtract. Give each difference in lowest terms.

1. $\frac{1}{2} - \frac{1}{4}$
2. $\frac{5}{8} - \frac{1}{2}$
3. $\frac{4}{5} - \frac{3}{10}$
4. $\frac{2}{3} - \frac{1}{6}$
5. $\frac{7}{8} - \frac{1}{4}$
6. $\frac{1}{2} - \frac{1}{3}$
7. $\frac{2}{3} - \frac{1}{4}$
8. $\frac{1}{2} - \frac{1}{5}$
9. $\frac{7}{10} - \frac{1}{2}$
10. $\frac{5}{9} - \frac{1}{3}$

Multiplication practice

Multiply.

1. 26 ×12	**2.** 84 ×19	**3.** 53 ×27	**4.** 48 ×39	**5.** 68 ×50
6. 58 ×26	**7.** 67 ×35	**8.** 49 ×48	**9.** 76 ×19	**10.** 85 ×92
11. 48 ×75	**12.** 93 ×68	**13.** 68 ×93	**14.** 56 ×74	**15.** 74 ×56
16. 374 ×65	**17.** 281 ×56	**18.** 519 ×74	**19.** 650 ×32	**20.** 942 ×60
21. 321 ×31	**22.** 746 ×92	**23.** 883 ×48	**24.** 609 ×85	**25.** 619 ×73
26. 344 ×76	**27.** 551 ×88	**28.** 668 ×26	**29.** 748 ×15	**30.** 526 ×65
31. \$2.53 ×24 10.12 50.60 \$60.72	**32.** \$.25 ×32	**33.** \$.69 ×50 3450	**34.** \$6.20 ×26	**35.** \$8.62 ×68
	36. \$2.74 ×62	**37.** \$3.58 ×79	**38.** \$5.06 ×82	**39.** \$9.53 ×91
40. \$6.27 ×34	**41.** \$8.36 ×15	**42.** \$4.98 ×25	**43.** \$5.08 ×42	**44.** \$7.69 ×83
45. \$2.96 ×39	**46.** \$7.09 ×64	**47.** \$8.81 ×69	**48.** \$9.20 ×53	**49.** \$2.37 ×68

One day a class kept a record of the number of people in each car that passed their school. They made this **picture graph** of what they found.

NUMBER OF PEOPLE IN CARS	
1 person	
2 people	
3 people	
4 people	
Each stands for 12 cars	

50. How many cars does each stand for?

51. How many cars had only 1 person?

52. How many had 2 people?

53. Can you tell exactly how many cars had 3 people? If not, estimate the number.

54. Estimate the number of cars that had 4 people.

55. About how many cars passed the school?

Computers at the catalog store

The Criser family is planning a camping vacation. They visited the Super Value Catalog Store to buy some camping equipment. They decided to buy several items. Mrs. Criser filled out the order form.

Name: Ruth Criser — Super Value Catalog Store

Catalog Item No.	Description	Quantity
42318	Alpine tent	2
37942	Sleeping bag	3
29005	Air mattress	3
12003	Can of heat	20
13216	Lantern	1

The clerk entered the item numbers and the quantity for each item purchased. The computer has the prices stored in its **memory.** The computer printed out this sales record:

Super Value Catalog Store

Catalog Item Number	Description	Quantity	Price for one	Total Price
42318	Alpine tent	2	$72.46	$144.92
37942	Sleeping bag	3	61.75	185.25
29005	Air mattress	3	8.96	26.88
12003	Can of heat	20	1.25	25.00
13216	Lantern	1	19.50	19.50
	TOTAL			401.55

Here are three completed order forms. Use the completed sales record on page 264 to find the unit prices the computer has stored in its memory.

What is the total cost of each order?

1\.

Name Juan Ortiz — Super Value Catalog Store

Catalog Item No.	Description	Quantity
37942	Sleeping bag	1
29005	Air mattress	1
13216	Lantern	2

2\.

Name Bill Higgins — Super Value Catalog Store

Catalog Item No.	Description	Quantity
42318	Alpine tent	1
29005	Air mattress	2
12003	Can of heat	5
13216	Lantern	1

3\.

Name Karen Clark — Super Value Catalog Store

Catalog Item No.	Description	Quantity
29005	Air mattress	3
12003	Can of heat	4
42318	Alpine tent	1

Problem solving

1. Buying Tickets

   Adult: $5.75
   Child: $2.35

   How much for 1 child and 2 adults?

2. Attendance

   Number of seats: 13,502
   Number of empty seats: 4185

   How many were at the basketball game?

3. Seats in Section

   Rows: 38
   Seats in each row: 26

   How many seats in all?

4. Free Throws, First Quarter
   Baskets made: 6
   Shots: 8

   What fraction of the free throws did they make? Give the fraction in lowest terms.

5. Field Goals, Second Quarter

   Baskets made: 12
   Shots: 15

   What fraction of the field goals did they make? Give your answer in lowest terms.

6. Season's Games

   Games played: 18
   Fraction of games won: $\frac{2}{3}$

   How many games did they win?

7. Home Games

   Games played: 18
   Fraction of games played at home: $\frac{5}{9}$

   How many home games did they play?

8. Leading Scorer

   Games played: 4
   Points scored: 92

   What was the average score per game?

9. Scoring by Center

   Average score per game: 19
   Games played: 17

   How many points did he score?

10. Players in Game

    Team members: 16
    Fraction that played: $\frac{7}{8}$

    How many players played?

11. Half Time

    Band members per row: 15
    Rows: 13

    How many people in the band?

12. Buying Snacks

    Hot dog: $.85
    Cold drink: $.40

    What was the total cost of 2 hot dogs and 3 cold drinks?

CHAPTER CHECKUP

Multiply. [pages 244–251]

1. $\begin{array}{r} 217 \\ \times 4 \\ \hline \end{array}$
2. $\begin{array}{r} 271 \\ \times 3 \\ \hline \end{array}$
3. $\begin{array}{r} 158 \\ \times 6 \\ \hline \end{array}$
4. $\begin{array}{r} 829 \\ \times 2 \\ \hline \end{array}$
5. $\begin{array}{r} 465 \\ \times 8 \\ \hline \end{array}$
6. $\begin{array}{r} 158 \\ \times 9 \\ \hline \end{array}$
7. $\begin{array}{r} 603 \\ \times 6 \\ \hline \end{array}$
8. $\begin{array}{r} 2167 \\ \times 7 \\ \hline \end{array}$
9. $\begin{array}{r} 1692 \\ \times 4 \\ \hline \end{array}$
10. $\begin{array}{r} 5938 \\ \times 5 \\ \hline \end{array}$

Multiply. [pages 254–257]

11. $\begin{array}{r} 32 \\ \times 10 \\ \hline \end{array}$
12. $\begin{array}{r} 417 \\ \times 10 \\ \hline \end{array}$
13. $\begin{array}{r} 68 \\ \times 40 \\ \hline \end{array}$
14. $\begin{array}{r} 82 \\ \times 80 \\ \hline \end{array}$
15. $\begin{array}{r} 735 \\ \times 60 \\ \hline \end{array}$

Multiply. [pages 258–263]

16. $\begin{array}{r} 17 \\ \times 54 \\ \hline \end{array}$
17. $\begin{array}{r} 86 \\ \times 27 \\ \hline \end{array}$
18. $\begin{array}{r} 65 \\ \times 95 \\ \hline \end{array}$
19. $\begin{array}{r} 142 \\ \times 18 \\ \hline \end{array}$
20. $\begin{array}{r} 674 \\ \times 39 \\ \hline \end{array}$

Solve. (pages 252–253, 264–267)

21. A small roll of film costs $2.35. A large roll of film costs $3.19. How much do one small roll and one large roll cost?
22. You have $5. You buy a large roll of film for $3.19. How much change should you get back?
23. You can take **36** pictures on a large roll of film. How many pictures can you take on **8** large rolls?
24. You can take **24** pictures on a small roll of film. How many pictures can you take on **20** small rolls?
25. A photo shop charges $.28 to develop and print a picture. How much would they charge to develop and print **36** pictures?
26. You have **96** pictures. You decide to put **6** pictures on each page of an album. How many pages will you need?

CHAPTER PROJECT

Napier's rods (named after John Napier, 1550–1617) can be used to multiply.

1. Get some graph paper and make a set of Napier's rods as shown. Notice that the numbers listed on each rod are the first 9 multiples of the "red" number.

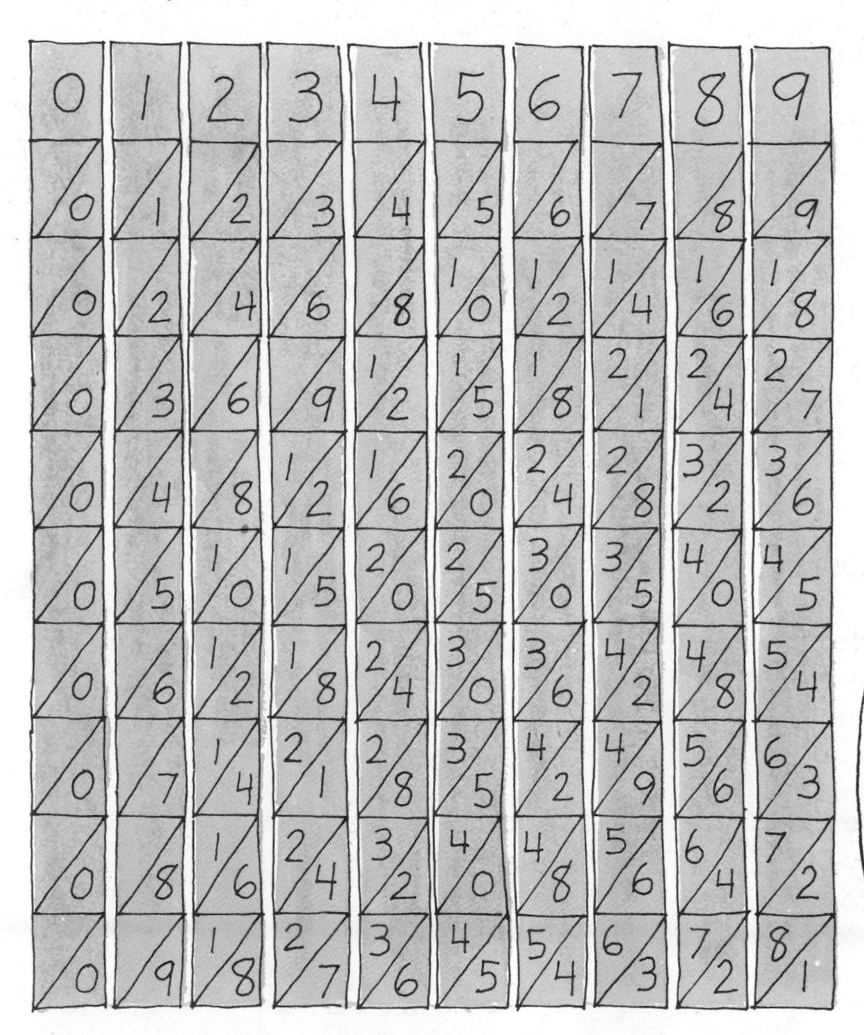

2. The example shows how to use your Napier's rods to find this product:

$$\begin{array}{r} 358 \\ \times 6 \\ \hline \end{array}$$

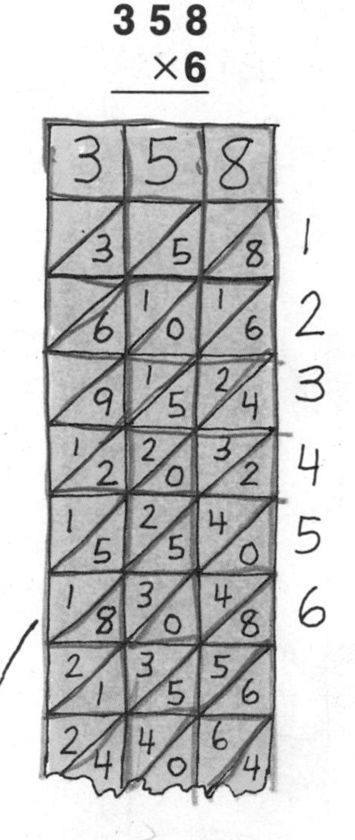

To find the product, add the numbers along the diagonals, as shown.

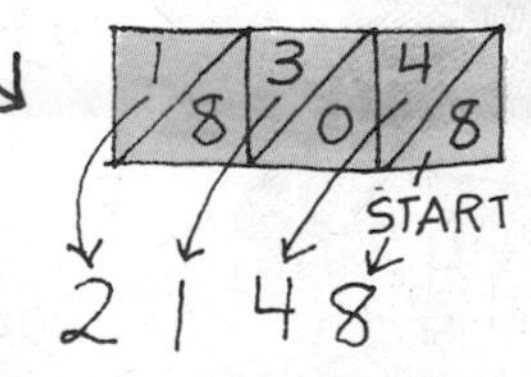

3. Use your set of rods to find these products.

a. $\begin{array}{r} 529 \\ \times 4 \\ \hline \end{array}$ b. $\begin{array}{r} 738 \\ \times 3 \\ \hline \end{array}$ c. $\begin{array}{r} 509 \\ \times 7 \\ \hline \end{array}$ d. $\begin{array}{r} 6831 \\ \times 6 \\ \hline \end{array}$ e. $\begin{array}{r} 5874 \\ \times 5 \\ \hline \end{array}$

CHAPTER REVIEW

2
427
×3
1281

Multiply.

1. 261 ×4	2. 134 ×7	3. 549 ×3	4. 528 ×5

1 2
5427
×3
16,281

Multiply.

5. 1643 ×5	6. 4072 ×8	7. 6183 ×4	8. 3516 ×7

427
×10
4270

Multiply.

9. 819 ×10	10. 73 ×10	11. 406 ×10	12. 520 ×10

427
×60
25,620

Multiply.

13. 561 ×20	14. 813 ×70	15. 96 ×40	16. 264 ×30

27
×63
81
1620
1701

Multiply.

17. 49 ×28	18. 83 ×71	19. 64 ×56	20. 72 ×39

427
×63
1281
25620
26,901

Multiply.

21. 618 ×42	22. 391 ×17	23. 804 ×39	24. 562 ×58

CHAPTER CHALLENGE

A story of long ago tells of a man whose horse needed shoes. He asked a blacksmith how much it would cost. The blacksmith explained that he had two ways of charging—either $1000 for the whole job or 2¢ for the first nail, 4¢ for the second nail, 8¢ for the third nail, and so on for the 16 nails needed to shoe the horse. The man decided to pay the second way.

1. How much did he have to pay for each nail to shoe his horse? To answer the question, copy and complete this table.

Nail	1	2	3	4	5	6	7	8	9	10	11	12	13	14	15	16
Charge	2¢	4¢	8¢	16¢	32¢	?	?	?	?	?	?	?	?	?	?	?

2. What was the total charge?
3. Did he pay more or less than $1000?
4. How much more or less?

MAJOR CHECKUP
STANDARDIZED FORMAT

Choose the correct letter.

1. 78,521 rounded to the nearest thousand is

- **a.** 78,000
- **b.** 78,500
- **c.** 79,000
- **d.** 80,000

2. Add.

$$\begin{array}{r} 2653 \\ +8797 \\ \hline \end{array}$$

- **a.** 11,340
- **b.** 11,450
- **c.** 10,450
- **d.** none of these

3. Subtract.

$$\begin{array}{r} 3205 \\ -1678 \\ \hline \end{array}$$

- **a.** 2473
- **b.** 1537
- **c.** 1627
- **d.** 1527

4. How many minutes from 9:45 A.M. to 10:25 A.M.?

- **a.** 50
- **b.** 30
- **c.** 70
- **d.** none of these

5. What is the value of 1 half-dollar, 1 quarter, 3 dimes, and 2 nickels?

- **a.** \$1.10
- **b.** \$1.15
- **c.** \$1.05
- **d.** none of these

6. Add.

$\frac{2}{9} + \frac{1}{3}$

- **a.** $\frac{5}{9}$
- **b.** $\frac{3}{12}$
- **c.** $\frac{1}{2}$
- **d.** none of these

7. Change to a fraction.

$2\frac{3}{4}$

- **a.** $\frac{5}{4}$
- **b.** $\frac{11}{4}$
- **c.** $\frac{10}{4}$
- **d.** none of these

8. What is the perimeter?

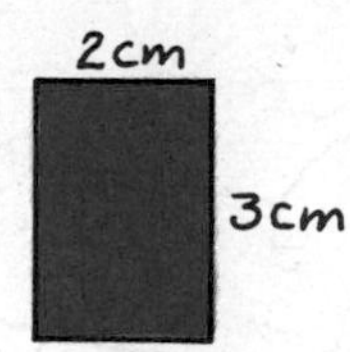

- **a.** 5 cm
- **b.** 6 cm
- **c.** 10 cm
- **d.** none of these

9. What is the area?

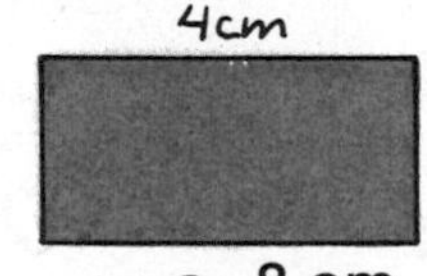

- **a.** 8 cm
- **b.** 6 square cm
- **c.** 8 square cm
- **d.** none of these

10. What is the volume?

3cm 4cm 3cm

- **a.** 10 cubic cm
- **b.** 36 cubic cm
- **c.** 36 square cm
- **d.** none of these

11. Each of 6 teams had 18 girls. Each of 8 teams had 24 boys. How many girls were there?

- **a.** 128
- **b.** 192
- **c.** 320
- **d.** none of these

12. Hot dog: \$.85
Milk: \$.45
What is the total cost of 2 hot dogs and 1 milk?

- **a.** \$1.30
- **b.** \$2.15
- **c.** \$2.60
- **d.** none of these

Geometry

EXERCISES

Name the shape of each numbered item.

A cube has 8 vertices (the plural of **vertex**), 12 edges, and 6 flat surfaces.

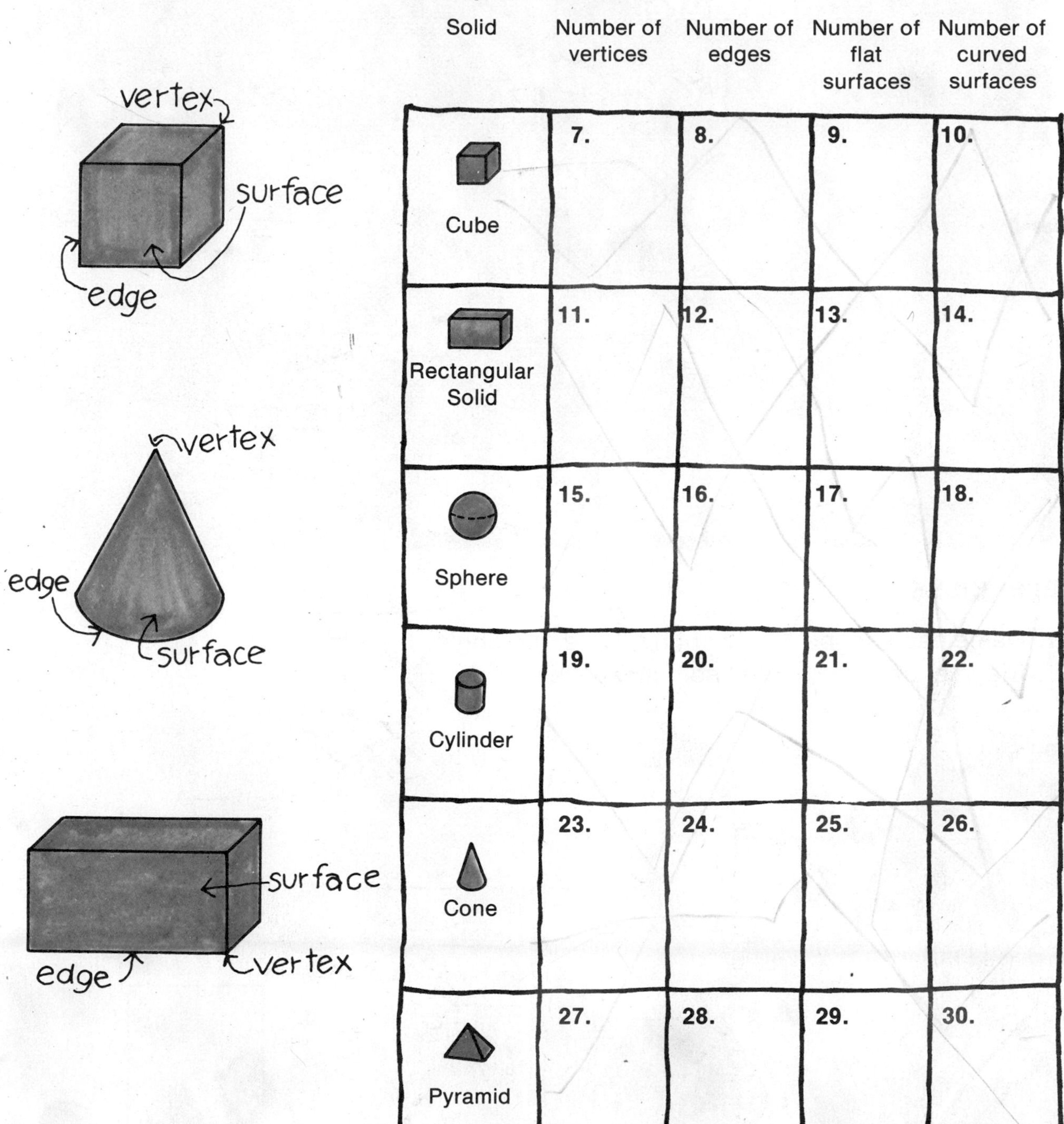

Complete this table.

Solid	Number of vertices	Number of edges	Number of flat surfaces	Number of curved surfaces
Cube	7.	8.	9.	10.
Rectangular Solid	11.	12.	13.	14.
Sphere	15.	16.	17.	18.
Cylinder	19.	20.	21.	22.
Cone	23.	24.	25.	26.
Pyramid	27.	28.	29.	30.

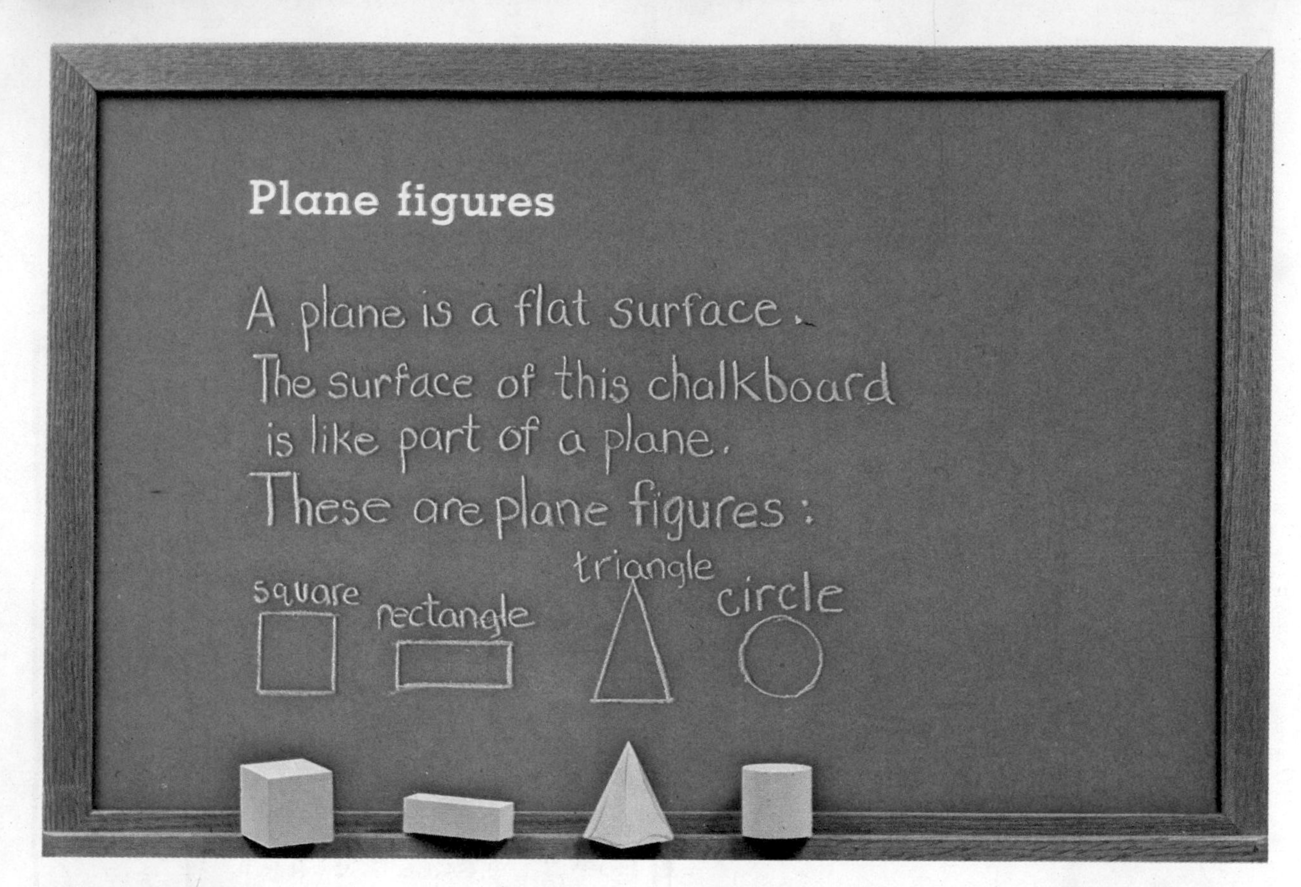

EXERCISES

1. Name each shape. If it is not a square, rectangle, triangle, or circle, write "something else."

a.

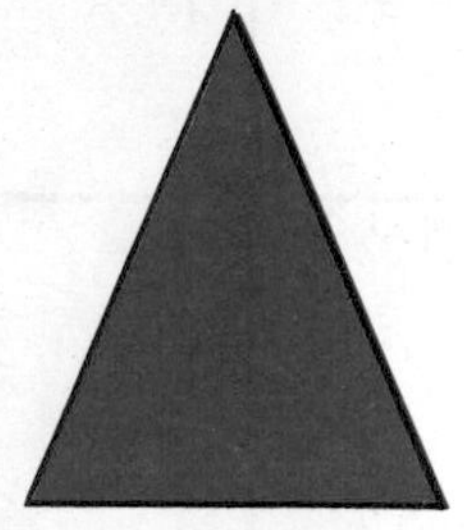

b.

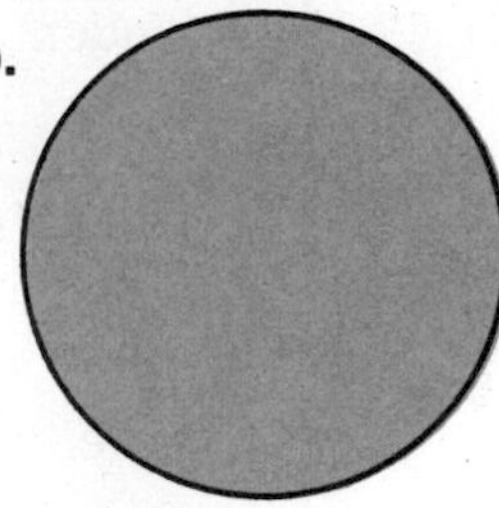

c.

d.

e.

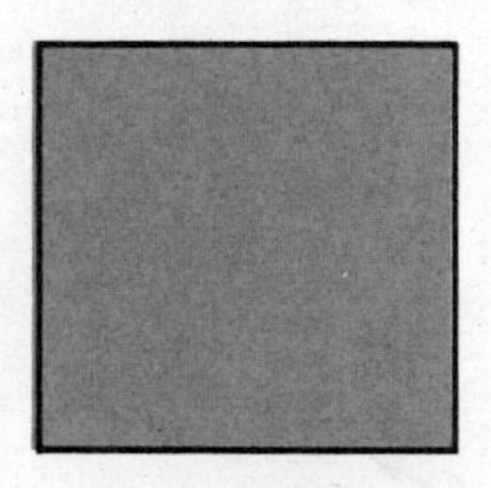

f.

g.

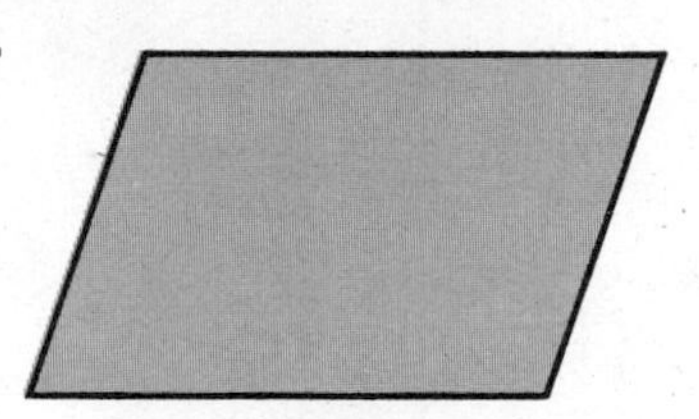

h.

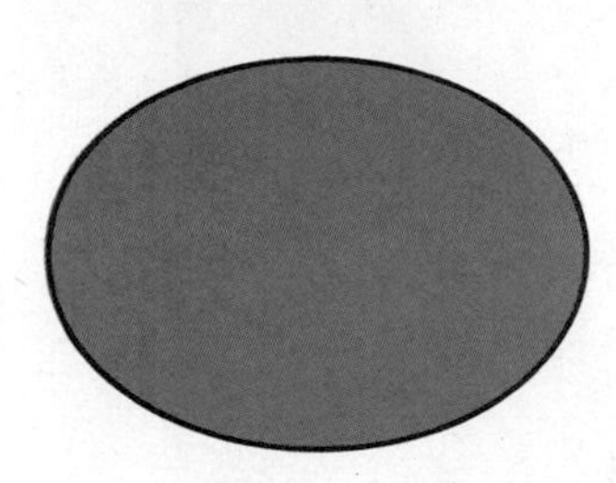

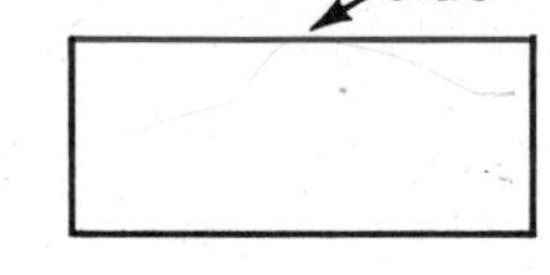

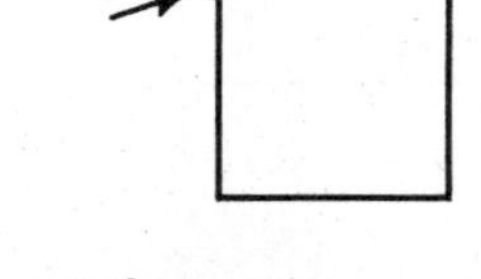

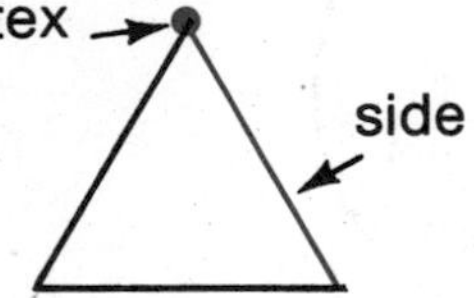

2. How many sides does a square have?

3. How many vertices (plural of *vertex*) does a rectangle have?

4. How many vertices does a triangle have?

5. Does a circle have a vertex?

Challenge!

6. How many triangles are shown here?

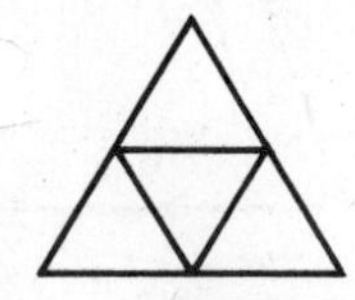

7. How many squares?

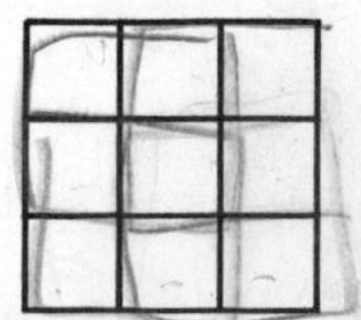

Points and segments

The sides of these plane figures are segments.

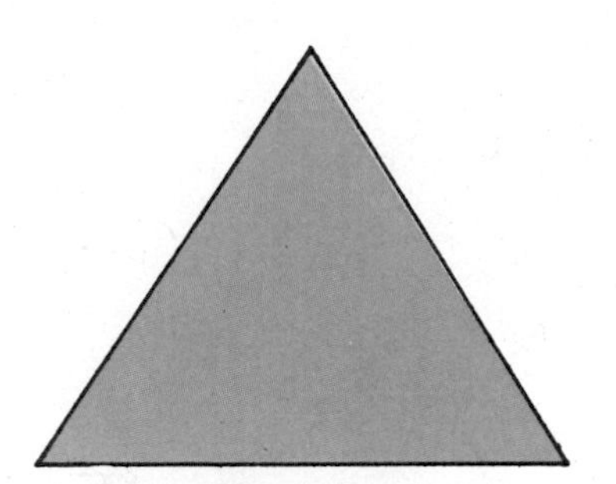

You can think of a segment as a straight path between two points.

Here is another segment.

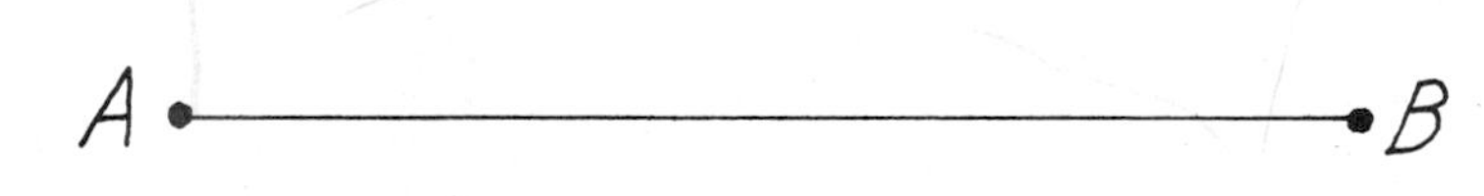

The **endpoints** are *A* and *B*.

You can call it either segment *AB* or segment *BA*. For short we write $\overline{AB}$ or $\overline{BA}$. $\overline{AB}$ is read as "segment *AB*."

EXERCISES

Which are segments?

1.

2.

3.

4. What are the endpoints of this segment?

5. What are two ways to name this segment?

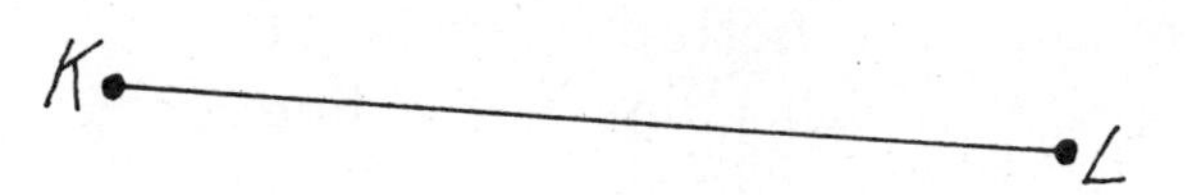

6. $\overline{AB}$ crosses $\overline{RS}$ at what point?

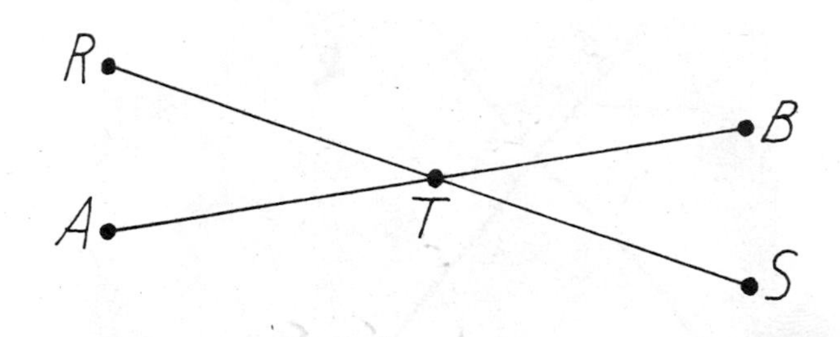

7. Give four segments that have point V as an endpoint.

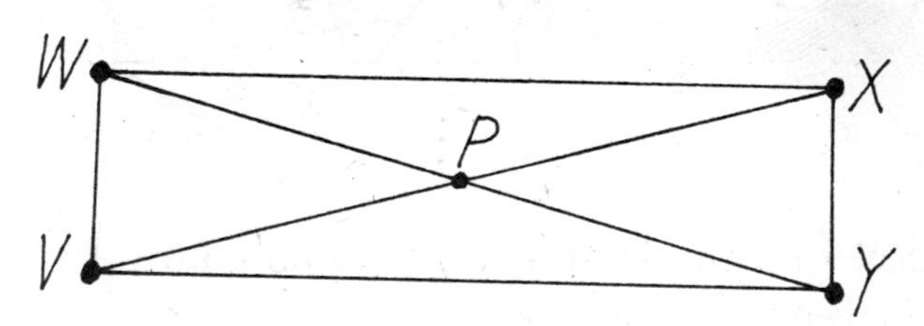

8. How many segments are shown?

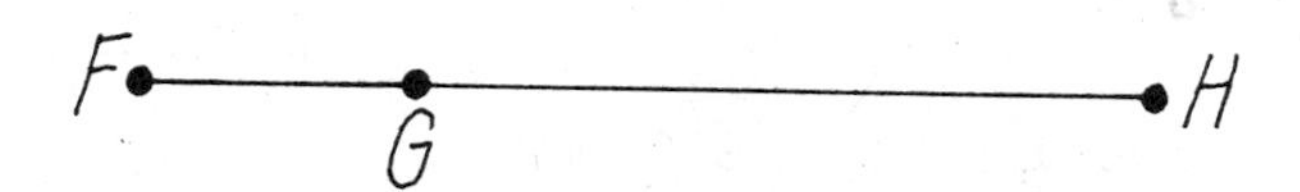

Draw a figure to go with each sentence. Be sure to write the letters for the endpoints and crossing points.

9. $\overline{RS}$ crosses $\overline{CD}$ at point P.

10. $\overline{AB}$ crosses $\overline{XY}$ at point X.

Challenge!

You can draw 6 segments using 4 points. Draw segments to complete the table.

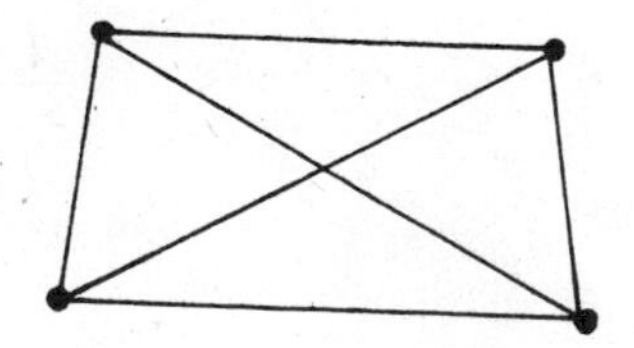

	11.	12.	13.	14.	15.
Number of points	2	3	4	5	6
Number of segments	?	?	6	?	?

Lines

A segment is part of a **line.** The arrows remind you that a line goes on and on in both directions. You can call it either line *RS* or line *SR*.

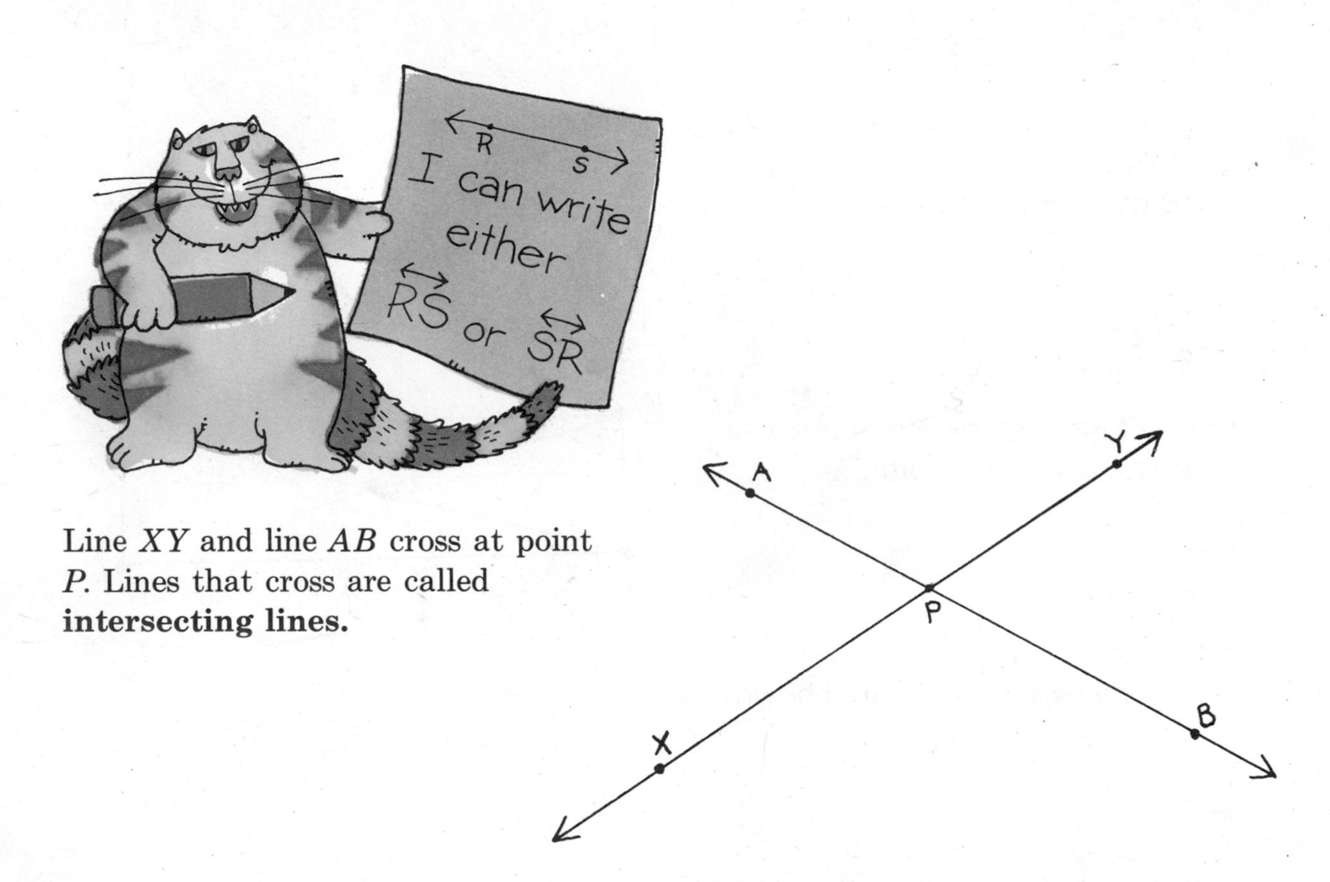

Line *XY* and line *AB* cross at point *P*. Lines that cross are called **intersecting lines.**

Line *CD* and line *EF* do not cross. Lines in a plane that do not cross are called **parallel lines.**

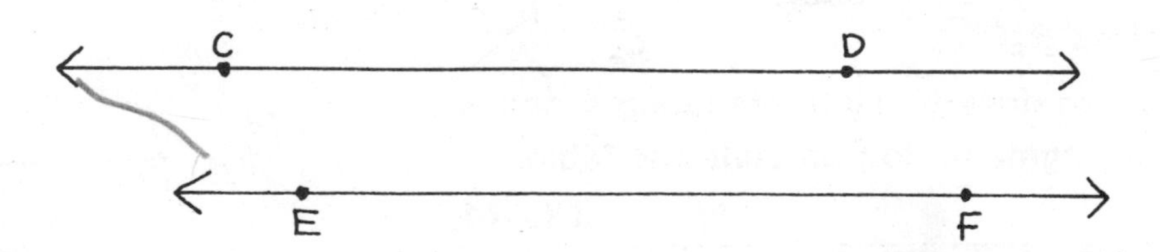

EXERCISES

The lines shown in each exercise were torn from a piece of paper. Do you think that the lines are intersecting or parallel?

1.

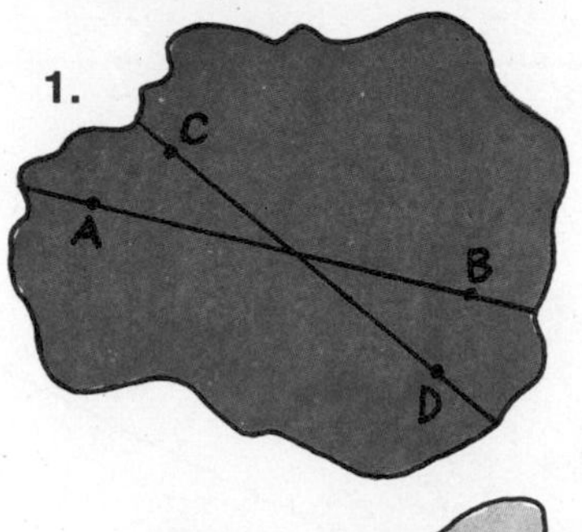

2.

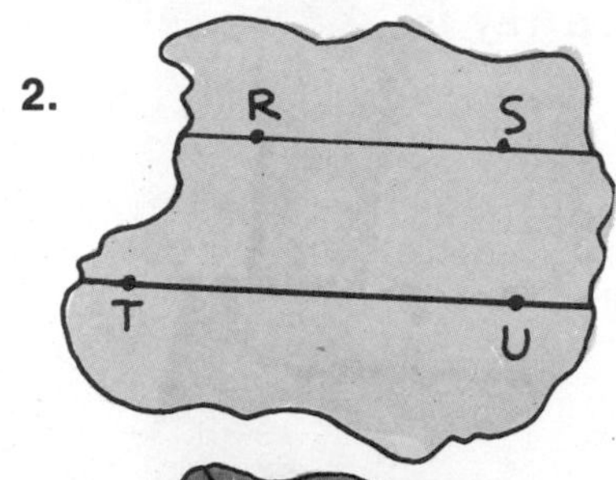

3.

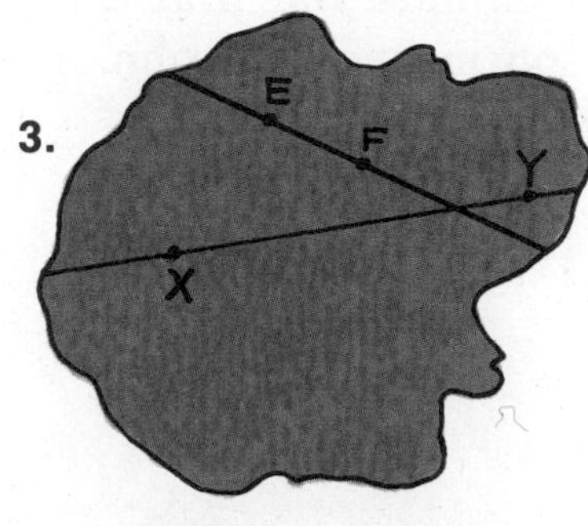

4.

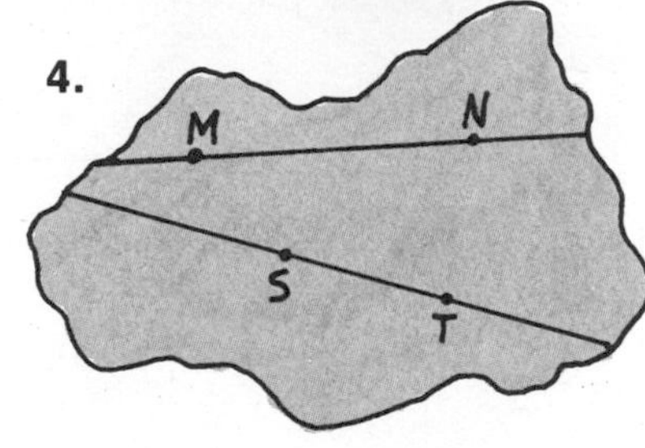

5.

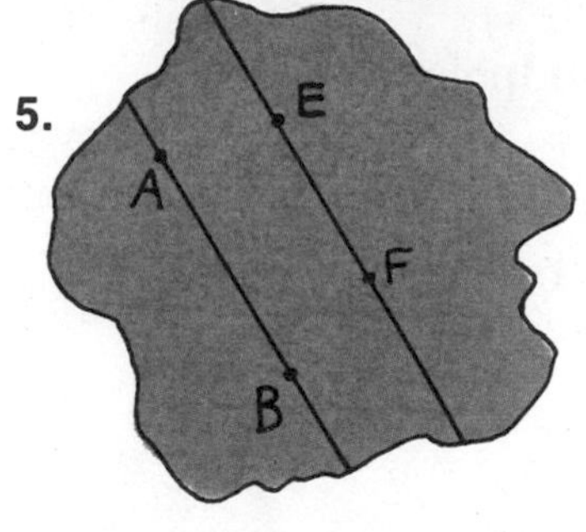

6.

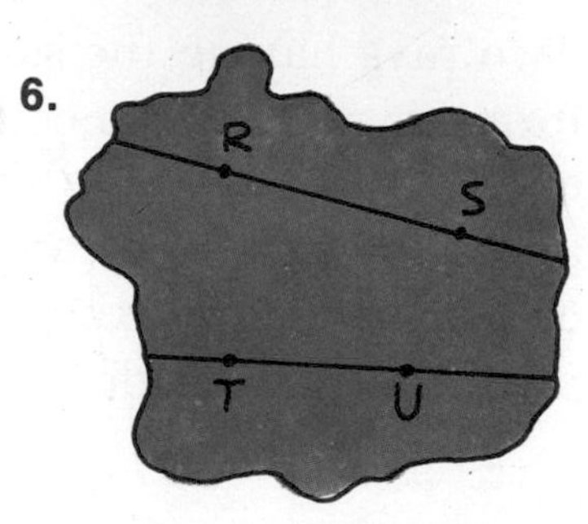

7. Trace these four points.
8. Draw $\overleftrightarrow{AB}$.
9. Draw $\overleftrightarrow{DC}$.
10. Do you think that line *AB* is parallel to line *DC*?

A B

D C

Draw a figure to go with each sentence.

11. $\overleftrightarrow{CD}$ intersects $\overleftrightarrow{RS}$ at point *T*.
12. $\overleftrightarrow{AB}$ is parallel to $\overleftrightarrow{UV}$, $\overleftrightarrow{RS}$ intersects $\overleftrightarrow{AB}$ at point *Y*, and $\overleftrightarrow{RS}$ intersects $\overleftrightarrow{UV}$ at point *Z*.

KEEPING SKILLS SHARP

1. 32×3
2. 46×2
3. 78×5
4. 153×4
5. 378×6
6. 528×9
7. 846×6
8. 935×8
9. 706×5
10. 931×9
11. 812×7
12. 777×9

Rays and angles

A **ray** is part of a line. The arrow reminds you that a ray goes on and on in one direction. The red ray is called ray RS.

Two rays having the same endpoint form an **angle.** The common endpoint is called the vertex of the angle. Ray AC and ray AB are called the sides of the angle. This angle is called angle A.

An angle that forms a square corner is called a **right angle.**

The part shown in blue is called angle A of the square. Which part is angle D?

EXERCISES

How many endpoints does each figure have?

1. segment

2. line

3. ray

4. Name the angle shown.

5. Give the vertex of the angle.

6. Give the sides.

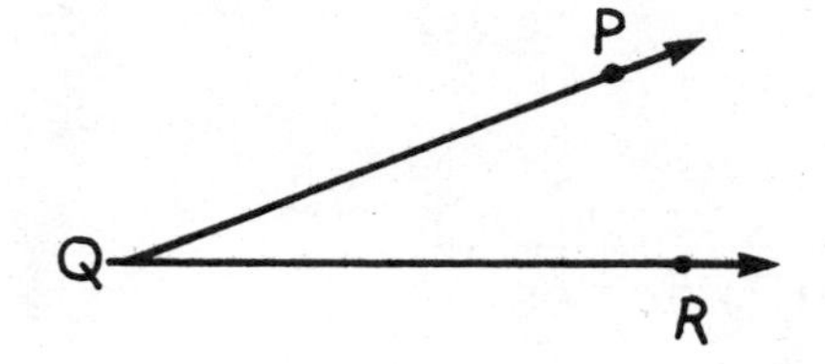

Is the angle a right angle?

7.

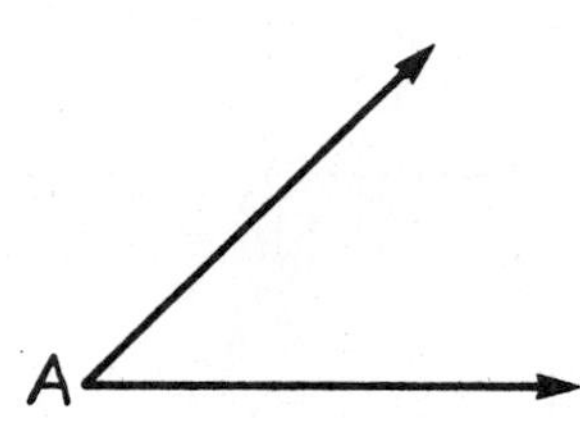

8.

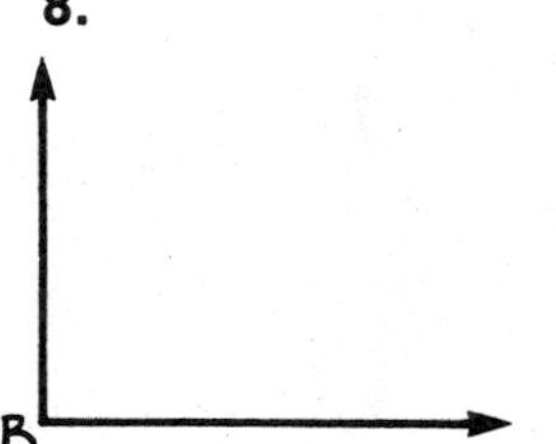

9.

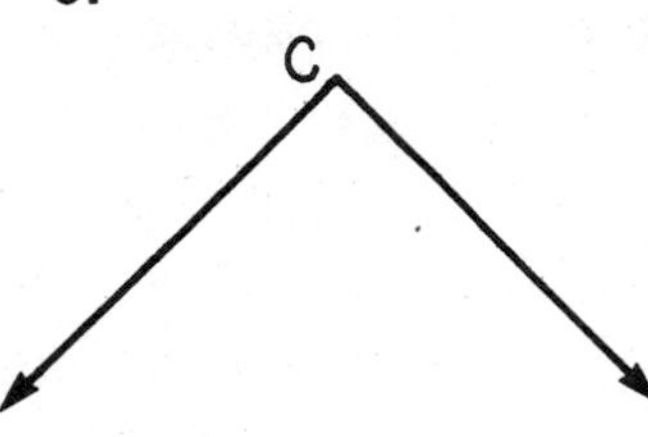

How many right angles does each of these figures have?

10.

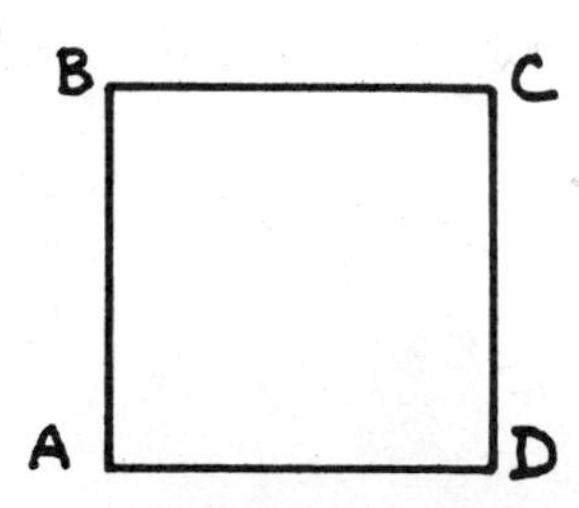

11.

12.

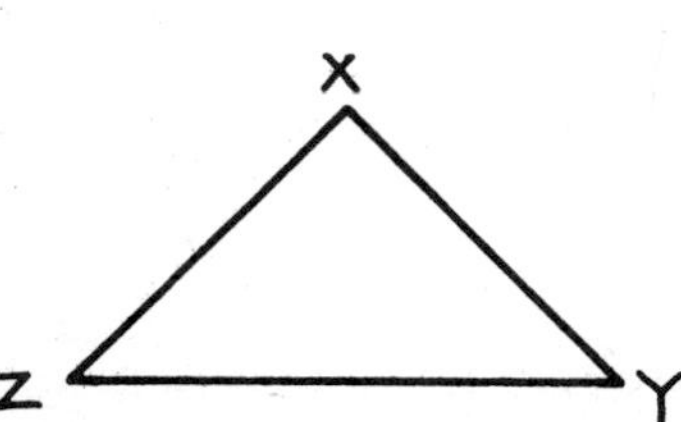

13. List some things in your classroom that form right angles.

14. Name two figures that have four right angles.

How many angles can you find in each figure?

15.

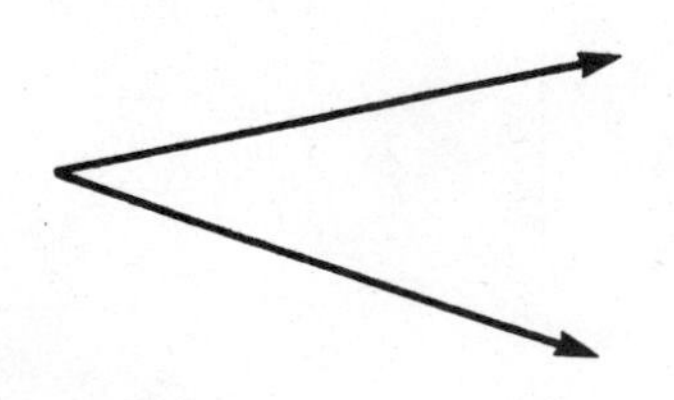

16.

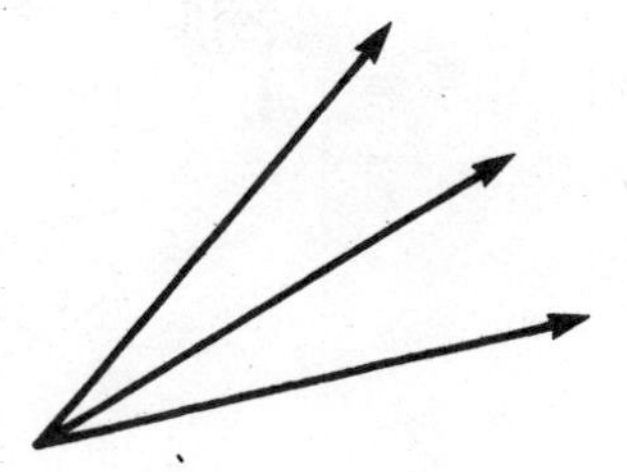

★ **17.**

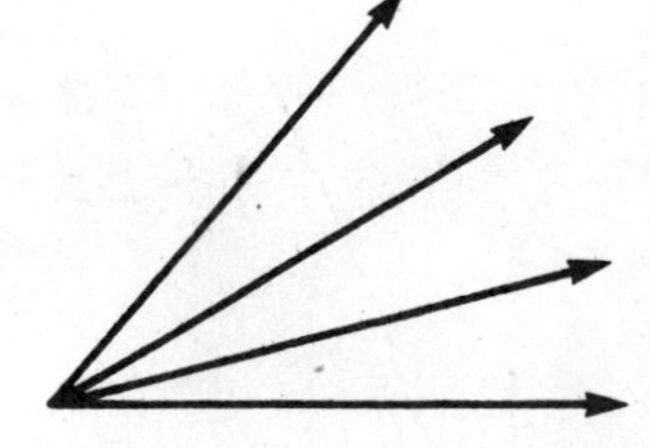

Similar figures and congruent figures

Figures that are the same shape are called **similar figures.**

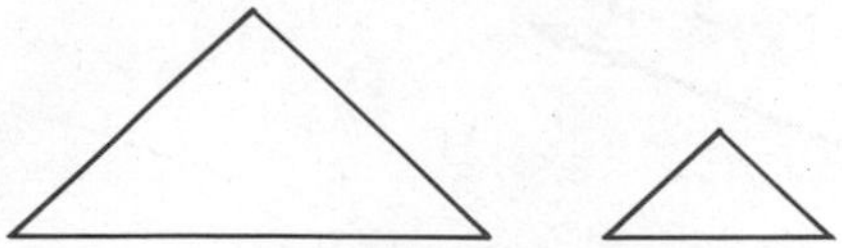

Similar Triangles

Figures that are the same size and shape are called **congruent figures.**

Congruent Rectangles

EXERCISES

Match similar figures.

1.

2.

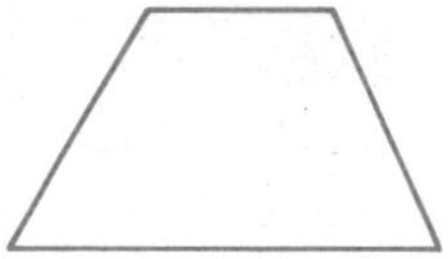

3.

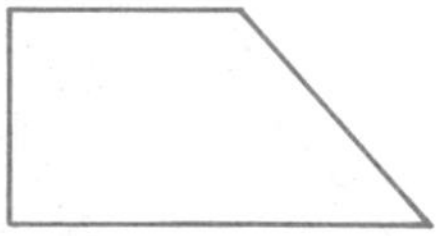

4.

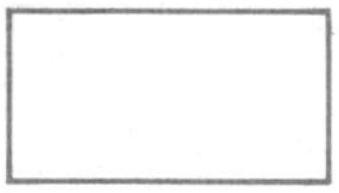

5.

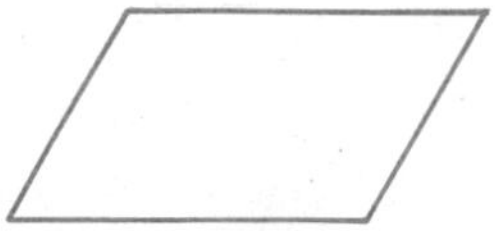

6.

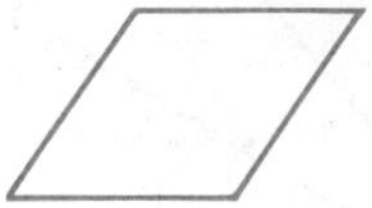

a.

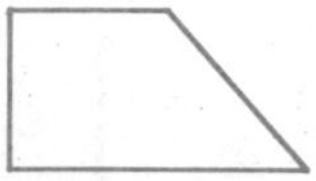

b.

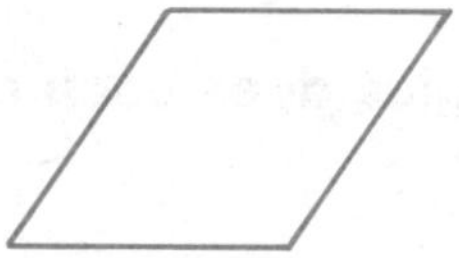

c.

d.

e.

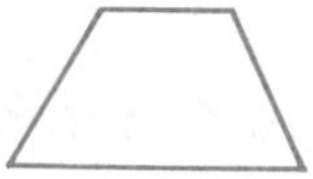

f.

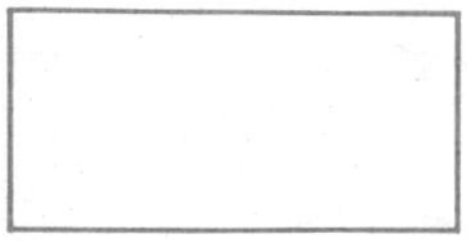

Which figure is congruent to the blue figure?

Step 1. **Trace the blue figure.**

Step 2. **See which red figure your tracing fits.**

7. **a.** **b.** **c.**

8.

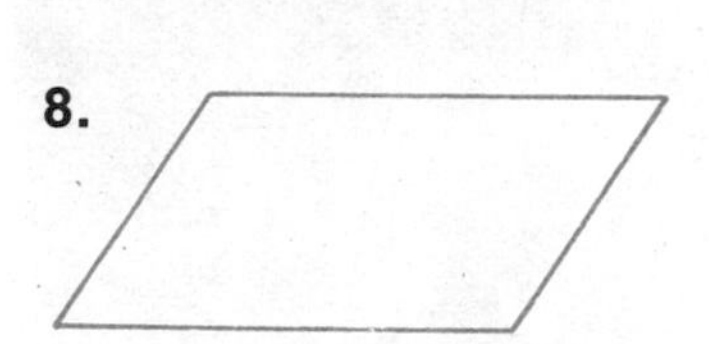

a.

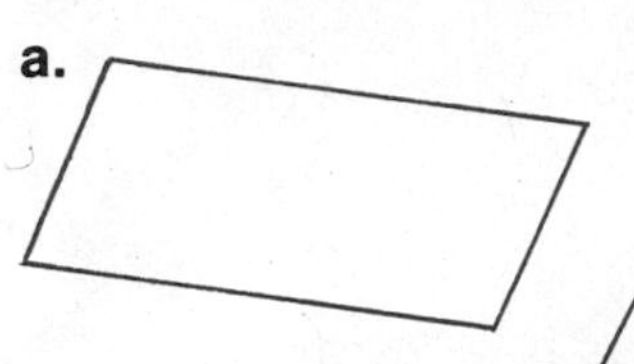

b. **c.**

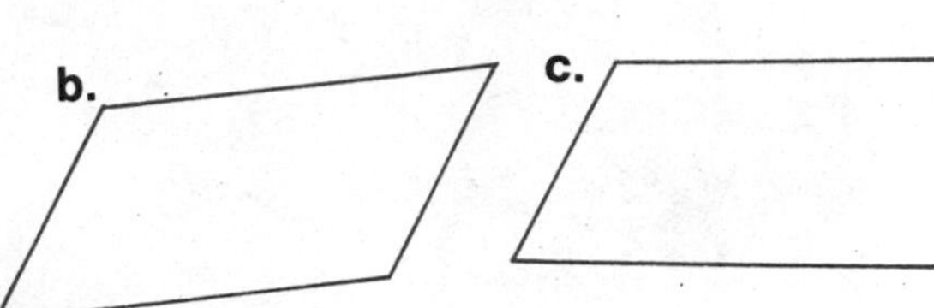

9.

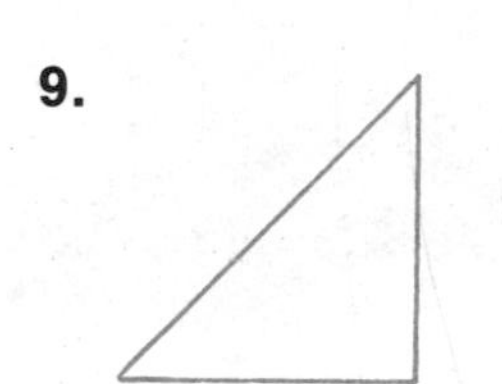

a.

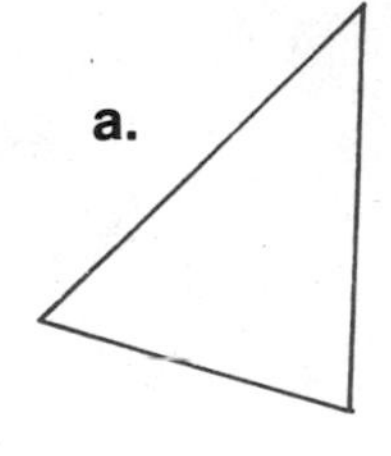

b. **c.**

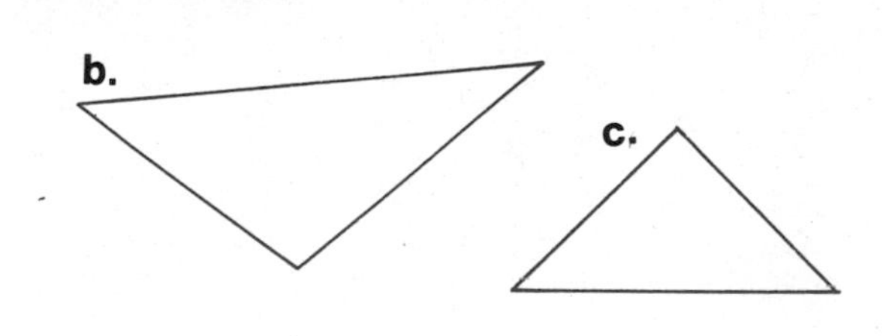

10.

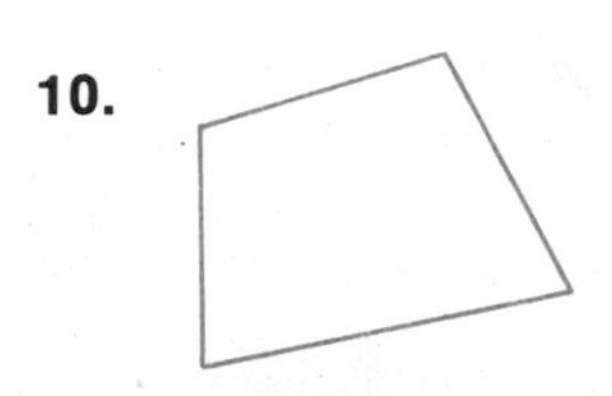

a.

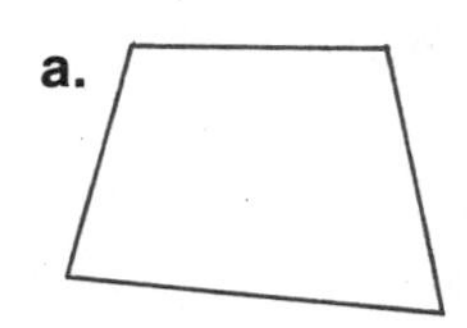

b.

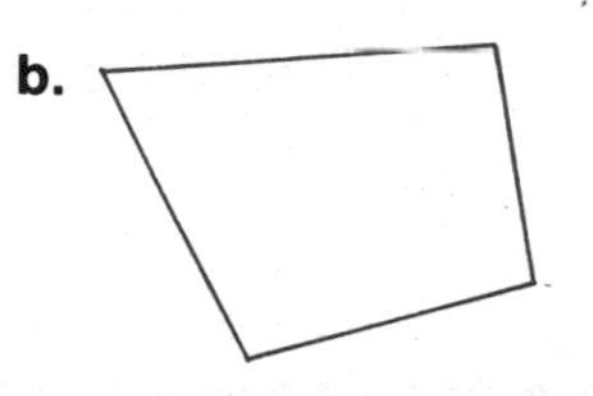

c.

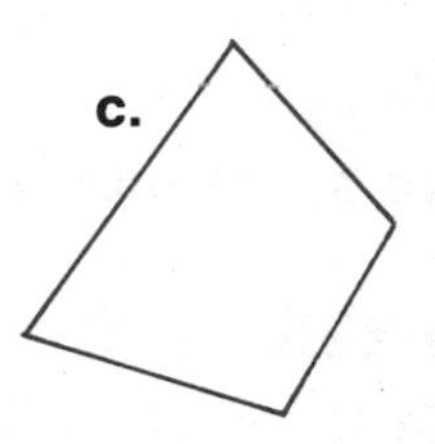

11. **a.** Is figure A the same shape as figure B?
b. Is figure A the same size as figure B?
c. Are they congruent?

12. **a.** Is figure A the same shape as figure C?
b. Is figure A the same size as figure C?
c. Are they congruent?

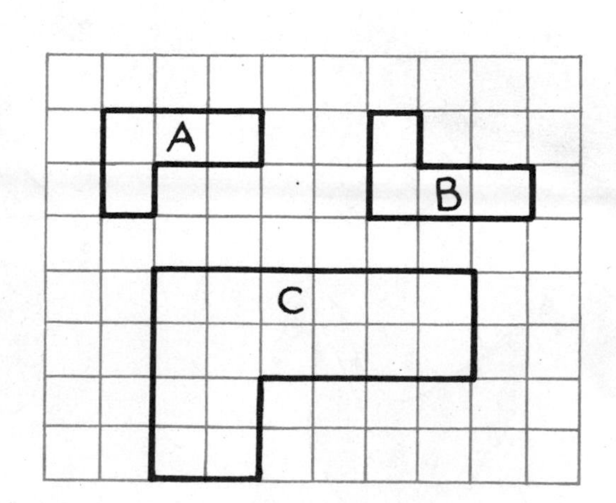

Line of symmetry

Sandy made this cutout from a folded piece of paper.

When she folded her cutout along the dashed line, the two halves fit. The dashed line is called a **line of symmetry.**

EXERCISES

Is the dashed line a line of symmetry?

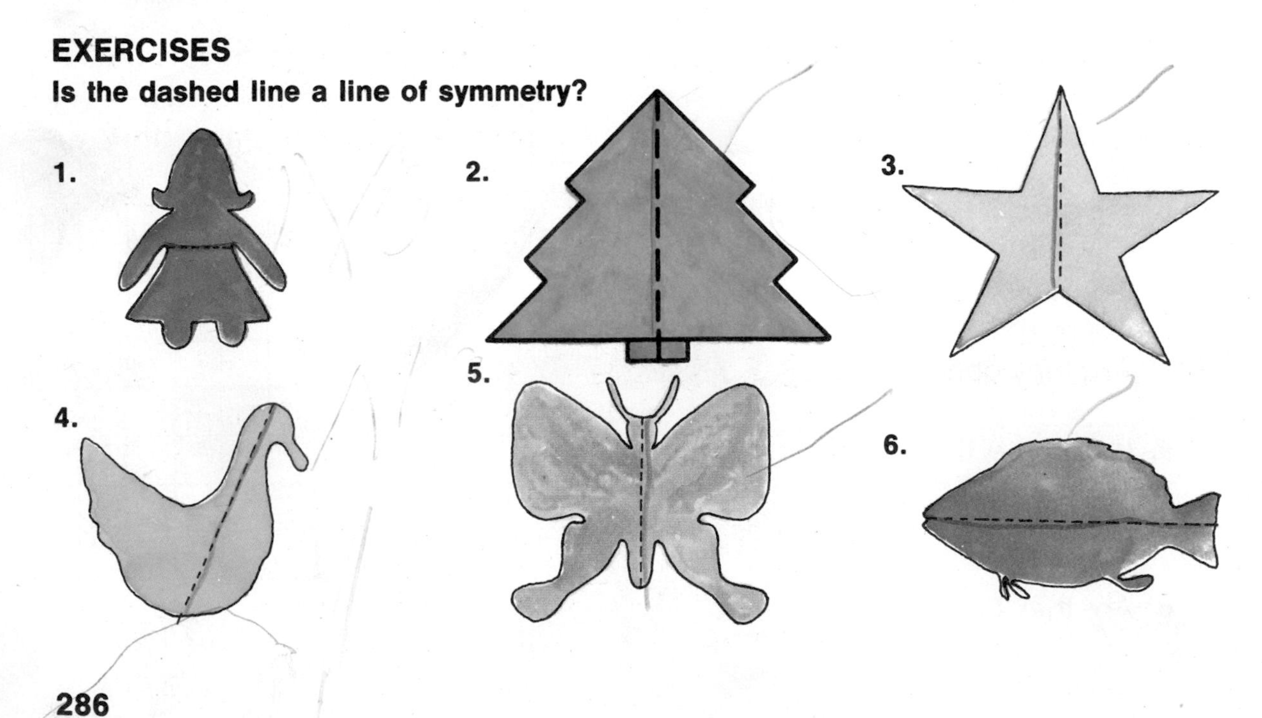

How many lines of symmetry?

7.

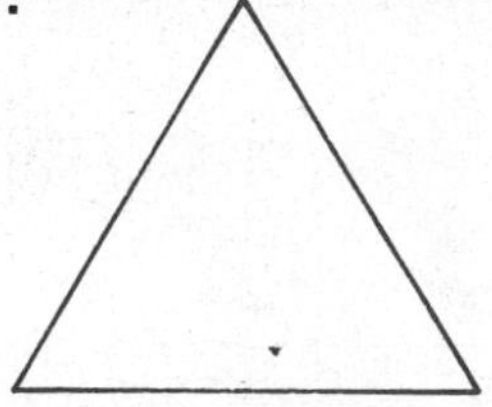

8.

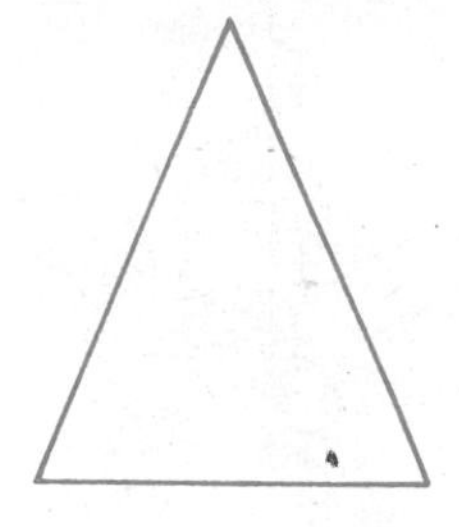

9.

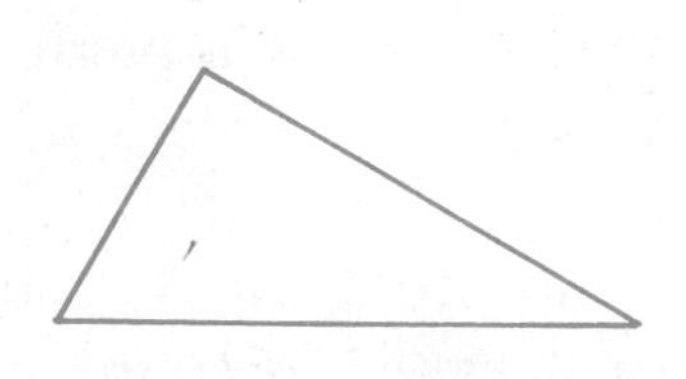

10.

11.

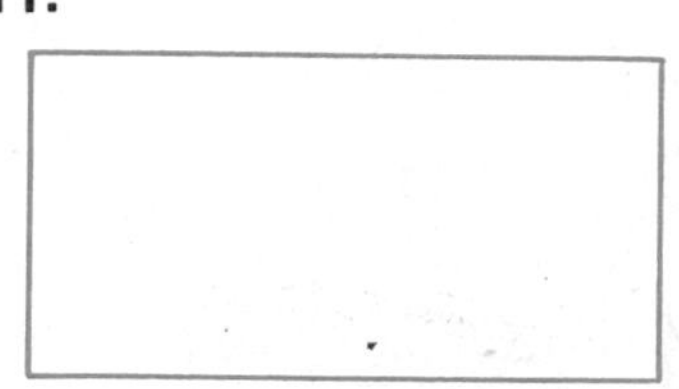

12.

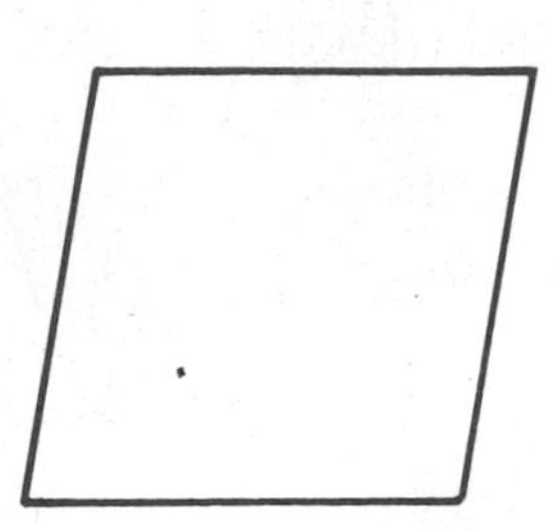

13.

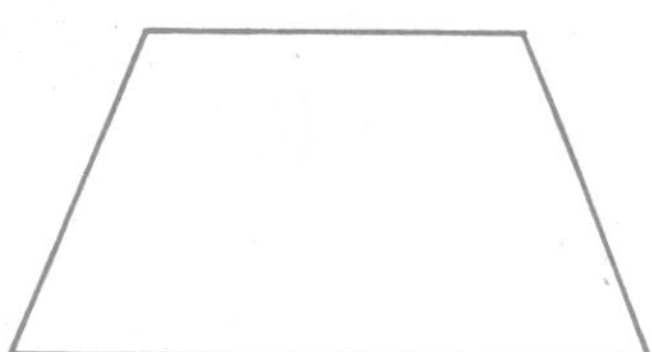

14.

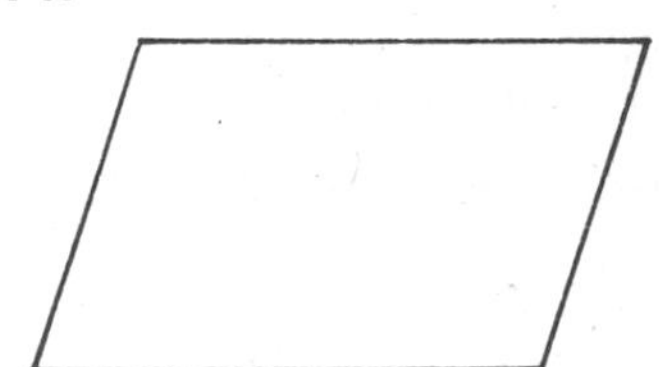

15.

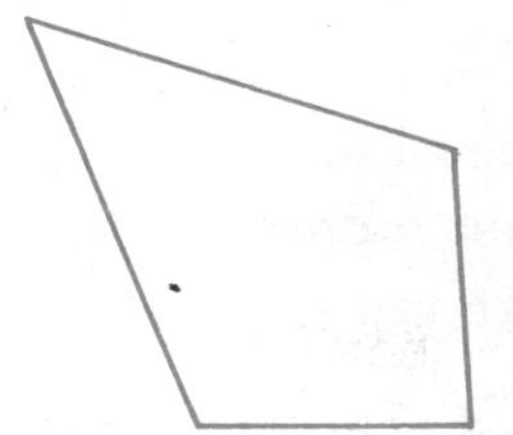

Challenge!

16. Fold a piece of paper. Cut out a figure from the folded edge. Draw the line of symmetry.

17. Can you find a way to fold the paper so you can cut out a figure that has **2** lines of symmetry?

18. Can you draw a triangle that has only **2** lines of symmetry?

19. Can you draw a **6**-sided figure that has only **1** line of symmetry?

KEEPING SKILLS SHARP

1. $2\overline{)68}$ **2.** $5\overline{)55}$ **3.** $3\overline{)72}$ **4.** $5\overline{)85}$ **5.** $4\overline{)92}$

6. $3\overline{)56}$ **7.** $5\overline{)90}$ **8.** $2\overline{)85}$ **9.** $4\overline{)74}$ **10.** $3\overline{)83}$

11. $4\overline{)90}$ **12.** $2\overline{)87}$ **13.** $5\overline{)83}$ **14.** $3\overline{)78}$ **15.** $6\overline{)98}$

Points on a grid

You can use a **number pair** to locate a point on a grid.

To locate point *A*, you start at 0 and count 7 units to the right and then count 5 units up.

The number pair (7, 5) locates point *A*.

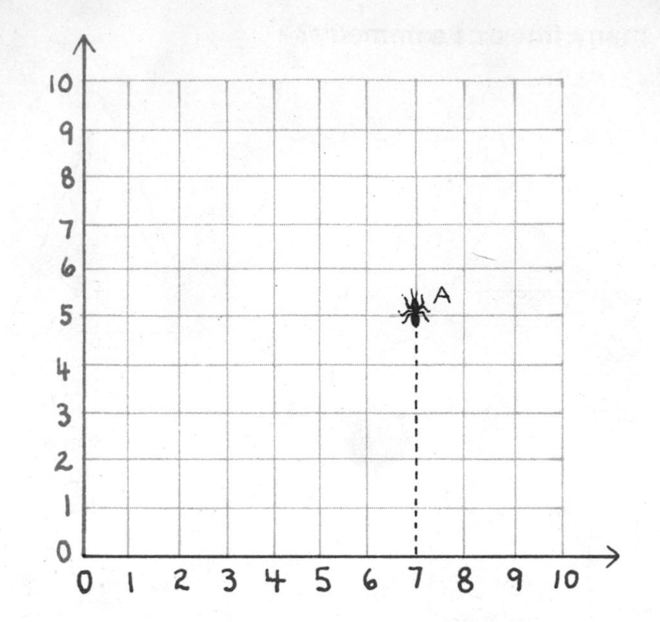

EXERCISES

Give the number pair for each point. *Remember:* The first number is the number of units to the right. The second number is the number of units up.

1. *D* (8,5)

2. *H* **3.** *A* **4.** *G*

5. *C* **6.** *I* **7.** *F*

8. *E* **9.** *J* **10.** *B*

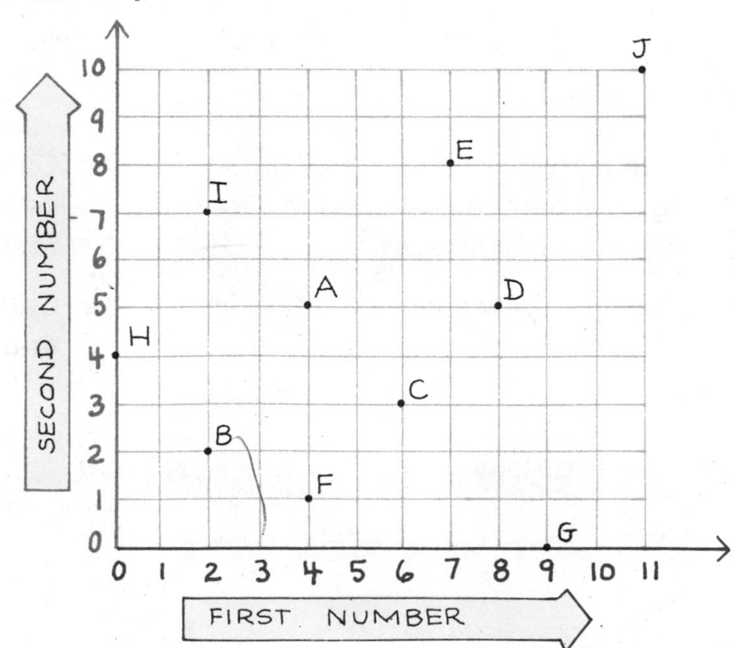

Give the point for each of these number pairs.

11. (2, 3)	**12.** (3, 2)
13. (5, 1)	**14.** (1, 7)
15. (8, 0)	**16.** (0, 8)
17. (6, 4)	**18.** (4, 6)
19. (4, 5)	**20.** (7, 3)

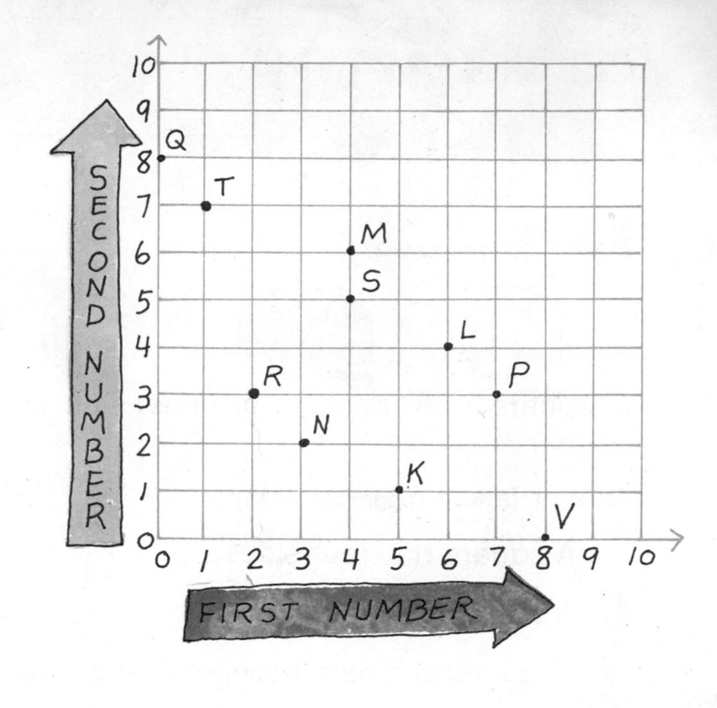

Challenge!

21. Kevin belonged to a secret club. Each member had a copy of this grid with all the letters of the alphabet. They used it to send secret messages. Here is a message that the president sent to each member. Use the grid to decode the message.

(8, 9) (6,2) (6,2) (8,2) (1,6) (8,2)
(8,2) (5,4) (6,2) (2,2) (6,8) (1,9) (3,8)
(5,4) (9,3) (1,9) (5,1) (6,2) (1,6) (8,2)
(2,5) (9,3) (9,3) (2,5)
(5,1) (1,6) (8,2) (1,9) (1,3) (3,1) (1,6) (10,5)

22. Use the grid to write a message of your own.

CHAPTER CHECKUP

Match. [pages 274–275]

1. 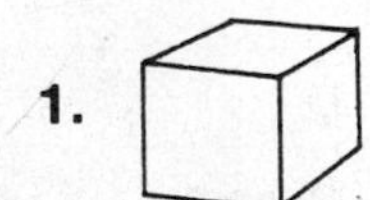 **2.** **3.** 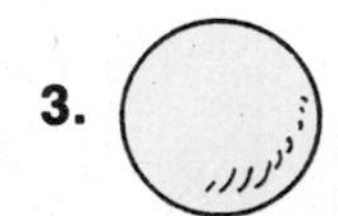 **4.**  **5.**  **6.** 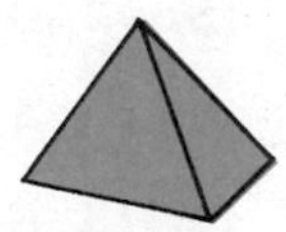

a. sphere **b.** cone **c.** cube **d.** cylinder **e.** pyramid **f.** rectangular solid

True or false? [pages 276–283]

7. A square has four sides.
8. A triangle has four vertices.
9. The shortest path between two points is a segment.
10. A line has two endpoints.
11. Parallel lines can intersect.
12. Any two rays form an angle.
13. A right angle forms a square corner.
14. A rectangle has four right angles.

15. How many lines of symmetry? [pages 284–287]

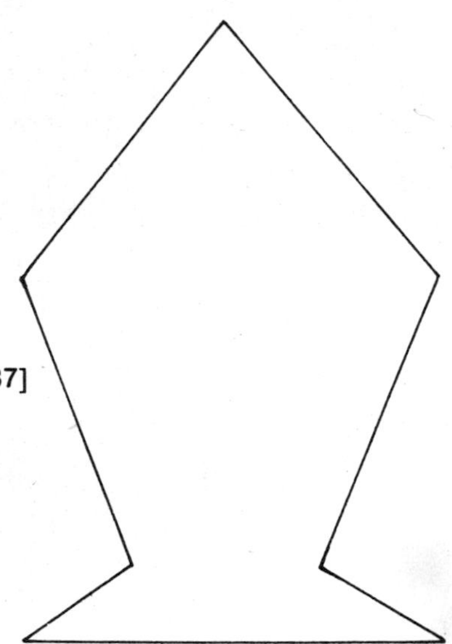

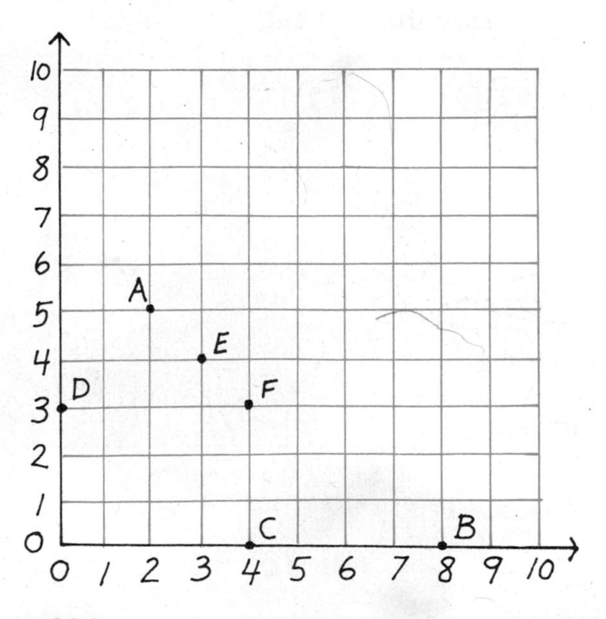

[pages 288–289]

16. Give the number pair for point *A*.
17. Give the number pair for point *B*.
18. What point is at (4, 3)?

CHAPTER PROJECT

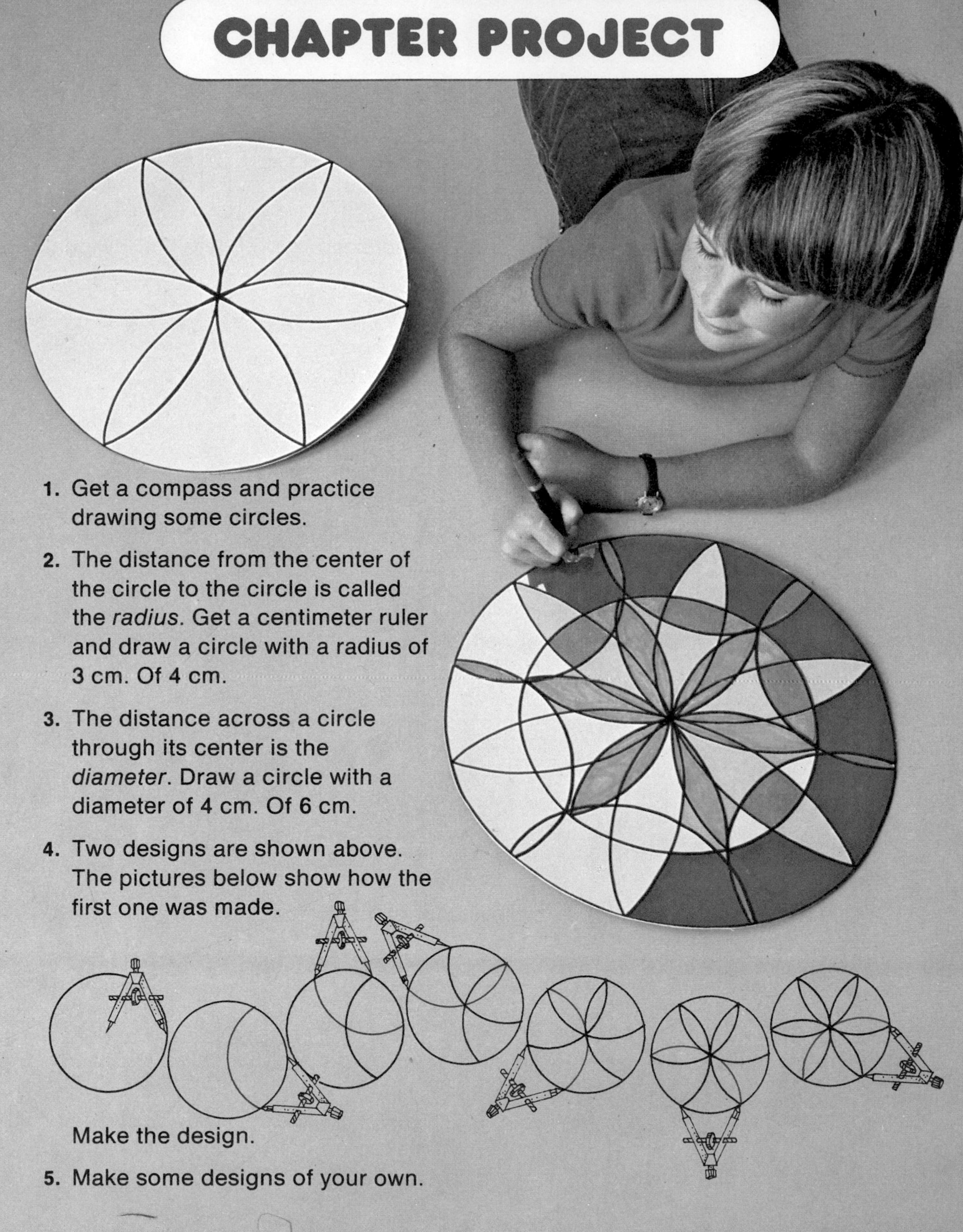

1. Get a compass and practice drawing some circles.
2. The distance from the center of the circle to the circle is called the *radius*. Get a centimeter ruler and draw a circle with a radius of 3 cm. Of 4 cm.
3. The distance across a circle through its center is the *diameter*. Draw a circle with a diameter of 4 cm. Of 6 cm.
4. Two designs are shown above. The pictures below show how the first one was made.

   Make the design.
5. Make some designs of your own.

CHAPTER REVIEW

Match.

1. 
2.
3. 
4. 
5. 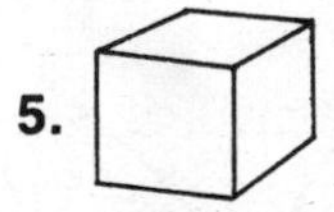
6. 

a. pyramid **b.** cube **c.** cone **d.** rectangular solid **e.** sphere **f.** cylinder

Match.

7. 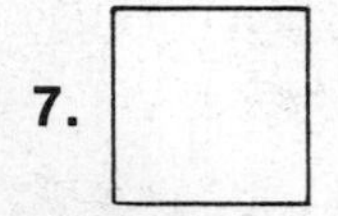
8.
9.
10. 

g. rectangle **h.** triangle **i.** square **j.** circle

True or false?

11. Point *A* and point *B* are called endpoints of segment *AB*.

12. Line *RS* and line *UV* intersect at point *P*.

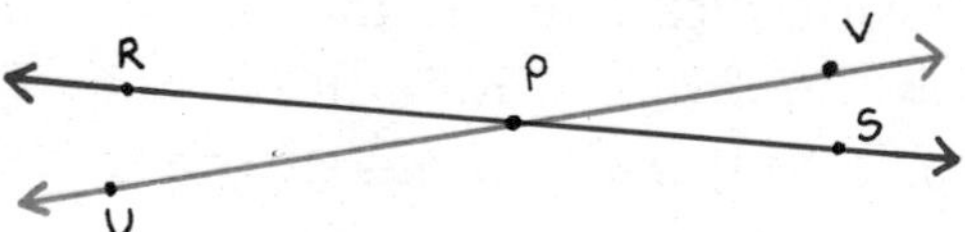

13. Point *D* is an endpoint of ray *CD*.

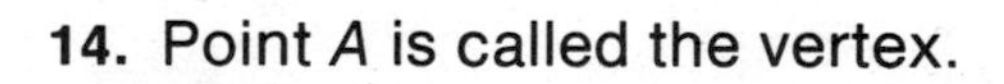

14. Point *A* is called the vertex.

15. Angle *A* is a right angle.

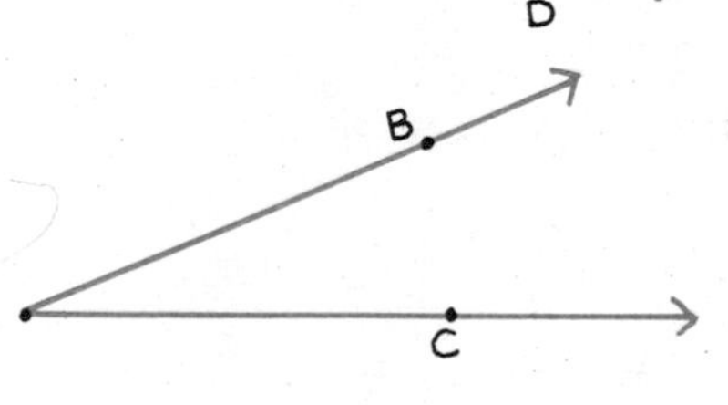

Complete.

16. Use a tracing to tell which triangle is congruent to the red triangle.

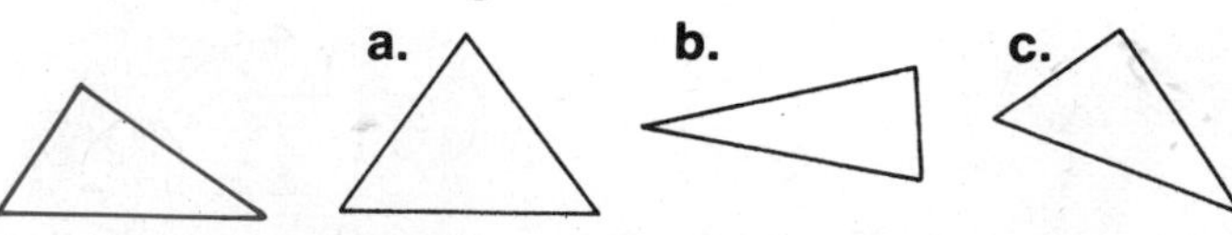

17. Is the dashed line a line of symmetry?

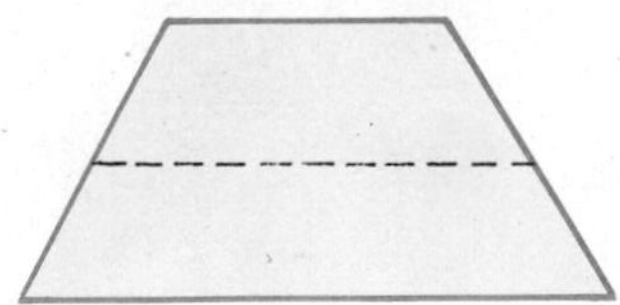

18. Give the number pair for point *A*.

CHAPTER CHALLENGE

Carla uses this computer. She gives it a rule and some starting numbers. The machine computes number pairs for her and prints them out. When she gave the computer the rule *Add 2* and the numbers 0 through 4, the printout looked like this:

Add 2

0	2
1	3
2	4
3	5
4	6

0+2=2

1+2=3

Here is how Carla graphed the number pairs:

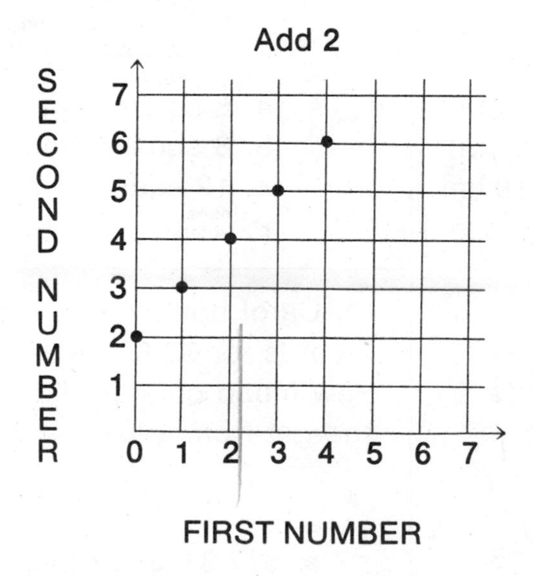

Complete the printouts. Then graph the number pairs.

1. Add 1

0	1
1	2
2	3
3	?
4	?

2. Subtract 2

2	0
3	1
4	?
5	?
6	?

3. Multiply by 2

0	0
1	2
2	?
3	?
4	?

MAJOR CHECKUP
STANDARDIZED FORMAT

Choose the correct letter.

1. Which is the smallest number?

a. 86,521
b. 769,431
c. 86,397
d. 86,400

2. 65,381 rounded to the nearest hundred is

a. 65,000
b. 65,400
c. 65,300
d. none of these

3. Subtract.

$$\begin{array}{r} 8204 \\ -3569 \\ \hline \end{array}$$

a. 3365
b. 4745
c. 4635
d. none of these

4. What time is shown?

a. quarter past one
b. quarter to one
c. quarter to two
d. quarter past two

5. Complete.

$\frac{2}{3}$ of 18 = ?

a. 27
b. 12
c. 15
d. none of these

6. Add.

$\frac{1}{2} + \frac{1}{6}$

a. $\frac{2}{3}$
b. $\frac{1}{3}$
c. $\frac{1}{4}$
d. none of these

7. Subtract.

$\frac{3}{8} - \frac{1}{4}$

a. $\frac{1}{8}$
b. $\frac{1}{4}$
c. $\frac{1}{2}$
d. none of these

8. What is the perimeter?

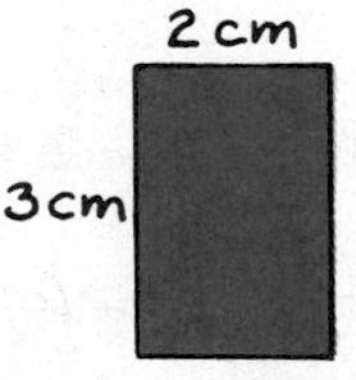

a. 6 cm
b. 5 cm
c. 10 cm
d. none of these

9. What is the area?

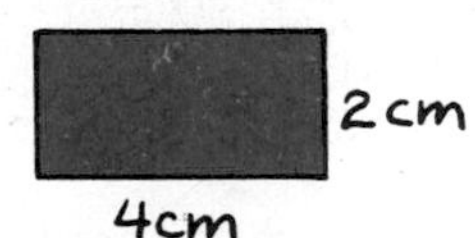

a. 6 square cm
b. 8 square cm
c. 12 square cm
d. none of these

10. Multiply.

$$\begin{array}{r} 758 \\ \times 6 \\ \hline \end{array}$$

a. 4548
b. 4208
c. 4248
d. none of these

11. Games played: 8
Points scored: 120
How many points did he average each game?

a. 10 **b.** 15
c. 64 **d.** none of these

12. Carol bought 5 records for $2.69 each. How much change did she get from $20?

a. $17.31 **b.** $13.45
c. $7.55 **d.** none of these

Division by 1- and 2-Digit Divisors

11

READY OR NOT?

Divide.

1. $5\overline{)25}$
2. $7\overline{)63}$
3. $5\overline{)40}$
4. $6\overline{)24}$
5. $4\overline{)28}$
6. $9\overline{)27}$
7. $3\overline{)24}$
8. $9\overline{)54}$
9. $5\overline{)30}$
10. $7\overline{)28}$
11. $4\overline{)36}$
12. $8\overline{)32}$
13. $3\overline{)18}$
14. $9\overline{)81}$
15. $4\overline{)32}$
16. $8\overline{)56}$
17. $3\overline{)21}$
18. $7\overline{)49}$
19. $4\overline{)16}$
20. $8\overline{)72}$
21. $2\overline{)14}$
22. $9\overline{)63}$
23. $5\overline{)20}$
24. $9\overline{)36}$
25. $3\overline{)27}$
26. $6\overline{)54}$
27. $5\overline{)15}$
28. $8\overline{)24}$
29. $2\overline{)16}$
30. $9\overline{)72}$
31. $4\overline{)20}$
32. $8\overline{)64}$
33. $8\overline{)16}$
34. $8\overline{)48}$
35. $7\overline{)21}$
36. $7\overline{)56}$
37. $6\overline{)30}$
38. $8\overline{)40}$
39. $9\overline{)9}$
40. $7\overline{)35}$

Division review

Philip works in King's supermarket. He puts fruits and vegetables into packages.

How many packages of 3 peppers can he make with 85 peppers?

85 ÷ 3 = ?

$3\overline{)85}$

Step 1. Divide tens and regroup.

$$\begin{array}{r} 2 \\ 3\overline{)85} \\ -6 \\ \hline 25 \end{array}$$

Step 2. Divide ones and write remainder.

$$\begin{array}{r} 28 \text{ R1} \\ 3\overline{)85}\phantom{\text{ R1}} \\ -6\phantom{\text{ R1}} \\ \hline 25\phantom{\text{ R1}} \\ -24\phantom{\text{ R1}} \\ \hline 1\phantom{\text{ R1}} \end{array}$$

Philip can make 28 packages of 3 peppers, with 1 pepper left over.

EXERCISES

Divide.

1. $3\overline{)36}$	2. $2\overline{)84}$	3. $6\overline{)66}$	4. $3\overline{)90}$	5. $4\overline{)48}$
6. $3\overline{)69}$	7. $2\overline{)26}$	8. $8\overline{)80}$	9. $2\overline{)68}$	10. $5\overline{)55}$
11. $3\overline{)48}$	12. $4\overline{)64}$	13. $6\overline{)90}$	14. $4\overline{)60}$	15. $5\overline{)65}$
16. $3\overline{)81}$	17. $5\overline{)95}$	18. $2\overline{)76}$	19. $7\overline{)91}$	20. $3\overline{)72}$
21. $3\overline{)74}$	22. $2\overline{)87}$	23. $6\overline{)80}$	24. $4\overline{)97}$	25. $5\overline{)48}$
26. $3\overline{)62}$	27. $7\overline{)88}$	28. $4\overline{)43}$	29. $5\overline{)81}$	30. $6\overline{)92}$

Solve.

31. a. How many packages of 4 could Philip make from this box of pears?

    b. If Philip had made packages of 6 pears, then how many packages would there have been?

32. a. How many packages of 4 could Philip make from this box of tomatoes? Would there be any tomatoes left over?

    b. If Philip had made packages of 6 tomatoes, then how many packages would there have been? Would there have been any tomatoes left over?

33. King's sells oranges in bags of 8. How many bags of 8 can be made from a box of 96 oranges?

34. King's sells apples in bags of 6 for $1.19 a bag. How much money will it cost to buy 12 bags of apples?

Dividing a 3-digit number

Think about dividing up the blocks.

3) 4 4 8

Step 1. Divide hundreds. Subtract.

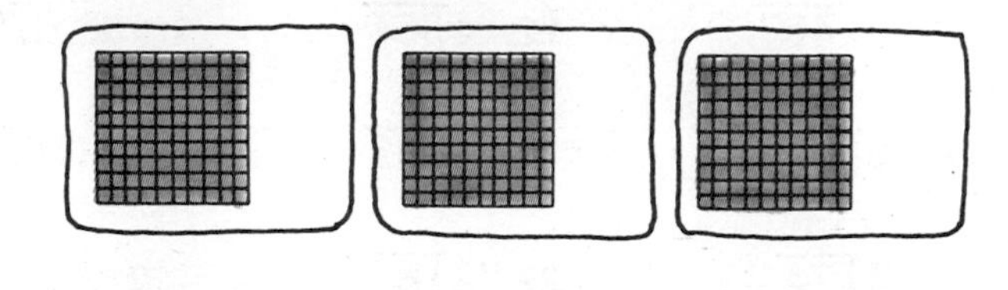

1
3) 4 4 8
-3
1

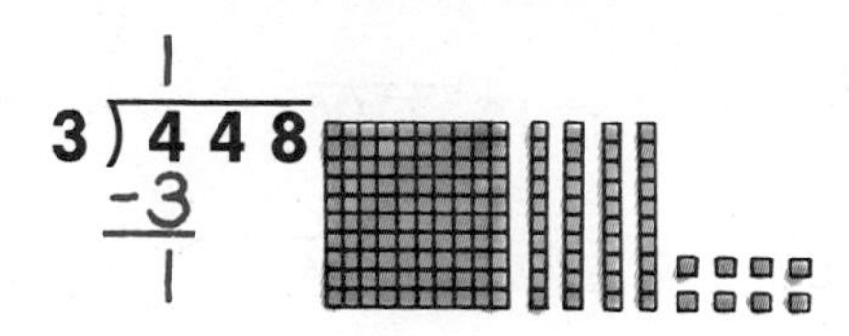

Step 2. Regroup 1 hundred for 10 tens.

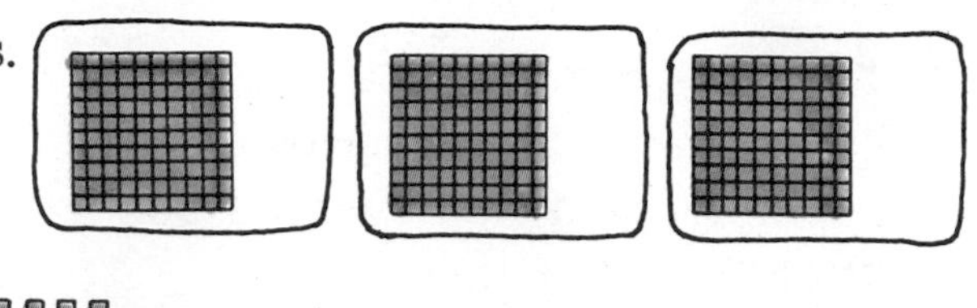

1
3) 4 4 8
-3
14

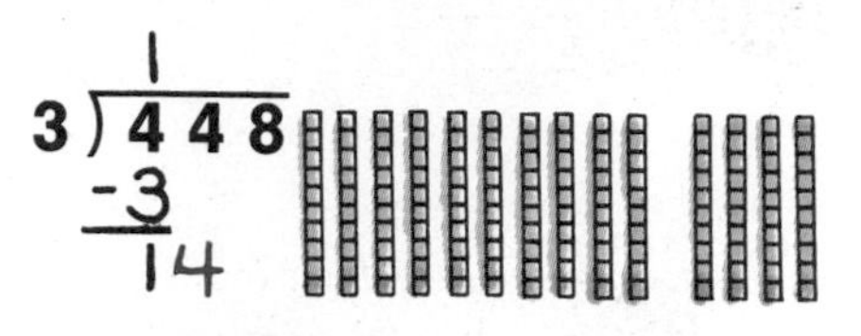

Step 3. Divide tens. Subtract.

14
3) 4 4 8
-3
14
-12
2

Step 4. Regroup 2 tens for 20 ones.

14
3) 4 4 8
-3
14
-12
28

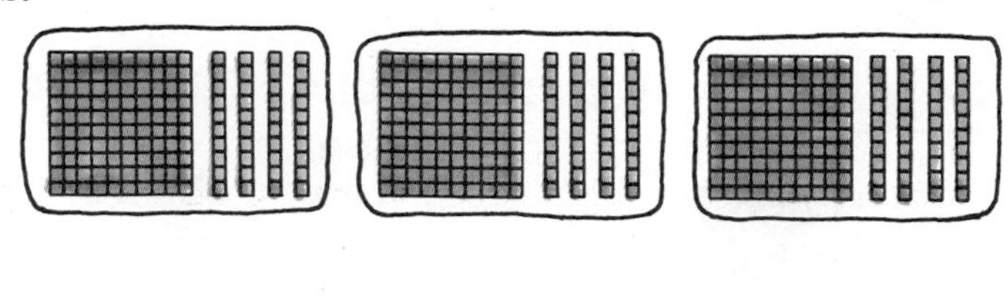

Step 5. Divide ones. Subtract.

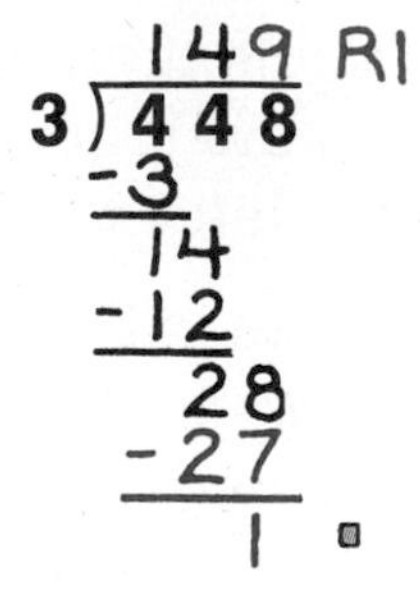

149 R1
3) 4 4 8
-3
14
-12
28
-27
1

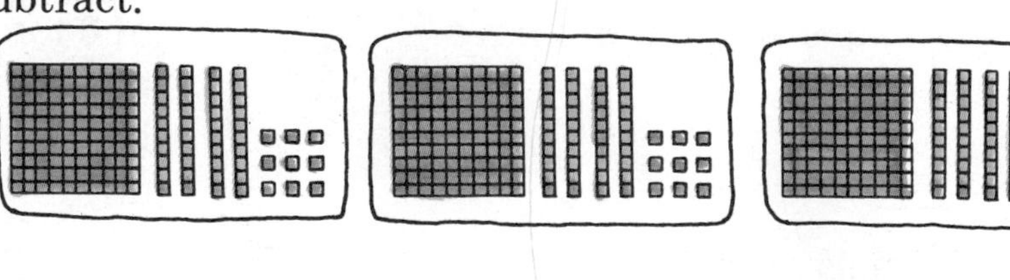

EXERCISES

Divide.

1. 3)527 2. 4)952 3. 2)504 4. 5)745 5. 2)963

6. 2)856 7. 5)742 8. 6)629 9. 4)859 10. 3)398

11. 7)749 12. 8)900 13. 2)378 14. 6)842 15. 4)726

16. 4)953 17. 9)974 18. 3)627 19. 2)700 20. 9)958

21. 2)$4.76 22. 7)$8.96 23. 8)$8.08 24. 3)$5.34 25. 5)$8.85

26. The Winn family drove 360 kilometers in 5 hours. How many kilometers did they average per hour?

27. The Winns' car averaged 8 kilometers per liter of gasoline. How many liters of gasoline did they use on the 360-kilometer trip?

Challenge!

Copy and complete these division problems.

28.
```
      □ 6 8 R1
  2)□ 3 □
    -4
     1 3
    -1 2
       1 □
     - □ □
         □
```

29.
```
      1 □ 5 R5
  6)□ □ □
   -□
    2 7
   -□ □
      3 □
     -3 □
        □
```

30.
```
       1 3 1
  □)□ □ □
   -□
    2 □
   -□ □
      □ □
       -□
        0
```

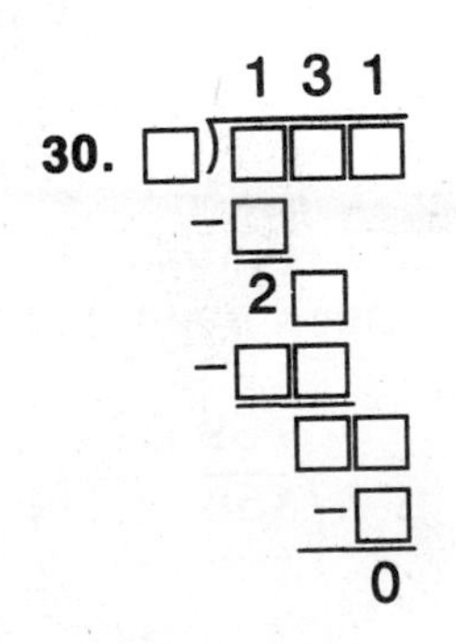

More about division

Sometimes you have to regroup before you start dividing.

Step 1. Not enough hundreds. So think about 26 tens.

$$3\overline{)268}$$

Step 2. Divide tens.

$$\begin{array}{r} 8 \\ 3\overline{)268} \\ -24 \\ \hline 2 \end{array}$$

Step 3. Regroup and divide ones.

$$\begin{array}{r} 89 \text{ R1} \\ 3\overline{)268} \\ -24 \\ \hline 28 \\ -27 \\ \hline 1 \end{array}$$

You can check your work.

Multiply the quotient by the divisor:

$$\begin{array}{r} 89 \\ \times 3 \\ \hline 267 \end{array}$$

Then add the remainder:

$$\begin{array}{r} 267 \\ +1 \\ \hline 268 \end{array}$$

It checks.

EXERCISES

Check each answer. Correct the wrong answers.

1. $4\overline{)158}$ 39 R2
2. $5\overline{)261}$ 50 R1
3. $3\overline{)174}$ 57 R2
4. $4\overline{)229}$ 57 R1
5. $6\overline{)361}$ 6 R1
6. $5\overline{)278}$ 53 R3
7. $3\overline{)258}$ 86
8. $5\overline{)227}$ 45 R1

Divide.

9. $6\overline{)312}$
10. $4\overline{)407}$
11. $3\overline{)148}$
12. $2\overline{)642}$
13. $6\overline{)357}$
14. $4\overline{)349}$
15. $5\overline{)409}$
16. $8\overline{)839}$
17. $3\overline{)745}$
18. $4\overline{)392}$
19. $7\overline{)314}$
20. $5\overline{)517}$
21. $6\overline{)236}$
22. $7\overline{)432}$
23. $5\overline{)819}$
24. $2\overline{)248}$
25. $3\overline{)731}$
26. $3\overline{)557}$
27. $7\overline{)603}$
28. $8\overline{)548}$
29. $2\overline{)726}$
30. $3\overline{)225}$
31. $7\overline{)837}$
32. $4\overline{)453}$
33. $9\overline{)329}$
34. $8\overline{)951}$
35. $8\overline{)626}$
36. $5\overline{)961}$
37. $5\overline{)969}$
38. $6\overline{)624}$

Solve.

39. Frank bought 3 records for \$4.35 each. What was the total cost?
40. Sarah had \$8.72. She spent \$3.79 for a record. How much money did she have left?
41. Al bought 2 records for \$2.85 each. He gave the clerk a \$10 bill. How much change did he receive?
42. Carol bought 4 records that cost \$2.85, \$3.74, \$2.94, and \$4.35. What was the average cost of each record?

43. What is the largest whole number you can multiply by 289 to get an answer between 4000 and 4500?
44. What is the largest whole number you can multiply by 357 to get an answer between 6500 and 7000?

Dividing a 4-digit number

Large numbers are divided in the same way as small numbers.
Remember that you can check division by multiplying.

The division tells us that there are 1776 blocks in each of 3 sets. So, we can multiply to see if we get 5328 blocks in all.

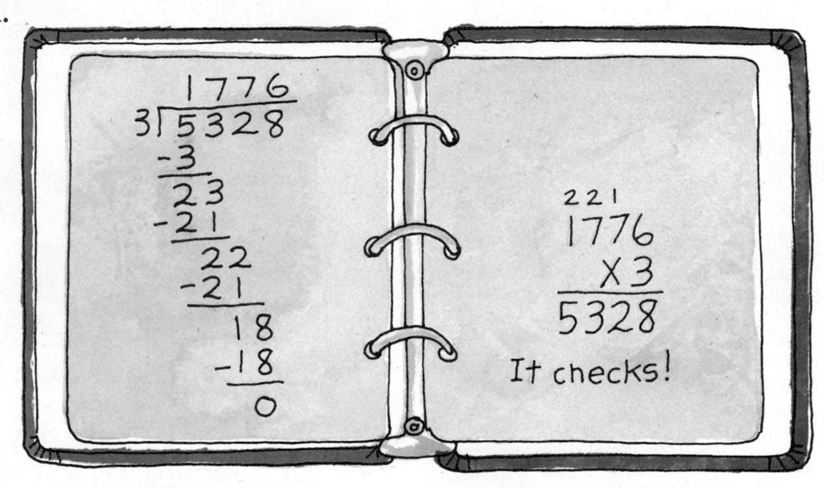

To check a division problem, multiply the quotient by the divisor. Then add the remainder.

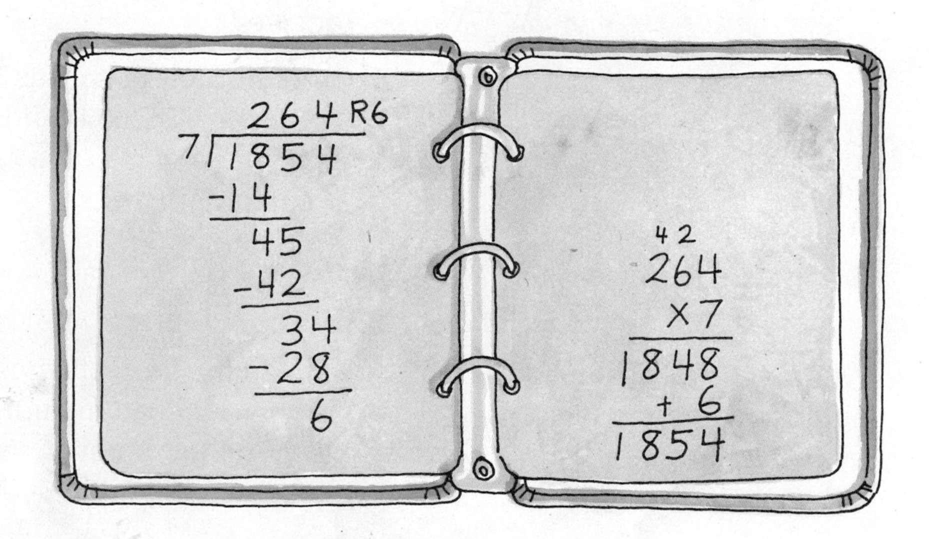

EXERCISES

First divide. Then check your answer.

1. $4\overline{)2248}$ 2. $5\overline{)1335}$ 3. $3\overline{)1506}$ 4. $6\overline{)4374}$

5. $7\overline{)3821}$ 6. $6\overline{)7342}$ 7. $9\overline{)5000}$ 8. $8\overline{)3967}$

Divide.

9. $3\overline{)75}$ 10. $2\overline{)54}$ 11. $5\overline{)78}$ 12. $6\overline{)93}$

13. $4\overline{)361}$ 14. $6\overline{)742}$ 15. $7\overline{)396}$ 16. $6\overline{)125}$

17. $9\overline{)2358}$ 18. $8\overline{)4695}$ 19. $5\overline{)7803}$ 20. $9\overline{)3016}$

21. $7\overline{)7213}$ 22. $4\overline{)4246}$ 23. $4\overline{)6381}$ 24. $4\overline{)5000}$

25. $8\overline{)4382}$ 26. $8\overline{)3005}$ 27. $9\overline{)2000}$ 28. $7\overline{)5378}$

29. $7\overline{)3999}$ 30. $5\overline{)2916}$ 31. $6\overline{)3742}$ 32. $5\overline{)9465}$

Challenge!

Guess my number.

33. If you divide my number by 6, you get a quotient of 248 and a remainder of 0.

34. If you divide my number by 8, you get a quotient of 429 and a remainder of 5.

KEEPING SKILLS SHARP

Subtract. Give each difference in lowest terms.

1. $\frac{5}{6} - \frac{1}{6}$ 2. $\frac{5}{9} - \frac{2}{9}$ 3. $\frac{3}{6} - \frac{1}{6}$ 4. $\frac{3}{4} - \frac{1}{4}$

5. $\frac{3}{8} - \frac{0}{8}$ 6. $\frac{5}{8} - \frac{1}{4}$ 7. $\frac{9}{16} - \frac{1}{8}$ 8. $\frac{5}{6} - \frac{1}{2}$

9. $\frac{5}{6} - \frac{2}{3}$ 10. $\frac{5}{6} - \frac{1}{3}$ 11. $\frac{2}{3} - \frac{1}{6}$ 12. $\frac{3}{4} - \frac{1}{2}$

Addition, subtraction, multiplication, and division

$242 + (756 \div 4) = 431$

$$\begin{array}{r} 189 \\ 4\overline{)756} \\ -4 \\ \hline 35 \\ -32 \\ \hline 36 \\ -36 \\ \hline 0 \end{array} \qquad \begin{array}{r} 242 \\ +189 \\ \hline 431 \end{array}$$

EXERCISES

Compute.

1. $(252 \div 2) - 28$
2. $681 + (396 \div 3)$
3. $753 - (400 \div 4)$
4. $(385 \div 5) - 19$
5. $(462 \div 3) \times 4$
6. $(507 - 287) \div 5$
7. $(246 \times 7) \div 7$
8. $(836 + 259) - 259$

Less than (<), equal to (=), or greater than (>)?

9. 382 + 167 ● 275 × 2
10. 974 − 395  4504 ÷ 8
11. 68 × 4 ● 1664 ÷ 6
12. 143 × 6 ● 1001 − 143

Challenge!

13. Which of these computer directions will give the same answer?
    a. PRINT (9 + 4) + 2 b. PRINT 9 + (4 + 2)
    c. PRINT (8 − 2) − 1 d. PRINT 8 − (2 − 1)
14. Put the parentheses in these print statements.

a.

b.

Which quotient is greater?

1. 2)4583   3)8254
2. 6)5872   8)7625
3. 4)7850   5)8407
4. 6)8364   4)6638

Play the game.

1. Choose a leader.
2. Make two cards for each of the digits.

3. Each player draws a table.

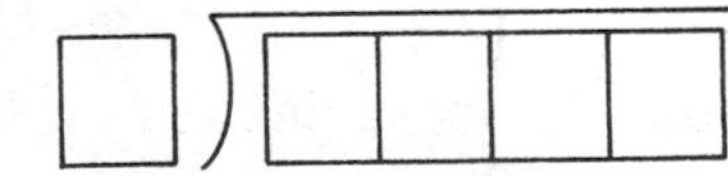

4. As the leader picks a digit, fill in your table.
5. Repeat step 4 until your table is filled in.
6. The player who builds the greatest quotient wins.
   Later you may wish to play that the lowest quotient wins.

Problem solving

Mr. Allen is a landscape gardener. He keeps lawns, bushes, and flowers healthy and growing.

Solve.

1. Mr. Allen planted 137 bushes on Monday and 127 bushes on Tuesday. How many bushes did he plant on those 2 days?

2. Mr. Allen delivered an apple tree on Wednesday. He left the nursery at 9:25 A.M. and reached the buyer's house at 10:40 A.M. How many minutes long was the trip?

3. Each can had 30 daffodils. How many daffodils were there in 6 cans?

4. Steve and Betty picked 264 carnations. They divided the carnations evenly into 8 cans. How many carnations were there in each can?

5. Mr. Allen grows tomato plants in long rows. There are 72 plants in each row. How many plants are there in 8 rows?

6. Mr. Allen's 3 helpers are planting bulbs. Each helper plants 2 bulbs in 1 minute. How many bulbs can they plant in 1 hour?

7. Betty mowed 3 lawns on Thursday. She mowed the Smiths' lawn, which is 1846 square feet. She mowed the Rodriguezes' lawn, which is 2306 square feet. She mowed the Kilpatricks' lawn, which is 3179 square feet. How many square feet of lawn did she mow?

8. Mr. Allen used Quik Gro plant food to feed 106 trees. He used 9 cups to feed each tree. How many cups of Quik Gro did he use?

9. There are 162 cups of Quik Gro plant food in a bag. To feed a tree, 9 cups of Quik Gro are used. How many trees can be fed from 1 bag?

10. Mr. Allen puts 6 marigold plants in a box. In 2 hours, he put 324 marigold plants in boxes. How many boxes of marigolds did he have?

11. Petunia plants are sold in boxes of 8 plants each. How many plants are there in 216 boxes?

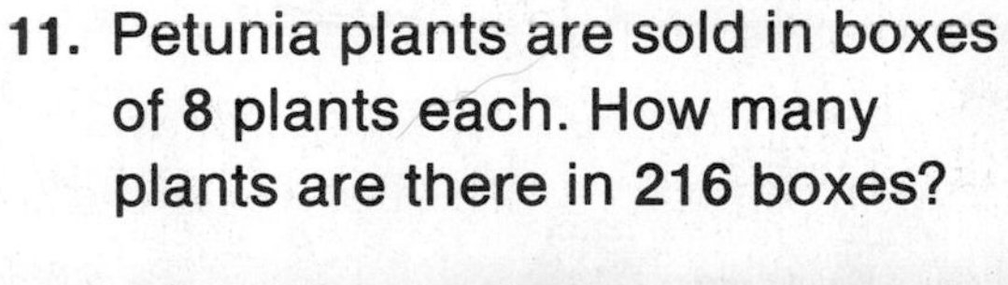

12. Would $10 be enough to buy these flowers?

Problem solving

Clerks at checkout counters have to enter the prices of items into the cash registers. Sometimes they have to solve a problem like this:

How much should the customer be charged for 1 can of soup?

To find the cost of 1 can, the clerk can divide:

$$\begin{array}{r} 26 \\ 3\overline{)79} \\ -6 \\ 19 \\ -18 \\ 1 \end{array}$$

Since there is a remainder, the clerk knows that the cost of 1 can is between 26¢ and 27¢. The rule used by most stores is always to round up to the next cent. Is this the same as rounding to the nearest cent?

EXERCISES

1. How much will 1 jar cost?

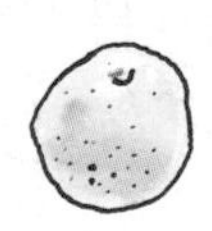

2. What is the cost of 1 grapefruit?

3. How much will 1 cost?

4. What will be the cost of 3 boxes?

   Hint: Add the costs of 2 boxes and 1 box.

5. If 2 cans of pears cost $1.89, what is the cost of 1 can?

6. If pudding is 4 packages for $.79, how much will 5 packages cost?

7. Terry took $1 to the store. He wanted 1 can of orange juice. The juice was 3 cans for $1.17. How much money did Terry have left?

8. Bread was on sale, 3 loaves for $1.89. Jessica bought a carton of milk for $.73 and 4 loaves of bread. What was the total cost?

9. Jam is on sale: 2 jars of grape for $1.75, 2 jars of raspberry for $1.89, and 3 jars of strawberry for $2.00. What is the cost of 3 jars of grape jam and 4 jars of strawberry jam?

KEEPING SKILLS SHARP

Add. Give each sum in lowest terms.

1. $\frac{2}{5} + \frac{1}{5}$
2. $\frac{2}{3} + \frac{1}{9}$
3. $\frac{2}{9} + \frac{1}{6}$
4. $\frac{1}{3} + \frac{1}{2}$
5. $\frac{1}{4} + \frac{1}{2}$
6. $\frac{5}{8} + \frac{1}{4}$
7. $\frac{3}{10} + \frac{1}{5}$
8. $\frac{1}{2} + \frac{1}{6}$
9. $\frac{1}{6} + \frac{2}{3}$
10. $\frac{3}{5} + \frac{1}{4}$
11. $\frac{2}{9} + \frac{1}{3}$
12. $\frac{5}{6} + \frac{1}{9}$
13. $\frac{1}{3} + \frac{2}{5}$
14. $\frac{1}{6} + \frac{3}{4}$
15. $\frac{1}{2} + \frac{2}{5}$

Dividing by tens

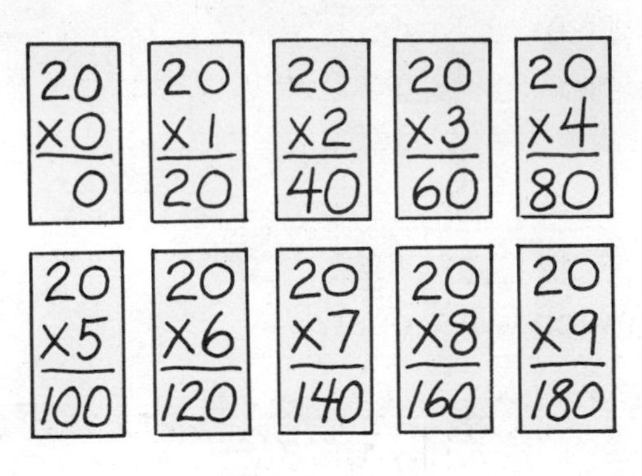

Dividing by a number that is 10 or greater is easy when you have the multiplication facts. Here are the multiplication facts for 20. We will use them to do this division.

20) 9 2 7

Step 1. Not enough hundreds.

20) 9 | 2 7

Step 2. Regroup and divide.
Think 92 tens.

```
      4|
20 ) 9 2|7
    -8 0
     1 2
```

20
×4
80

Step 3. Regroup and divide.

```
      4 6| R7
20 ) 9 2 7|
    -8 0
     1 2 7
    -1 2 0
         7
```

20
×6
120

EXERCISES

Use the given multiplication facts to help you divide.

10	10	10	10	10
×0	×1	×2	×3	×4
0	10	20	30	40

10	10	10	10	10
×5	×6	×7	×8	×9
50	60	70	80	90

1. $10\overline{)72}$ **2.** $10\overline{)162}$ **3.** $10\overline{)720}$

4. $10\overline{)882}$ **5.** $10\overline{)990}$ **6.** $10\overline{)342}$

7. $10\overline{)702}$ **8.** $10\overline{)684}$ **9.** $10\overline{)738}$

40	40	40	40	40
×0	×1	×2	×3	×4
0	40	80	120	160

40	40	40	40	40
×5	×6	×7	×8	×9
200	240	280	320	360

10. $40\overline{)853}$ **11.** $40\overline{)675}$ **12.** $40\overline{)347}$

13. $40\overline{)921}$ **14.** $40\overline{)792}$ **15.** $40\overline{)296}$

16. $40\overline{)578}$ **17.** $40\overline{)729}$ **18.** $40\overline{)907}$

Challenge!

First list the multiplication facts for 32. Then divide.

32	32	32	32	32
×0	×1	×2	×3	×4

32	32	32	32	32
×5	×6	×7	×8	×9

19. $32\overline{)674}$ **20.** $32\overline{)829}$ **21.** $32\overline{)283}$

22. $32\overline{)981}$ **23.** $32\overline{)569}$ **24.** $32\overline{)994}$

25. $32\overline{)482}$ **26.** $32\overline{)873}$ **27.** $32\overline{)296}$

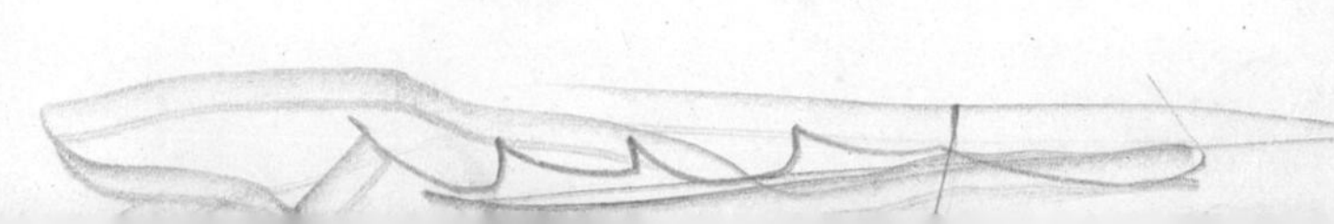

Reviewing rounding and multiplication

Before you learn more about division, you will need to review some skills.

ORAL EXERCISES

Round to the nearest ten. *Remember:* If the number is halfway between, round up.

1. 69	**2.** 42	**3.** 63	**4.** 79	**5.** 37	**6.** 47	**7.** 62
8. 53	**9.** 89	**10.** 56	**11.** 82	**12.** 59	**13.** 14	**14.** 54
15. 86	**16.** 61	**17.** 71	**18.** 51	**19.** 46	**20.** 65	**21.** 72
22. 74	**23.** 66	**24.** 92	**25.** 45	**26.** 78	**27.** 55	**28.** 88
29. 58	**30.** 87	**31.** 52	**32.** 76	**33.** 44	**34.** 83	**35.** 77
36. 84	**37.** 41	**38.** 94	**39.** 57	**40.** 73	**41.** 39	**42.** 68
43. 43	**44.** 64	**45.** 67	**46.** 36	**47.** 38	**48.** 16	**49.** 33
50. 48	**51.** 75	**52.** 85	**53.** 34	**54.** 81	**55.** 49	**56.** 35

ORAL EXERCISES

Give each product.

57. 50 ×3	**58.** 60 ×2	**59.** 40 ×4	**60.** 30 ×3	**61.** 80 ×6	**62.** 30 ×5
63. 60 ×4	**64.** 50 ×2	**65.** 50 ×8	**66.** 90 ×7	**67.** 40 ×2	**68.** 90 ×6
69. 70 ×4	**70.** 40 ×5	**71.** 80 ×9	**72.** 40 ×9	**73.** 80 ×2	**74.** 80 ×5
75. 60 ×9	**76.** 90 ×4	**77.** 60 ×5	**78.** 40 ×6	**79.** 50 ×4	**80.** 70 ×5
81. 80 ×8	**82.** 50 ×7	**83.** 80 ×3	**84.** 90 ×3	**85.** 30 ×4	**86.** 50 ×9
87. 70 ×3	**88.** 70 ×6	**89.** 50 ×6	**90.** 40 ×7	**91.** 60 ×6	**92.** 20 ×8
93. 80 ×4	**94.** 90 ×5	**95.** 30 ×6	**96.** 90 ×8	**97.** 90 ×2	**98.** 30 ×2
99. 60 ×8	**100.** 40 ×8	**101.** 70 ×8	**102.** 50 ×5	**103.** 60 ×7	**104.** 70 ×7

Dividing by a 2-digit number

When you divide, you can guess which multiplication facts you will need by rounding the divisor to the nearest ten.

EXAMPLE 1. **23) 1 1 7**

Step 1. Think about dividing by 20.

23) 1 1 7

Step 2. *Think:* $20 \times 4 = 80$
$20 \times 5 = 100$
$20 \times 6 = 120$ ← Too big, so try 5.

5 R2
23) 1 1 7
-1 1 5
2

23
×5
115

EXAMPLE 2. **27) 7 2 5**

Step 1. Think about dividing by 30.

30

27) 7 2 5

Step 2. *Think:* $30 \times 2 = 60$
$30 \times 3 = 90$ ← Too big, so try 2.

2
27) 7 2 5
-54
18

27
×2
54

Step 3. Regroup.

2
27) 7 2 5
-54
185

Step 4. *Think:* $30 \times 5 = 150$
$30 \times 6 = 180$
$30 \times 7 = 210$

26 R23
27) 7 2 5
-54
185
-162
23

27
×6
162

EXERCISES

Divide.

1. $31\overline{)175}$	**2.** $18\overline{)123}$	**3.** $37\overline{)172}$	**4.** $33\overline{)169}$	**5.** $21\overline{)178}$
6. $18\overline{)681}$	**7.** $42\overline{)698}$	**8.** $23\overline{)576}$	**9.** $38\overline{)296}$	**10.** $12\overline{)496}$
11. $32\overline{)983}$	**12.** $19\overline{)682}$	**13.** $28\overline{)753}$	**14.** $43\overline{)778}$	**15.** $18\overline{)822}$
16. $21\overline{)173}$	**17.** $28\overline{)194}$	**18.** $19\overline{)127}$	**19.** $38\overline{)293}$	**20.** $34\overline{)176}$
21. $31\overline{)968}$	**22.** $29\overline{)742}$	**23.** $42\overline{)942}$	**24.** $20\overline{)860}$	**25.** $36\overline{)526}$
26. $21\overline{)873}$	**27.** $43\overline{)895}$	**28.** $38\overline{)940}$	**29.** $32\overline{)582}$	**30.** $19\overline{)678}$

Solve.

31. There were 768 cookies in all. There were 24 cookies in each box. How many boxes were there?

32. There were 492 boxes in all. There were 12 rolls in each box. How many rolls were there?

33. There were 96 loaves of bread. The baker baked 24 more loaves. How many loaves were there then?

34. There were 96 ounces of dough. The dough was used to make loaves of bread. Each loaf weighed 24 ounces. How many loaves of bread were made?

More on dividing by a 2-digit number

Sometimes your first guess needs to be changed.

EXAMPLE 1. $27\overline{)167}$

Step 1. Think about dividing by 30.

(30) $27\overline{)167}$

Step 2. *Think:* $30 \times 4 = 120$
$30 \times 5 = 150$
$30 \times 6 = 180$ ← Too big, so try 5.

$$\begin{array}{r} 5 \\ 27\overline{)167} \\ -135 \\ \hline 32 \end{array} \qquad \begin{array}{r} 27 \\ \times 5 \\ \hline 135 \end{array}$$

Remainder is greater than 27, so 5 is too small.

Step 3. Try 6.

$$\begin{array}{r} 6\text{ R5} \\ 27\overline{)167} \\ -162 \\ \hline 5 \end{array} \qquad \begin{array}{r} 27 \\ \times 6 \\ \hline 162 \end{array}$$

EXAMPLE 2. $24\overline{)817}$

Step 1. Think about dividing by 20.

(20) $24\overline{)817}$

Step 2. *Think:* $20 \times 3 = 60$
$20 \times 4 = 80$
$20 \times 5 = 100$ ← Too big, so try 4.

$$\begin{array}{r} 4 \\ 24\overline{)817} \\ -96 \\ \hline \end{array} \qquad \begin{array}{r} 24 \\ \times 4 \\ \hline 96 \end{array}$$

You can't subtract, so 4 is too big.

Step 3. Try 3.

$$\begin{array}{r} 3 \\ 24\overline{)817} \\ -72 \\ \hline 9 \end{array} \qquad \begin{array}{r} 24 \\ \times 3 \\ \hline 72 \end{array}$$

Step 4. Regroup and divide.

$$\begin{array}{r} 34\text{ R1} \\ 24\overline{)817} \\ -72 \\ \hline 97 \\ -96 \\ \hline 1 \end{array} \qquad \begin{array}{r} 24 \\ \times 4 \\ \hline 96 \end{array}$$

EXERCISES

Divide.

1. $32\overline{)184}$
2. $32\overline{)386}$
3. $32\overline{)736}$
4. $32\overline{)465}$
5. $32\overline{)158}$
6. $38\overline{)217}$
7. $38\overline{)502}$
8. $38\overline{)824}$
9. $38\overline{)651}$
10. $38\overline{)964}$
11. $24\overline{)718}$
12. $24\overline{)173}$
13. $24\overline{)396}$
14. $24\overline{)987}$
15. $24\overline{)583}$
16. $36\overline{)784}$
17. $47\overline{)496}$
18. $18\overline{)683}$
19. $23\overline{)870}$
20. $51\overline{)682}$
21. $42\overline{)900}$
22. $19\overline{)748}$
23. $27\overline{)918}$
24. $49\overline{)401}$
25. $37\overline{)765}$
26. $22\overline{)851}$
27. $16\overline{)810}$
28. $53\overline{)864}$
29. $13\overline{)689}$
30. $44\overline{)916}$

To raise money for summer camp, 6 children collected newspapers. The table shows how much each collected the first 5 days.

Pounds of Newspaper Collected

	Mon.	*Tues.*	*Wed.*	*Thur.*	*Fri.*
Betty	78	92	84	70	81
Craig	53	74	96	59	88
George	64	59	42	60	75
Nancy	53	68	71	65	93
Steven	42	75	83	46	94
Terry	58	36	51	40	55

Solve.

31. How many pounds of newspaper did Betty collect the first 5 days?
32. How many pounds did Betty average per day during the first 5 days?
33. How many pounds did Craig average per day?
34. Who had the greater average per day, Nancy or Steven? How much greater?
35. Who had the greater average per day, George or Terry? How much greater?
36. What was the average number of pounds collected on Monday?

Checking division

The students collected 985 kilograms of paper for the school paper drive. If they divide the paper into stacks of 36 kilograms each, then how many stacks of paper will there be? How many kilograms of paper will be left over?

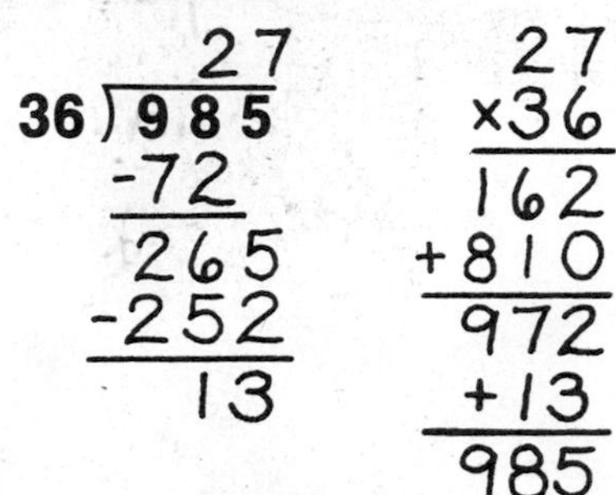

There will be 27 stacks of paper.
13 kilograms of paper will be left over.

EXERCISES

Divide. Then check your answer.

1. 23)873
2. 42)749
3. 36)963
4. 43)805
5. 18)542
6. 18)784
7. 24)862
8. 15)705
9. 29)690
10. 32)800

Divide.

11. 16)$4.48 — $ .28; -32; 128; -128; 0
12. 30)$9.60
13. 26)$9.88
14. 21)$9.87
15. 18)$9.00
16. 38)596
17. 17)809
18. 40)999
19. 24)862
20. 19)832
21. 37)782
22. 20)593
23. 45)900
24. 17)783
25. 41)865
26. 33)875
27. 28)965

Some children in a class weighed themselves on a kilogram scale. They wrote their names and weights on slips of paper.

28. How many weigh 24 kilograms?
29. How many weigh more than 25 kilograms?
30. Who weighs the least?
31. Who weighs the same as Terry?
32. What is their total weight?
33. What is their average weight?
34. How many weigh more than the average?
35. How many weigh less than the average?

Problem solving—not enough information

1. Study and understand.
2. Plan and do.
3. Answer and check.

Sometimes not enough facts are given in a problem. You need to find a fact so you can solve the problem.

EXAMPLE 1.

Teresa said we needed to make 360 sandwiches for the picnic. How many slices of bread did we need?

Missing fact: the number of slices of bread in each sandwich.

360 × (missing fact)

Do you know the missing fact? If so, solve the problem.

EXAMPLE 2.

We needed 720 slices of bread to make sandwiches for the picnic. How many loaves of bread did we need?

Missing fact: the number of slices of bread in each loaf.

720 ÷ (missing fact)

Do you know the missing fact? If so, solve the problem.

What is the missing fact?
If you know the missing fact, solve the problem.

1. Jennifer said we needed 80 pints of cider for the picnic. How many gallon jugs of cider did we need?

2. There were 180 people going to the picnic. How many buses were needed?

3. We each had to pay $1.50 for bus fare. Arthur had his fare in nickels. How many nickels did he have?

4. The driver of the last bus counted 2 empty seats when she closed the door. How many passengers were on the last bus?

5. Carl played football from 9:15 to 11:25 in the morning. How many minutes did he play?

6. David scored 3 touchdowns in the football game. How many points was that?

7. Barbara made 4 bull's-eyes in an archery game. How many points was that?

8. The water in one part of the lake was 12 feet deep. How many yards was that?

9. A man in a sailboat told Linda the water on the other side of the lake was 5 fathoms deep. How many yards was that?

10. Patricia said there were 20 paper plates left after the picnic. How many paper plates had been used?

11. Heather collected 24 empty quart bottles after the picnic. How much money did she get for the deposits?

★12. Joseph said he got a dollar's worth of dimes and quarters for the empty bottles he collected. How many dimes did he get? How many quarters did he get?

CHAPTER CHECKUP

Divide. [pages 296–305]

1. $8\overline{)941}$ 2. $6\overline{)927}$ 3. $3\overline{)948}$ 4. $9\overline{)826}$ 5. $8\overline{)964}$

6. $6\overline{)359}$ 7. $7\overline{)903}$ 8. $4\overline{)637}$ 9. $5\overline{)849}$ 10. $9\overline{)652}$

11. $4\overline{)1828}$ 12. $5\overline{)1927}$ 13. $8\overline{)2721}$ 14. $5\overline{)9420}$ 15. $6\overline{)3945}$

Divide. [pages 310–318]

16. $11\overline{)793}$ 17. $32\overline{)959}$ 18. $24\overline{)590}$ 19. $15\overline{)675}$ 20. $27\overline{)758}$

21. $34\overline{)268}$ 22. $32\overline{)703}$ 23. $33\overline{)681}$ 24. $25\overline{)605}$ 25. $17\overline{)963}$

26. $20\overline{)874}$ 27. $14\overline{)861}$ 28. $22\overline{)789}$ 29. $72\overline{)645}$ 30. $31\overline{)785}$

Solve. [pages 306–309, 319–321]

31. In one orchard Mr. Butler has 108 rows of trees with 36 trees in each row. How many trees does he have in that orchard?

32. From one orchard Donald got 572 boxes of apples. He stored the same number of boxes in each of 4 sheds. How many boxes did he put in each shed?

33. Katherine ordered 234 apple trees. She wants to plant them in 9 equal rows. How many should she plant in each row?

34. A fruit-stand owner bought 27 boxes of apples. She paid \$6.75 for each box. How much did she pay Mr. Butler?

CHAPTER PROJECT

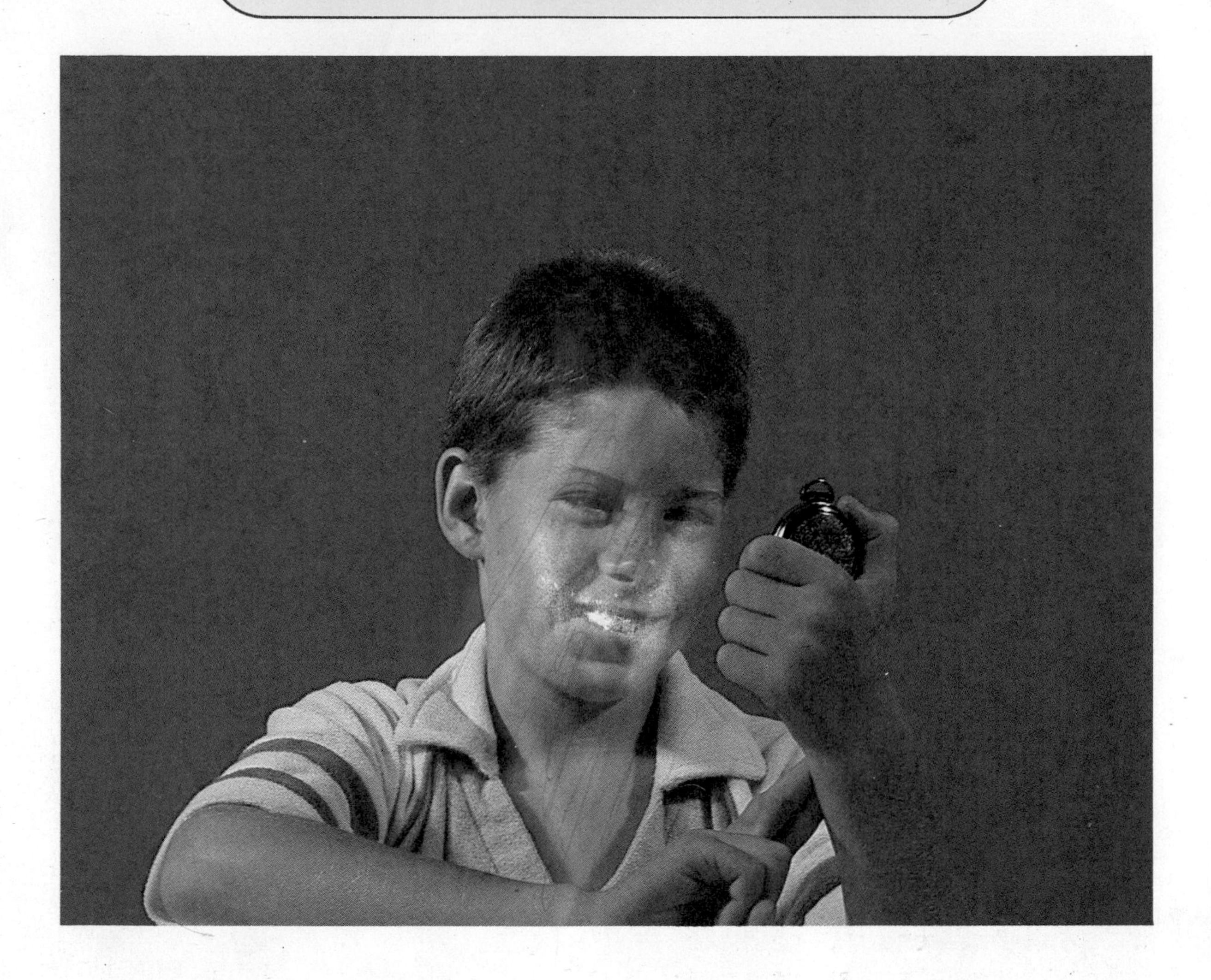

1. Find out how to take your pulse.
2. Count the number of times your heart beats in a minute.
3. How many times does it beat in an hour?
4. How many times in a day?
5. How many times does your heart beat in 30 days?
6. About how many times does it beat in a year?

CHAPTER REVIEW

4)349

87
4)349
-32
29

87 R1
4)349
-32
29
-28
1

Divide.

1. 2)506	2. 4)859	3. 3)397
4. 3)625	5. 6)940	6. 7)750
7. 7)395	8. 6)500	9. 8)621
10. 9)3723	11. 5)4380	12. 7)1601

24)432

18
24)432
-24
192
-192
0

Divide.

13. 24)936	14. 24)648	15. 24)149
(30) 16. 33)597	(20) 17. 22)784	(40) 18. 38)936
19. 43)543	20. 27)781	21. 39)960
22. 21)794	23. 47)381	24. 17)804

CHAPTER CHALLENGE

If the remainder is 0 when you divide a number by 2, the number is **divisible** by 2.

Can it be 0? 2? 4? 6? 8?

A number is divisible by 2 if its last digit is divisible by 2.

Is the number divisible by 2?

1. 578
2. 639
3. 507
4. 7405
5. 6384
6. 3310

A number is divisible by 3 if the sum of its digits is divisible by 3.

Is the number divisible by 3?

7. 147
8. 352
9. 601
10. 723
11. 537
12. 888

A number is divisible by 4 if its last two digits are divisible by 4.

Is the number divisible by 4?

13. 7132
14. 8404
15. 3915
16. 7726
17. 5384
18. 6032

19. See if you can find a rule to tell whether a number is divisible by 5.

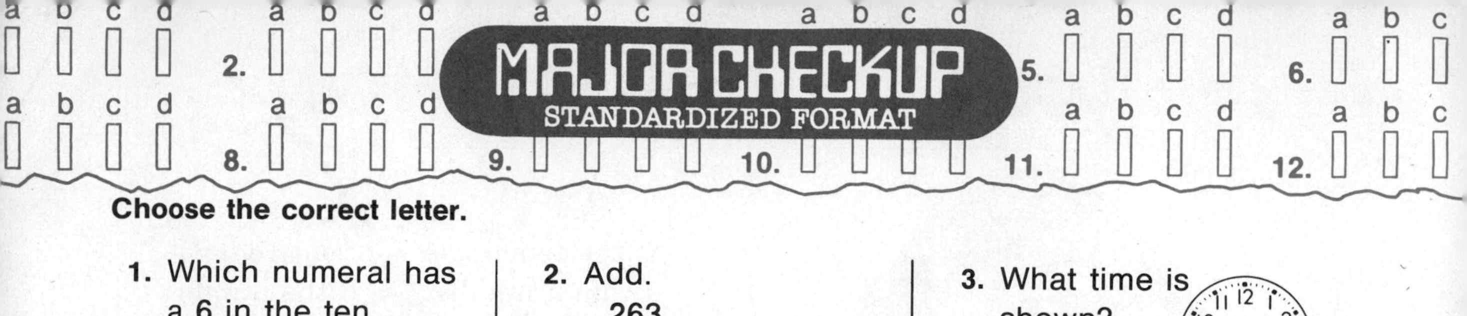

Choose the correct letter.

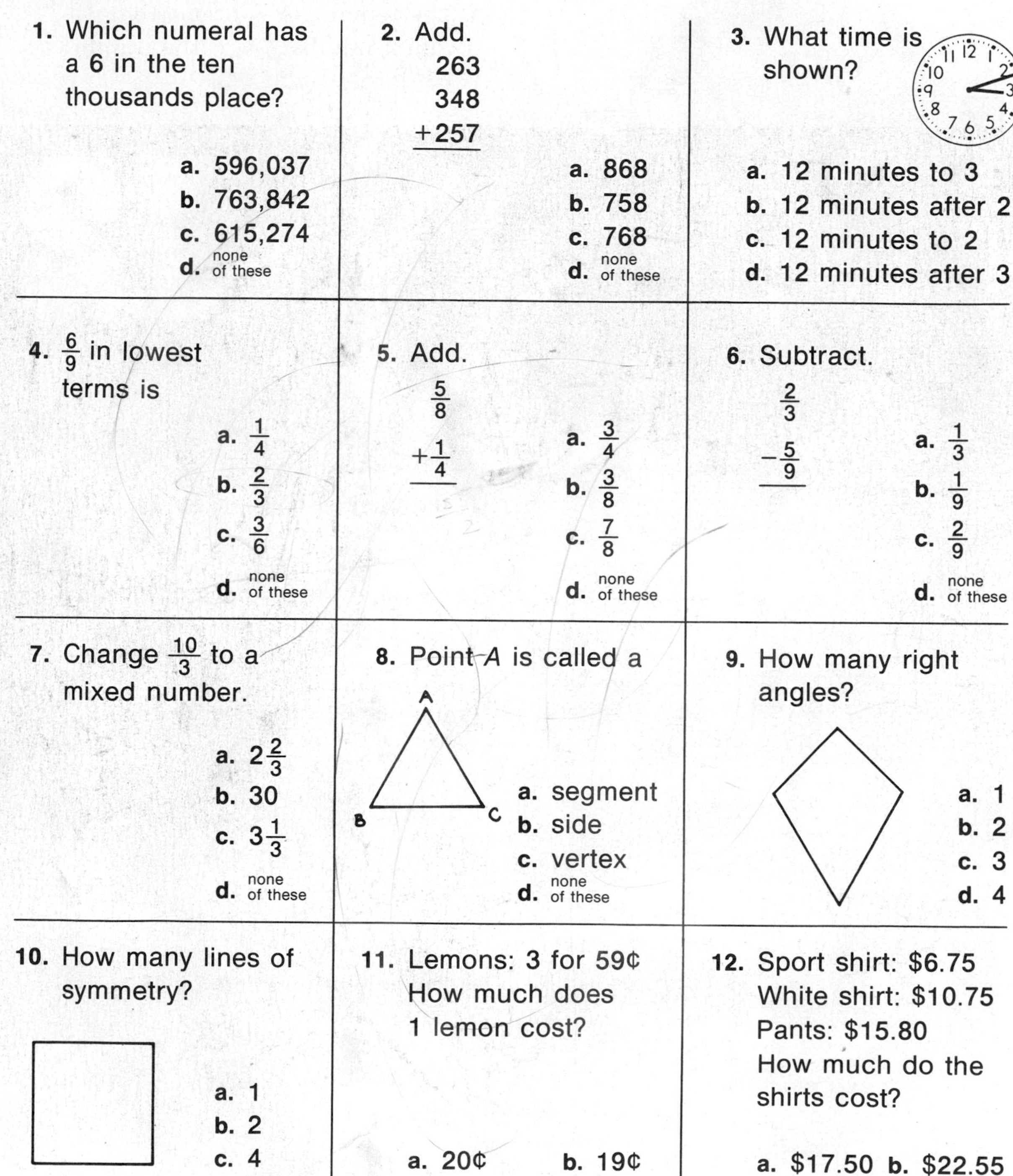

1. Which numeral has a 6 in the ten thousands place?
   a. 596,037
   b. 763,842
   c. 615,274
   d. none of these

2. Add.
   263
   348
   +257
   a. 868
   b. 758
   c. 768
   d. none of these

3. What time is shown?
   a. 12 minutes to 3
   b. 12 minutes after 2
   c. 12 minutes to 2
   d. 12 minutes after 3

4. $\frac{6}{9}$ in lowest terms is
   a. $\frac{1}{4}$
   b. $\frac{2}{3}$
   c. $\frac{3}{6}$
   d. none of these

5. Add.
   $\frac{5}{8}$
   $+\frac{1}{4}$
   a. $\frac{3}{4}$
   b. $\frac{3}{8}$
   c. $\frac{7}{8}$
   d. none of these

6. Subtract.
   $\frac{2}{3}$
   $-\frac{5}{9}$
   a. $\frac{1}{3}$
   b. $\frac{1}{9}$
   c. $\frac{2}{9}$
   d. none of these

7. Change $\frac{10}{3}$ to a mixed number.
   a. $2\frac{2}{3}$
   b. 30
   c. $3\frac{1}{3}$
   d. none of these

8. Point *A* is called a
   a. segment
   b. side
   c. vertex
   d. none of these

9. How many right angles?
   a. 1
   b. 2
   c. 3
   d. 4

10. How many lines of symmetry?
    a. 1
    b. 2
    c. 4
    d. none of these

11. Lemons: 3 for 59¢
    How much does 1 lemon cost?
    a. 20¢
    b. 19¢
    c. $1.77
    d. none of these

12. Sport shirt: $6.75
    White shirt: $10.75
    Pants: $15.80
    How much do the shirts cost?
    a. $17.50
    b. $22.55
    c. $26.55
    d. $33.30

Decimals

12

Tenths

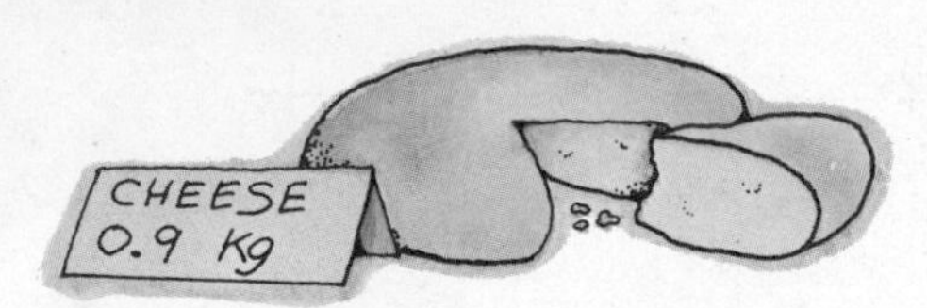

These numerals are called **decimal fractions** or just **decimals.**
To read decimals, think about unit squares that have been colored.

EXAMPLE 1.

$1\frac{4}{10}$ unit squares have been colored.

We can write the number in a place-value table like this:

Ones	Tenths
1	4

or without a table
like this: **1.4**

Read "1.4" as "one and four tenths."

EXAMPLE 2.

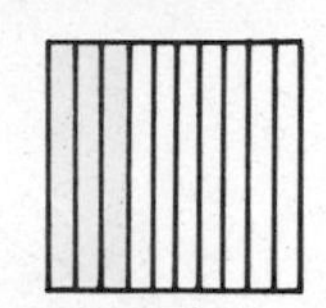

0.3

Read "0.3" as "three tenths."

EXAMPLE 3.

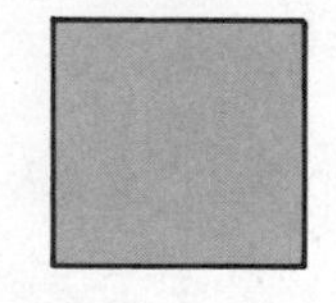 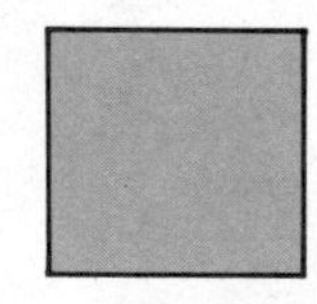 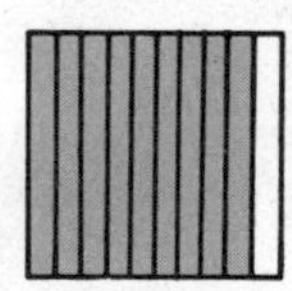

2.9

Read "2.9" as "two and nine tenths."

EXERCISES

Multiple choice. Choose the correct letter.

1.   

a. 3.2 **b.** 2.3 **c.** 2.5

2. 

a. 6.1 **b.** 2.6 **c.** 1.6

3. 

a. 1.7 **b.** 0.7 **c.** 0.3

Give a decimal for the number of unit squares that have been painted.

4. 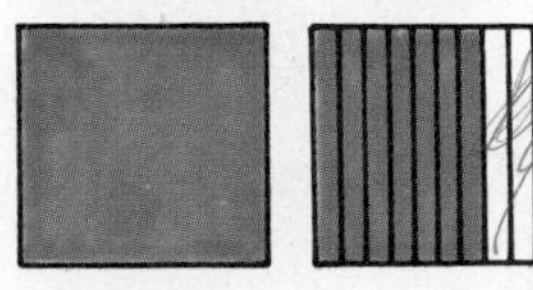

5. 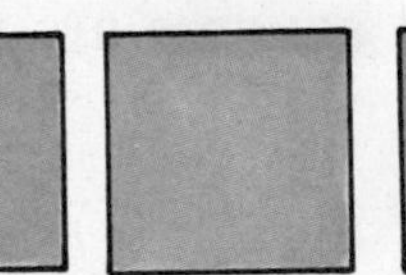 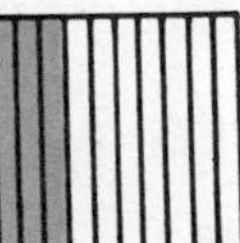

6. 

7. 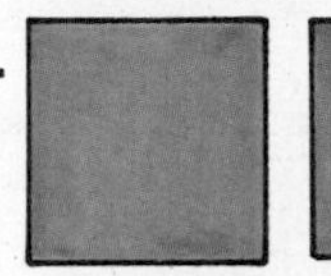  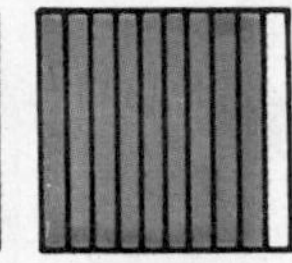

8. 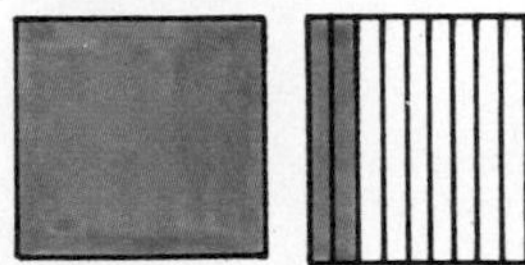

9.   

Match.

10. 2.1	**a.** two and three tenths
11. 1.2	**b.** three and two tenths
12. 3.2	**c.** two and one tenth
13. 2.3	**d.** one and two tenths

Read these decimals.

14. 3.6	**15.** 4.8	**16.** 2.9	**17.** 5.0	**18.** 6.8
19. 5.9	**20.** 10.3	**21.** 15.4	**22.** 0.8	**23.** 21.8

Challenge!

Use a decimal to answer each question.

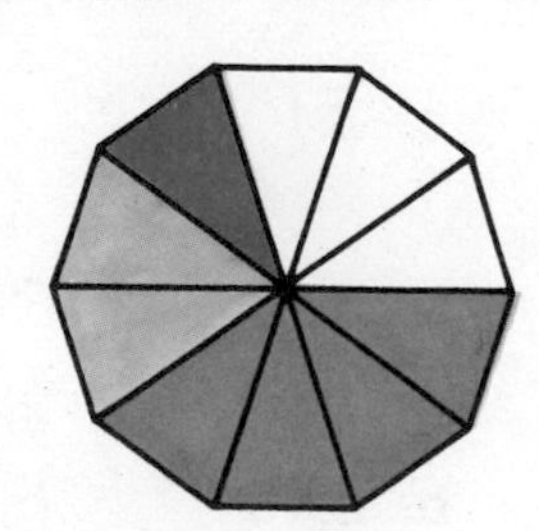

What part of the figure is

24. white? **25.** red?

26. yellow? **27.** yellow or blue?

Hundredths

How many unit squares are painted?

EXAMPLE 1.

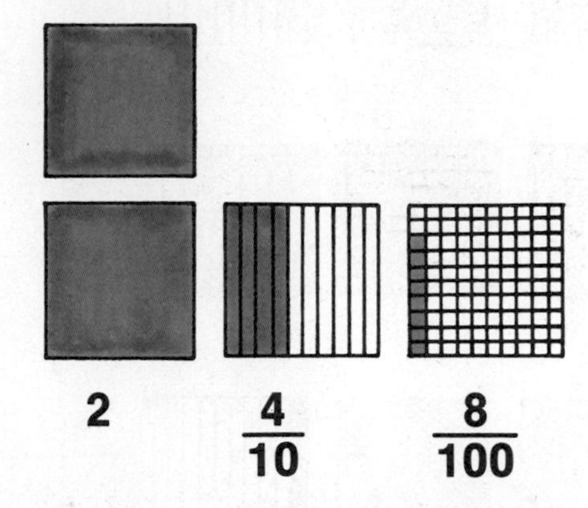

2 $\quad \frac{4}{10} \quad \frac{8}{100}$

Ones	Tenths	Hundredths
2	4	8

2.48

two and forty-eight hundredths

EXAMPLE 2.

1.06

one and six hundredths

EXAMPLE 3.

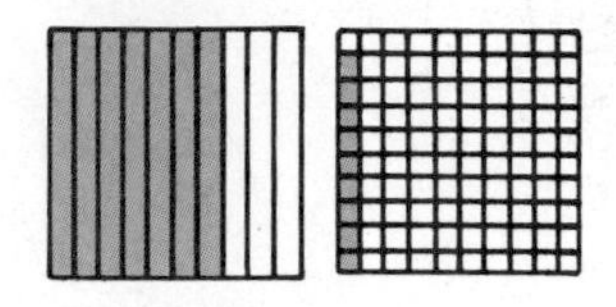

0.79

seventy-nine hundredths

EXERCISES

Multiple choice. Choose the correct letter.

1.   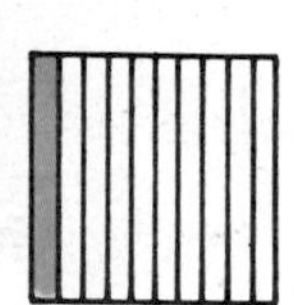

a. 2.01 **b.** 2.1 **c.** 1.02

2.  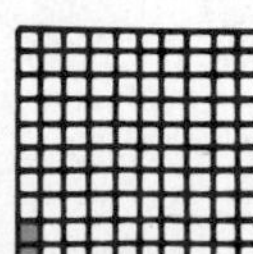

a. 2.01 **b.** 2.1 **c.** 1.02

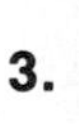

3.   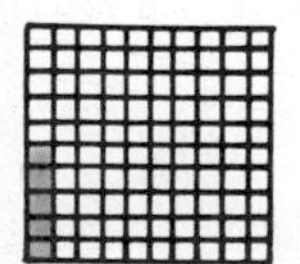

a. 1.25 **b.** 2.15 **c.** 2.05

How many unit squares are painted?
Give answers as decimals.

4. 5.

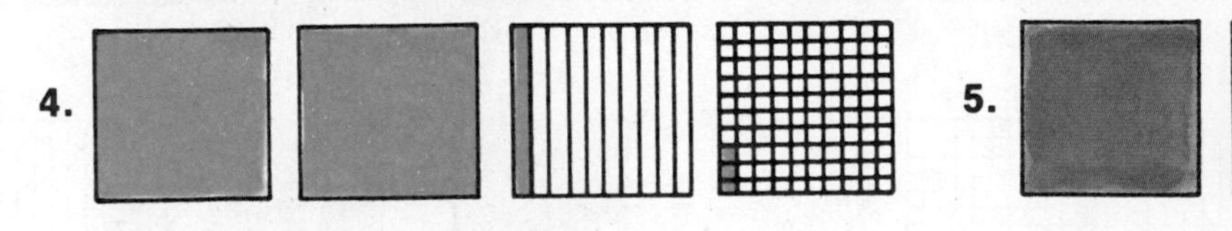

6.

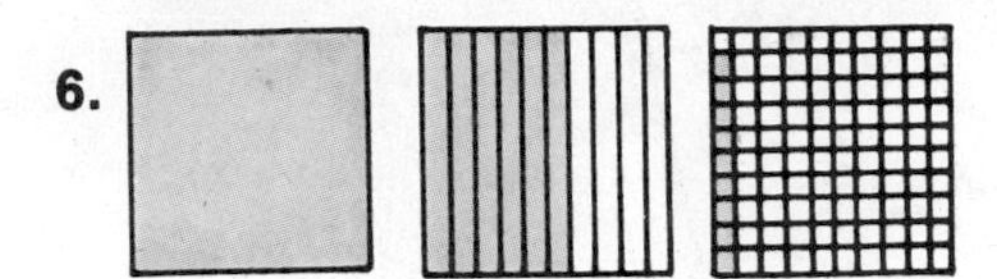

7.

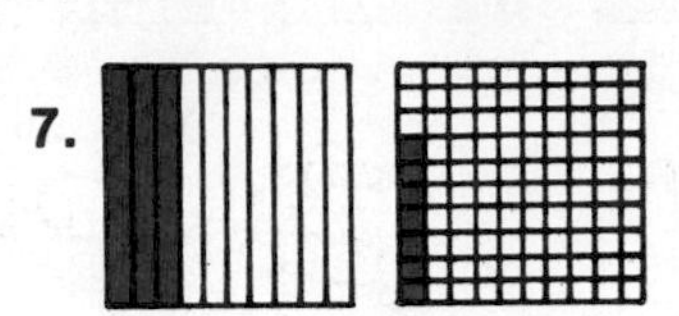

What digit is in the

8. tens place?
9. hundredths place?
10. hundreds place?
11. tenths place?

Write the decimal.

12. five and eight tenths
13. twenty-three and five tenths
14. nine and thirty-two hundredths
15. forty and twelve hundredths
16. thirty-four and six hundredths
17. fifty-eight hundredths

Build a decimal.

18. **4** in the **tenths** place
    2 in the **ones** place

19. **3** in the **tens** place
    6 in the **tenths** place
    0 in the **ones** place

20. **5** in the **tens** place
    2 in the **ones** place
    1 in the **tenths** place
    4 in the **hundreds** place
    3 in the **hundredths** place

21. **9** in the **tenths** place
    3 in the **ones** place
    8 in the **hundreds** place
    7 in the **tens** place
    1 in the **hundredths** place

Ordering and comparing decimals

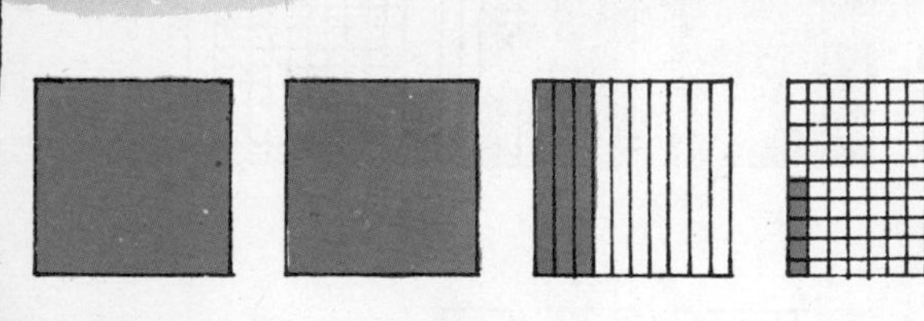

EXERCISES

Less than (<) or greater than (>)?

1. 1 ● 0.1	**2.** 0.3 ● 0.7	**3.** 0.8 ● 0.81
4. 0.9 ● 1	**5.** 2 ● 2.1	**6.** 0.2 ● 2
7. 2.35 ● 2.26	**8.** 3.46 ● 3.4	**9.** 5.37 ● 5.4
10. 17.8 ● 1.78	**11.** 0.14 ● 13	**12.** 15.02 ● 15.2

Give the next three numbers.

13. 2.3, 2.4, 2.5, ?, ?, ?	**14.** 15.8, 15.9, 16.0, ?, ?, ?
15. 8.6, 8.7, 8.8, ?, ?, ?	**16.** 6.02, 6.03, 6.04, ?, ?, ?
17. 5.95, 5.96, 5.97, ?, ?, ?	**18.** 9.96, 9.97, 9.98, ?, ?, ?

Give a decimal that is between the two numbers.

19. 3, 4	**20.** 5, 6	**21.** 9, 10	**22.** 11, 12
23. 2.1, 2.3	**24.** 15.4, 15.6	**25.** 19.5, 19.6	**26.** 17.8, 18.0
27. 23.9, 24.0	**28.** 25.37, 25.40	**29.** 28.99, 29.01	★ **30.** 40.98, 40.99

Complete.

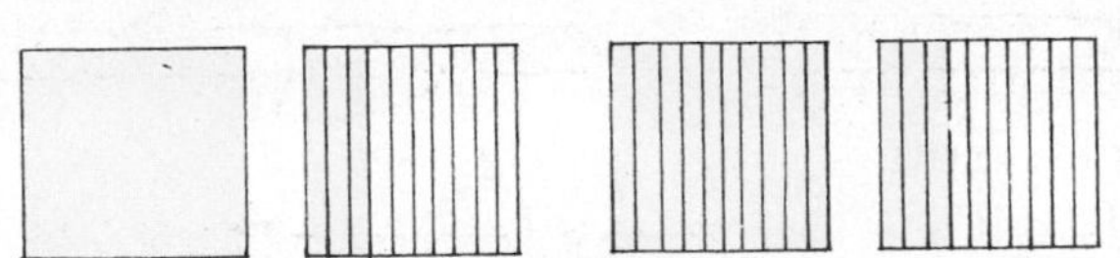

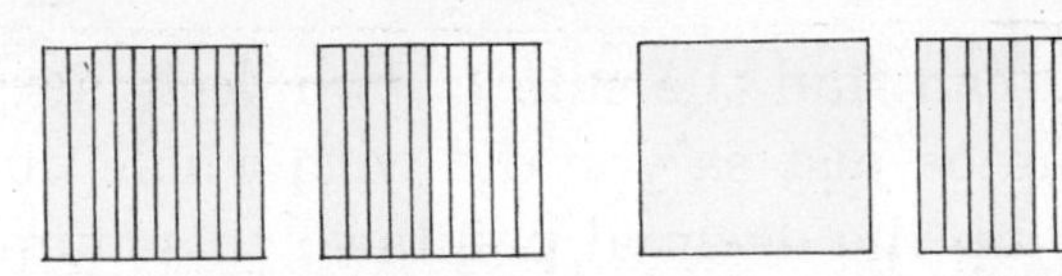

31. 1 one and 3 tenths = ? tenths

32. 15 tenths = ? one and ? tenths

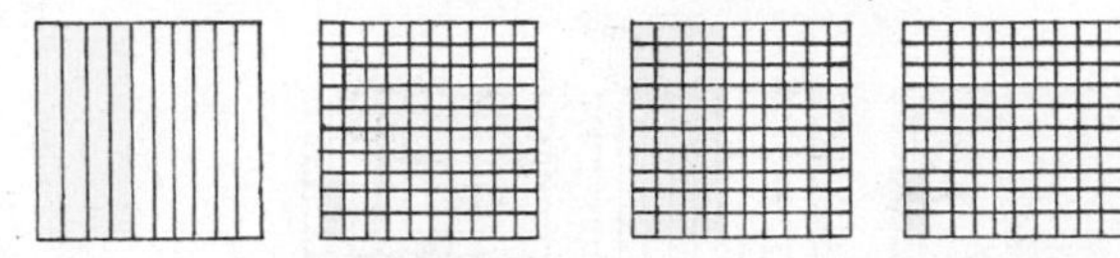

33. 4 tenths and 3 hundredths = ? hundredths

Challenge!

Forgetful Freddie wrote these "facts" about himself. Of course he forgot the decimal points.

Copy each numeral, and place the decimal point so the "fact" makes sense.

34. I am 105 years old.
35. My height is 1206 centimeters.
36. My weight is 2725 kilograms.
37. I drank 25 liters of milk for breakfast.
38. I took 245 minutes to eat breakfast.
39. I can high-jump 756 centimeters.

KEEPING SKILLS SHARP

1. 62×28
2. 59×42
3. 74×29
4. 68×50
5. 93×39
6. 118×16
7. 205×24
8. 400×48
9. 326×39
10. 742×26

Adding decimals

You can find the sum of two decimals by adding in columns just as you did with whole numbers. You line up the decimal points so that numerals with the same place value are lined up.

EXAMPLE 1.

1.43 + **1.13**

Step 1.
Line up the decimal points.

$$\begin{array}{r} 1.43 \\ +1.13 \\ \hline \end{array}$$

Step 2.
Add hundredths.

$$\begin{array}{r} 1.43 \\ +1.13 \\ \hline 6 \end{array}$$

Step 3.
Add tenths.

$$\begin{array}{r} 1.43 \\ +1.13 \\ \hline .56 \end{array}$$

Step 4.
Add ones.

$$\begin{array}{r} 1.43 \\ +1.13 \\ \hline 2.56 \end{array}$$

Put in the decimal point.

When a sum is 10 or greater, you will need to regroup.

EXAMPLE 2.

2.36 + **1.87**

Step 1.
Line up the decimal points to add.

$$\begin{array}{r} 2.36 \\ +1.87 \\ \hline \end{array}$$

Step 2.
Add hundredths and regroup.

$$\begin{array}{r} \scriptstyle 1 \\ 2.36 \\ +1.87 \\ \hline 3 \end{array}$$

Step 3.
Add tenths and regroup.

$$\begin{array}{r} \scriptstyle 1\;\;1 \\ 2.36 \\ +1.87 \\ \hline .23 \end{array}$$

Step 4.
Add ones.

$$\begin{array}{r} \scriptstyle 1\;\;1 \\ 2.36 \\ +1.87 \\ \hline 4.23 \end{array}$$

Put in the decimal point.

EXERCISES

Add.

1. $\begin{array}{r} 2.6 \\ +3.1 \\ \hline \end{array}$
2. $\begin{array}{r} 5.3 \\ +2.5 \\ \hline \end{array}$
3. $\begin{array}{r} 0.74 \\ +0.21 \\ \hline \end{array}$
4. $\begin{array}{r} 0.53 \\ +0.42 \\ \hline \end{array}$
5. $\begin{array}{r} 1.5 \\ +0.4 \\ \hline \end{array}$
6. $\begin{array}{r} 3.5 \\ +2.8 \\ \hline \end{array}$
7. $\begin{array}{r} 0.39 \\ +0.67 \\ \hline \end{array}$
8. $\begin{array}{r} 0.48 \\ +0.75 \\ \hline \end{array}$
9. $\begin{array}{r} 9.2 \\ +6.3 \\ \hline \end{array}$
10. $\begin{array}{r} 4.8 \\ +7.6 \\ \hline \end{array}$
11. $\begin{array}{r} 32.8 \\ +53.6 \\ \hline \end{array}$
12. $\begin{array}{r} 2.95 \\ +4.83 \\ \hline \end{array}$
13. $\begin{array}{r} 69.1 \\ +78.2 \\ \hline \end{array}$
14. $\begin{array}{r} 5.76 \\ +2.18 \\ \hline \end{array}$
15. $\begin{array}{r} 3.49 \\ +2.38 \\ \hline \end{array}$
16. $\begin{array}{r} 58.4 \\ +26.8 \\ \hline \end{array}$
17. $\begin{array}{r} 3.95 \\ +3.17 \\ \hline \end{array}$
18. $\begin{array}{r} 76.2 \\ +29.8 \\ \hline \end{array}$
19. $\begin{array}{r} 9.43 \\ +1.57 \\ \hline \end{array}$
20. $\begin{array}{r} 75.1 \\ +85.9 \\ \hline \end{array}$

Give each sum.
Line up the decimal points!

21. 56.3 + 38.5
22. 7.96 + 2.87
23. 18.6 + 9.7
24. 28.92 + 3.74
25. 38.29 + 6.54
26. 72.36 + 9.57
27. 70.6 + 8.9
28. 6.23 + 57.85
29. 4.8 + 37.6
30. 7.59 + 5.38
31. 6.55 + 1.34
32. 88.8 + 21.6
33. 3.74 + 7.43
34. 92.1 + 56.9
35. 20.6 + 9.8
36. 43.18 + 7.83
37. 52.6 + 34.7
38. 300.4 + 75.8

Solve.

39. Ms. Weaver drove 236.8 kilometers before lunch and 348.5 kilometers after lunch. How far did she drive that day?
40. She bought 42.6 liters of gas at one gas station and 37.9 liters of gas at another. How much gas did she buy in all?

Subtracting decimals

You can find the difference of two decimals by subtracting in columns just as you did with whole numbers.

EXAMPLE 1.

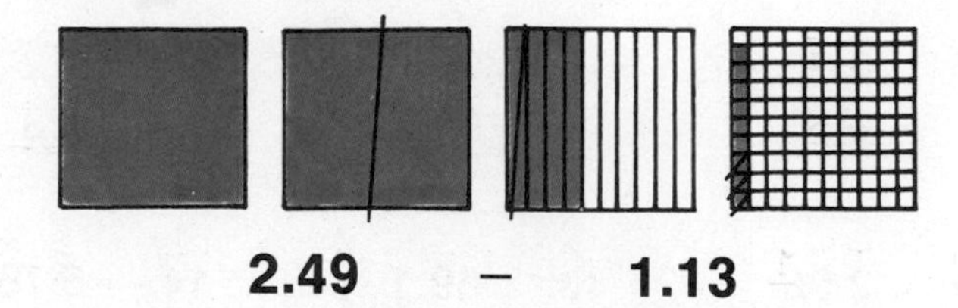

2.49 − 1.13

Step 1. Line up the decimal points.

$$\begin{array}{r} 2.49 \\ -1.13 \\ \hline \end{array}$$

Step 2. Subtract hundredths.

$$\begin{array}{r} 2.49 \\ -1.13 \\ \hline 6 \end{array}$$

Step 3. Subtract tenths.

$$\begin{array}{r} 2.49 \\ -1.13 \\ \hline .36 \end{array}$$

Step 4. Subtract ones.

$$\begin{array}{r} 2.49 \\ -1.13 \\ \hline 1.36 \end{array}$$

Remember to put in the decimal point.

Sometimes you have to regroup.

EXAMPLE 2.

3.36 − 1.49

Step 1. Line up the decimal points to subtract.

$$\begin{array}{r} 3.36 \\ -1.49 \\ \hline \end{array}$$

Step 2. Regroup and subtract hundredths.

$$\begin{array}{r} 2\ 16 \\ 3.\not{3}\not{6} \\ -1.49 \\ \hline 7 \end{array}$$

Step 3. Regroup and subtract tenths.

$$\begin{array}{r} 12\ \ \\ 2\ \not{2}\ 16 \\ \not{3}.\not{3}\not{6} \\ -1.49 \\ \hline .87 \end{array}$$

Step 4. Subtract ones.

$$\begin{array}{r} 12\ \ \\ 2\ \not{2}\ 16 \\ \not{3}.\not{3}\not{6} \\ -1.49 \\ \hline 1.87 \end{array}$$

EXERCISES

Subtract.

1. $\begin{array}{r} 0.83 \\ -0.21 \\ \hline \end{array}$ 2. $\begin{array}{r} 7.5 \\ -6.2 \\ \hline \end{array}$ 3. $\begin{array}{r} 0.98 \\ -0.15 \\ \hline \end{array}$ 4. $\begin{array}{r} 6.9 \\ -2.4 \\ \hline \end{array}$ 5. $\begin{array}{r} 0.87 \\ -0.33 \\ \hline \end{array}$

6. $\begin{array}{r} 7.2 \\ -3.7 \\ \hline \end{array}$ 7. $\begin{array}{r} 0.81 \\ -0.39 \\ \hline \end{array}$ 8. $\begin{array}{r} 6.3 \\ -2.8 \\ \hline \end{array}$ 9. $\begin{array}{r} 9.2 \\ -4.7 \\ \hline \end{array}$ 10. $\begin{array}{r} 0.54 \\ -0.23 \\ \hline \end{array}$

11. $\begin{array}{r} 42.5 \\ -23.8 \\ \hline \end{array}$ 12. $\begin{array}{r} 93.1 \\ -45.6 \\ \hline \end{array}$ 13. $\begin{array}{r} 30.7 \\ -21.8 \\ \hline \end{array}$ 14. $\begin{array}{r} 5.43 \\ -1.97 \\ \hline \end{array}$ 15. $\begin{array}{r} 7.85 \\ -1.39 \\ \hline \end{array}$

Give each difference.
Line up the decimal points.

16. 18.3 − 12.4
17. 6.73 − 2.89
18. 65.2 − 28.0
19. 5.64 − 2.39
20. 93.0 − 76.4
21. 3.12 − 1.26
22. 18.6 − 9.4
23. 10.3 − 6.7
24. 89.6 − 24.8
25. 15.23 − 5.74
26. 34.8 − 9.9
27. 3.25 − 1.46

Solve.

28. John high-jumped 1.13 meters and Paul high-jumped 1.06 meters. How much higher did John jump?

29. Jane swam across the pool in 9.6 seconds and then swam back in 10.8 seconds. How long did it take her to swim across and back?

KEEPING SKILLS SHARP

Add or subtract. Give answers in lowest terms.

1. $\frac{1}{8} + \frac{3}{8}$
2. $\frac{7}{10} - \frac{1}{10}$
3. $\frac{7}{8} - \frac{1}{2}$
4. $\frac{5}{6} - \frac{4}{9}$
5. $\frac{3}{10} + \frac{2}{5}$
6. $\frac{2}{3} + \frac{1}{6}$
7. $\frac{2}{3} + \frac{1}{4}$
8. $\frac{2}{5} + \frac{1}{3}$
9. $\frac{1}{3} + \frac{2}{9}$
10. $\frac{5}{6} - \frac{1}{4}$
11. $\frac{1}{2} + \frac{1}{5}$
12. $\frac{9}{10} - \frac{1}{2}$

More about adding and subtracting decimals

To add or subtract, line up the decimal points.

EXERCISES

Add.

1. 55.8 + 22.0	**2.** 38.0 + 42.9	**3.** 6.38 + 2.50	**4.** 7.30 + 3.59	**5.** 5.00 + 2.63
6. 82.6 + 23	**7.** 5.78 + 2.4	**8.** 39.7 + 25	**9.** 2.54 + 3.2	**10.** 4.93 + 2

11. 38.6 + 22 **12.** 1.8 + 2.54 **13.** 34.9 + 26

14. 27 + 5.08 **15.** 32.6 + 0.38 **16.** 0.06 + 24.5

Subtract.

17. 9.80 − 5.27
18. 6.50 − 2.74
19. 3.34 − 1.6
20. 82.0 − 14.5
21. 25.9 − 19
22. 37 − 24.65
23. 21 − 17.38
24. 48.1 − 26.35
25. 75.32 − 35.8
26. 93 − 25.32
27. 53.2 − 27
28. 78 − 21.6
29. 42 − 38.5
30. 2.36 − 1.5
31. 5.28 − 2.6
32. 9.3 − 0.05

Solve.

33. A tennis racket is on sale for $13. A can of balls is on sale for $2.79. How much will both cost?

34. John bought 8 fishhooks for $1.84. How much did one fishhook cost?

35. Mary had a ten-dollar bill. She bought a baseball glove for $8.67. How much money was left?

36. Tennis balls are on sale for $.86 each. How much will 12 cost?

37. Joanna got $20 for her birthday. She wanted to buy tennis shoes for $29.65. How much more money did she need?

38. Tennis socks are on sale for $1.89 a pair. How much will 6 pairs cost?

39. Diana bought 3 cans of balls for $2.79 each and a racket for $16.38. What was the total cost?

40. Terry bought 2 pairs of tennis shorts for $9.42 each. She gave the clerk $20. How much change should she get back?

KEEPING SKILLS SHARP

1. 20)400
2. 21)903
3. 19)912
4. 32)992
5. 29)725
6. 33)853
7. 24)964
8. 34)755
9. 25)800
10. 23)776
11. 42)803
12. 30)900
13. 22)599
14. 18)653
15. 41)943

Problem solving—interpreting a table

Mary and her father went to the auto races. The results of the race are shown in the table.

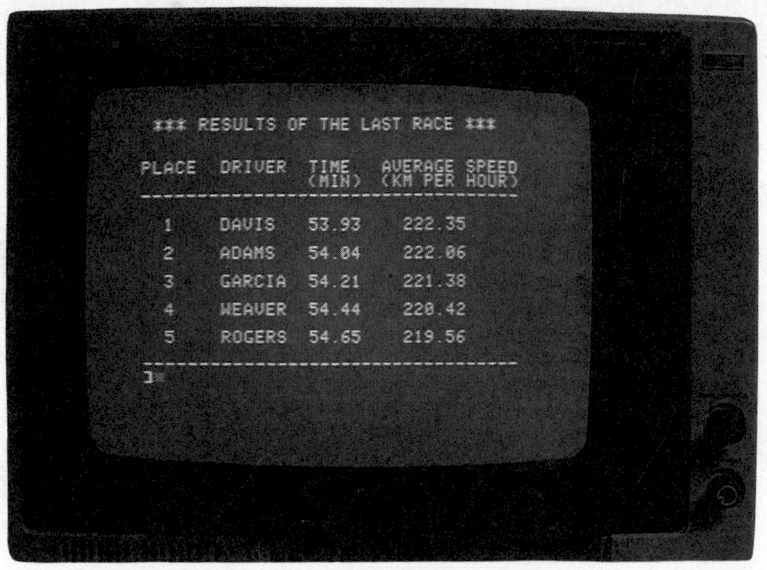

*** RESULTS OF THE LAST RACE ***

PLACE	DRIVER	TIME (MIN)	AVERAGE SPEED (KM PER HOUR)
1	DAVIS	53.93	222.35
2	ADAMS	54.04	222.06
3	GARCIA	54.21	221.38
4	WEAVER	54.44	220.42
5	ROGERS	54.65	219.56

1. Who was the winner?
2. How many minutes did it take the winning car to run the race?
3. What was the average speed of the fourth-place car?
4. The last-place car took how much longer than the first-place car?
5. What was the difference in the average speed of the first- and last-place cars?
6. How many minutes behind the winning car was Adams?
7. Which driver averaged 1.82 kilometers per hour less than Garcia?
8. Which driver took 0.4 min longer than Adams?

Computers at the grocery store

A small computer has many uses in a grocery store. The meat and produce departments use the computer to figure prices. The computer performs the necessary computation rapidly and accurately. A clerk places the item on the computer's scale and enters the price per pound, and the computer prints a label with the total price.

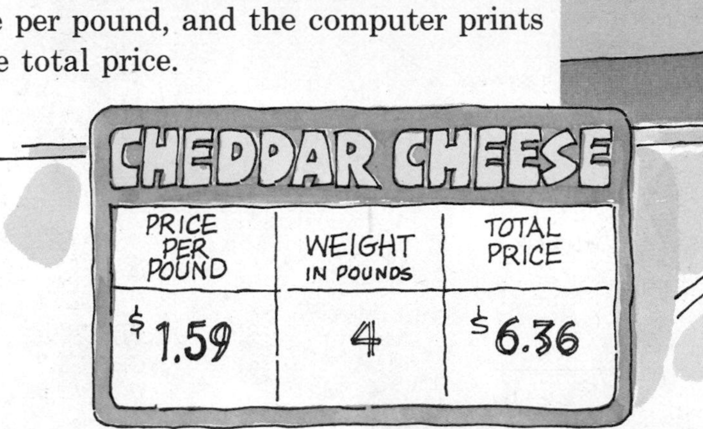

EXERCISES

Find the total price.

1. CHEESE

PRICE PER POUND	WEIGHT IN POUNDS	TOTAL PRICE
$1.59	3	

2. CARROTS

PRICE PER POUND	WEIGHT IN POUNDS	TOTAL PRICE
$ .49	4	

3. BANANAS

PRICE PER POUND	WEIGHT IN POUNDS	TOTAL PRICE
$ .47	3	

4. HAMBURGER

PRICE PER POUND	WEIGHT IN POUNDS	TOTAL PRICE
$1.95	3	

5. MUSHROOMS

PRICE PER POUND	WEIGHT IN POUNDS	TOTAL PRICE
$1.42	2	

6. POTATOES

PRICE PER POUND	WEIGHT IN POUNDS	TOTAL PRICE
$ .16	7	

The computer at the checkout counter uses the prices in its memory to compute sales tax, the total due, and the amount of change.

Solve.

7. Find the total of the grocery items purchased.
8. How many bakery items were purchased? Find the bakery total.
9. Find the total of the meat items purchased.
10. How much was the total?
11. How much change was returned?

GROCERY	.50
GROCERY	1.09
GROCERY	1.09
BAKERY	.25
GROCERY	.25
GROCERY	.30
MEAT	2.03
MEAT	2.78
BAKERY	.95
BAKERY	.78
GROCERY	1.19
GROCERY	.47
GROCERY	.43
VEGETABLE	.35
FRUIT	.75
GROCERY	.79
GROCERY	.59
GROCERY	.49
GROCERY	.97
VEGETABLE	.49
GROCERY	1.99
TOTAL	18.53
CASH RECEIVED	20.00
CHANGE	1.47

THANK YOU - COME AGAIN

CHAPTER CHECKUP

[pages 328–333]

In 534.27, what digit is in the

534.27

1. tens place?
2. hundredths place?
3. ones place?
4. tenths place?
5. hundreds place?

Add. [pages 334–335, 338–339]

6. 3.8 + 2.1	**7.** 0.46 + 0.23	**8.** 7.8 + 1.4	**9.** 5.9 + 6.8	**10.** 0.34 + 0.68
11. 35.2 + 16.5	**12.** 7.58 + 2.97	**13.** 52.6 + 39.4	**14.** 78.84 + 34.2	**15.** 5.9 + 3.09

Subtract. [pages 336–339]

16. 5.8 − 2.4	**17.** 0.76 − 0.33	**18.** 8.2 − 5.8	**19.** 7.0 − 5.3	**20.** 0.92 − 0.84
21. 76.3 − 48.5	**22.** 9.02 − 3.75	**23.** 5.72 − 2.8	**24.** 32 − 16.5	**25.** 5.8 − 2.37

Solve. [pages 340–343]

26.

How much for both?

27.

How much more does the large record cost?

CHAPTER PROJECT

Work with a group of classmates.

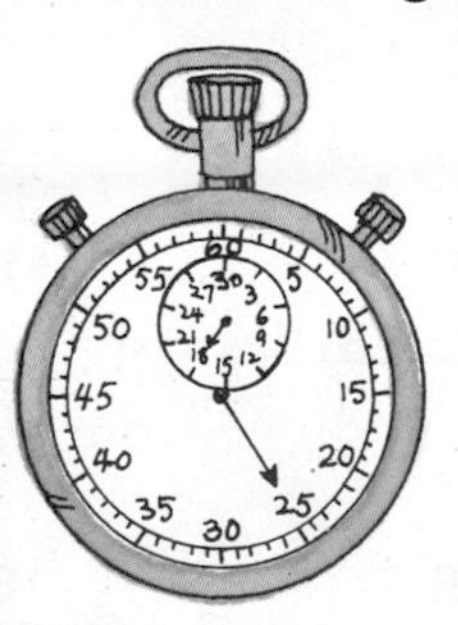

1. Measure off 40 meters in the schoolyard.
2. Get a stopwatch.
3. Have each classmate run a 40-meter dash twice. Record the better time.
4. Show the times on a graph.
5. Tell some things that your graph shows.

CHAPTER REVIEW

What digit is in the

1. ones place?
2. hundredths place?
3. tenths place?

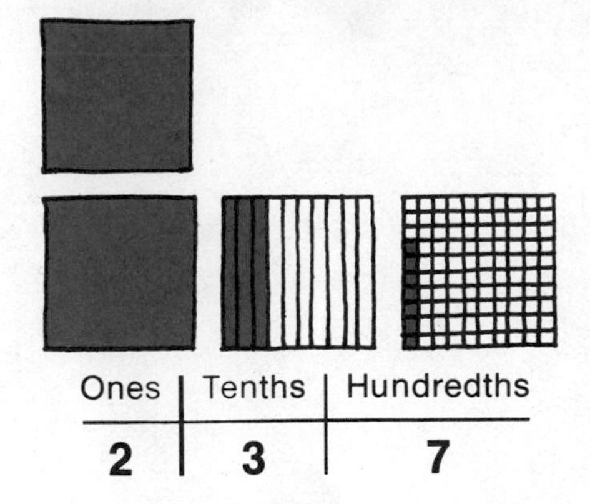

2.37

Use a decimal to tell how many squares are painted.

4.  

5.  

6\.

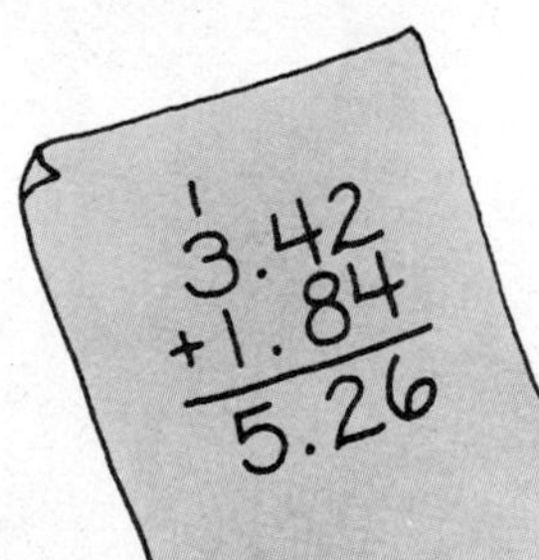

Add.

7. $2.3 + 4.2$
8. $6.8 + 2.6$
9. $0.57 + 0.69$
10. $5.38 + 4.95$
11. $2.5 + 9.75$
12. $65.4 + 32$

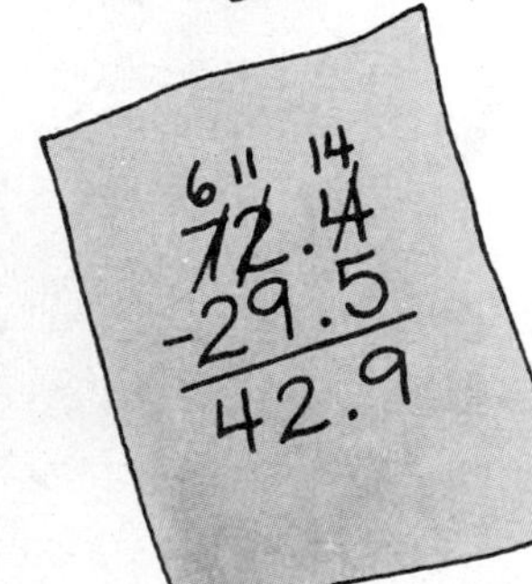

Subtract.

13. $5.9 - 2.6$
14. $0.74 - 0.28$
15. $8.4 - 3.7$
16. $4.26 - 1.59$
17. $6.2 - 1.75$
18. $8 - 7.65$

CHAPTER CHALLENGE

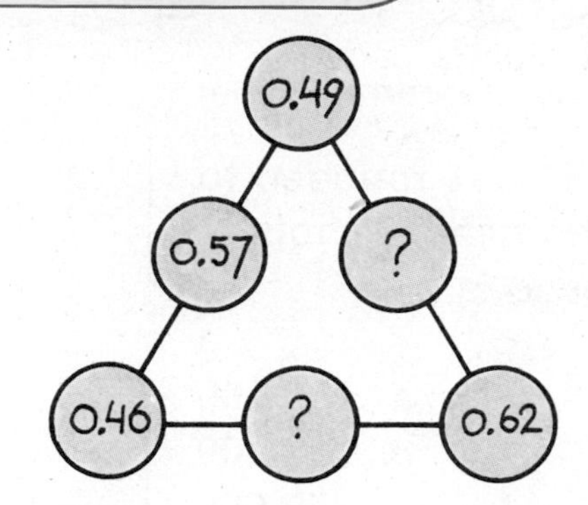

1. Copy and fill in the circles so that the sums along the three sides of the triangle are the same.

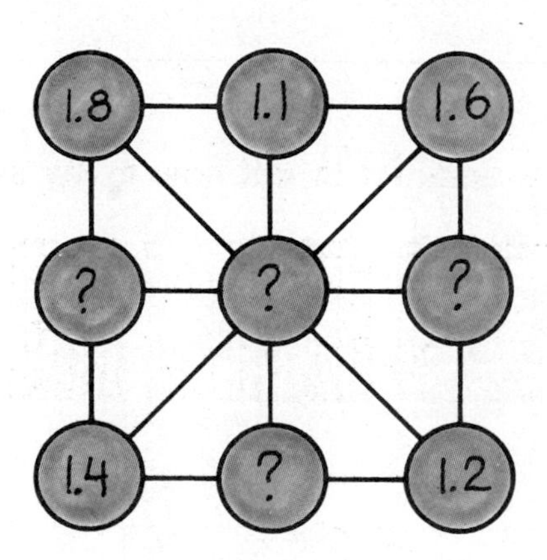

2. Copy and fill in the circles so that the sums along each row, column, and diagonal are the same.

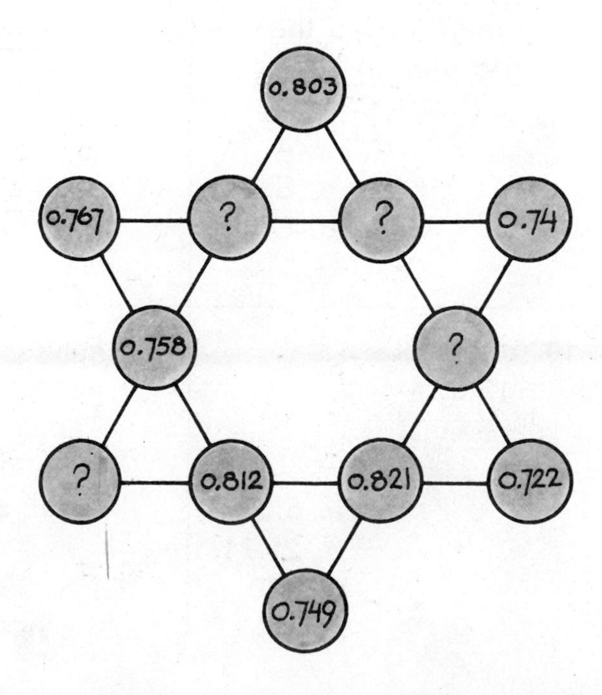

3. Copy and fill in the circles so that the sums along the rows are all the same.

Choose the correct letter.

1. 752,831 rounded to the nearest thousand is
   a. 750,000
   b. 752,000
   c. 753,000
   d. none of these

2. Add.
$$\begin{array}{r} 26 \\ 34 \\ 52 \\ 87 \\ +54 \\ \hline \end{array}$$
   a. 253
   b. 233
   c. 243
   d. none of these

3. $\frac{6}{12}$ in lowest terms is
   a. $\frac{3}{6}$
   b. $\frac{2}{4}$
   c. $\frac{1}{2}$
   d. none of these

4. Add.
$\frac{1}{3} + \frac{2}{9}$
   a. $\frac{5}{9}$
   b. $\frac{1}{4}$
   c. $\frac{1}{3}$
   d. none of these

5. Subtract.
$\frac{3}{4} - \frac{3}{8}$
   a. $\frac{0}{8}$
   b. $\frac{0}{4}$
   c. $\frac{3}{8}$
   d. none of these

6. Multiply.
$$\begin{array}{r} 726 \\ \times 48 \\ \hline \end{array}$$
   a. 34,848
   b. 34,648
   c. 34,508
   d. none of these

7. Which triangle is congruent to the red triangle?

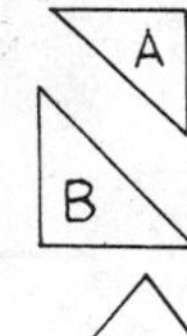

   a. A
   b. B
   c. C
   d. none of these

8. What is the remainder?
$32\overline{)583}$
   a. 18
   b. 9
   c. 7
   d. none of these

9. Add.
$$\begin{array}{r} 18.62 \\ +9.5 \\ \hline \end{array}$$
   a. 27.12
   b. 28.12
   c. 9.12
   d. none of these

10. Subtract.
$$\begin{array}{r} 13.05 \\ -8.66 \\ \hline \end{array}$$
   a. 5.39
   b. 21.71
   c. 439
   d. none of these

11. Records: \$3.75 each
Tapes: \$4.50 each
How much do 12 tapes cost?
   a. \$99 b. \$54
   c. \$45 d. none of these

12. Bob ran the race in 8.3 seconds. Eve ran it in 7.8 seconds. How many seconds faster was Eve's time?
   a. 1.5 b. 0.5
   c. 16.1 d. none of these

RESOURCES

GETTING READY FOR COMPUTER PROGRAMMING

Using commands in programs

You write a **program** to tell the moon mobile where to move. The directions you write in the program are called **commands.**

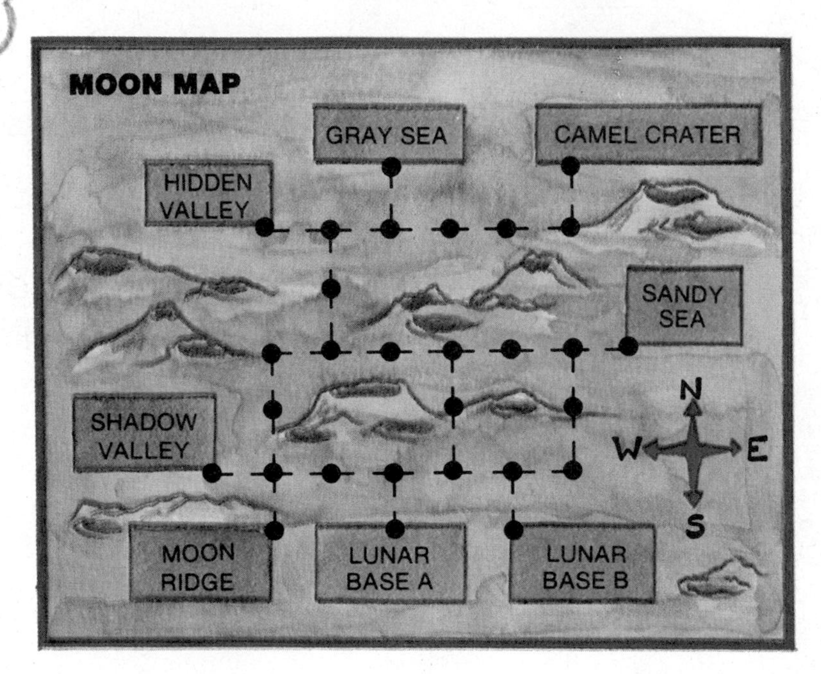

1. Use your finger to follow the path on the map.

PROGRAM BASE A/HIDDEN
BEGIN LUNAR BASE A
N1 ... Go north 1 space.
E1
N2 ... Go north 2 spaces.
W2
N2
W1
END HIDDEN VALLEY

a. Where did you begin?

b. Where did you end?

Use your finger to follow the path on the map. Complete each program.

2. PROGRAM HIDDEN/ ____a____
   BEGIN HIDDEN VALLEY
   E1
   S2
   W1
   S3
   END ____b____

3. PROGRAM MOON/ ____a____
   BEGIN MOON RIDGE
   N1
   E5
   N2
   E1
   END ____b____

There is a mistake in each problem. Rewrite each program correctly.

4. PROGRAM SHADOW/SANDY
   BEGIN SHADOW VALLEY
   E6
   N3
   E1
   END SANDY SEA

5. PROGRAM SHADOW/CAMEL
   BEGIN SHADOW VALLEY
   E2
   N2
   E1
   N2
   E4
   N1
   END CAMEL CRATER

Complete each program.

6. PROGRAM BASE B/GRAY
   BEGIN LUNAR BASE B
   N1
   W4
   N2
   ____a____
   ____b____
   ____c____
   ____d____
   END GRAY SEA

★7. There are 3 different ways to get from Lunar Base B to Camel Crater using 9 commands. Write a program for one of these ways.

New commands—take and leave

You can pick up a rock by using the command TAKE ROCK. You can put down the rock by using the command LEAVE ROCK.

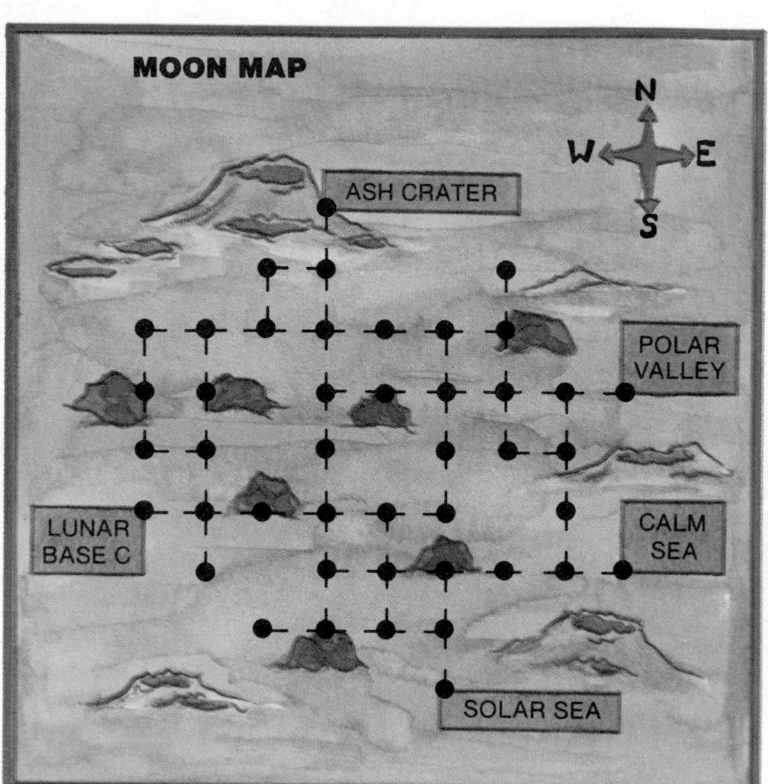

1. **Use your finger to follow the path on the map.**

   PROGRAM CALM/SOLAR
   BEGIN CALM SEA
   W3
   TAKE ROCK ... Use this command to pick up a rock.
   S2
   LEAVE ROCK ... Use this command to leave a rock.
   END SOLAR SEA

   a. Where did you begin?
   b. Where did you leave the rock?
   c. Where did you end?

Use your finger to follow the path on the map.
Complete the program.

2. PROGRAM SOLAR/ ___a___
BEGIN SOLAR SEA
N1
W2
TAKE ROCK
N7
LEAVE ROCK
END ___b___

3. PROGRAM ASH/ ___a___
BEGIN ASH CRATER
S3
E1
TAKE ROCK
E4
LEAVE ROCK
END ___b___

There is a mistake in each program. Rewrite the program correctly.

4. PROGRAM POLAR/BASE C
BEGIN POLAR VALLEY
W5
N1
W2
S3
TAKE ROCK
S2
W1
LEAVE ROCK
END LUNAR BASE C

5. PROGRAM SOLAR/ASH
BEGIN SOLAR SEA
N1
W1
N2
W2
TAKE ROCK
E1
N5
LEAVE ROCK
END POLAR VALLEY

Complete each program.

6. PROGRAM BASE C/ASH
BEGIN LUNAR BASE C
E1
N1
W1
N1
TAKE ROCK
___a___
___b___
___c___
LEAVE ROCK
___d___

★7. Write a program to go from Solar Sea to Polar Valley that does not go past any rocks.

New commands—jump and go back

You can use the command JUMP to make the moon mobile jump over a crater in its path. You can use the command GO BACK to make the moon mobile return to Lunar Storage along the same path.

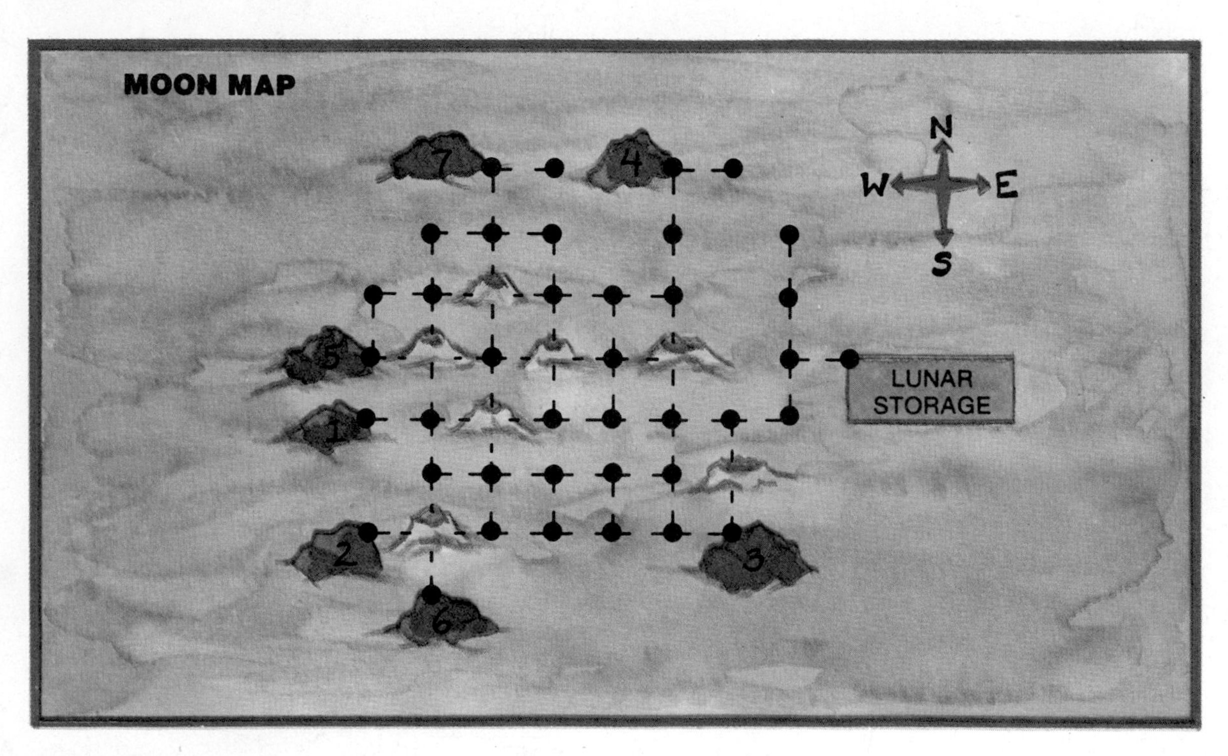

1. Use your finger to follow the path on the map.

PROGRAM ROCK 1
BEGIN LUNAR STORAGE
W1
S1
W4
JUMP W . . . This command tells the moon mobile to jump west and land on the next dot.
W1
TAKE ROCK
GO BACK . . . Go back along the same path.
LEAVE ROCK
END LUNAR STORAGE

a. Where did you begin?
b. Which rock did you pick up?
c. Where did you leave the rock?
d. Where did you end?

Use your finger to follow the path on the map.
Complete each program.

2. PROGRAM ROCK 2
   BEGIN LUNAR STORAGE
   W1
   S1
   W2
   S2
   W3
   JUMP __a__
   TAKE ROCK
   GO BACK
   LEAVE ROCK
   END __b__

3. PROGRAM ROCK __a__
   BEGIN LUNAR STORAGE
   W1
   S1
   W1
   JUMP S
   TAKE ROCK
   __b__
   LEAVE ROCK
   END LUNAR STORAGE

4. PROGRAM ROCK __a__
   BEGIN LUNAR STORAGE
   W1
   S1
   W2
   S1
   W4
   JUMP __b__
   TAKE ROCK
   __c__
   LEAVE ROCK
   END LUNAR STORAGE

5. PROGRAM ROCK __a__
   BEGIN LUNAR STORAGE
   W1
   S1
   W2
   JUMP N
   N2
   TAKE ROCK
   __b__
   __c__
   END __d__

★6.

Procedures

In this lesson, the moon mobile will always begin at Lunar Storage. After the moon mobile takes a rock back to Lunar Storage, it must go to the Solar Station to recharge its battery. You can use the command RECHARGE to tell the moon mobile to recharge its battery.

Since the path from Lunar Storage to the Solar Station is always the same, we will write a procedure for the path.

A procedure is a list of commands that is used often. Each procedure has a name. You can use the name of a procedure as a command in any program.

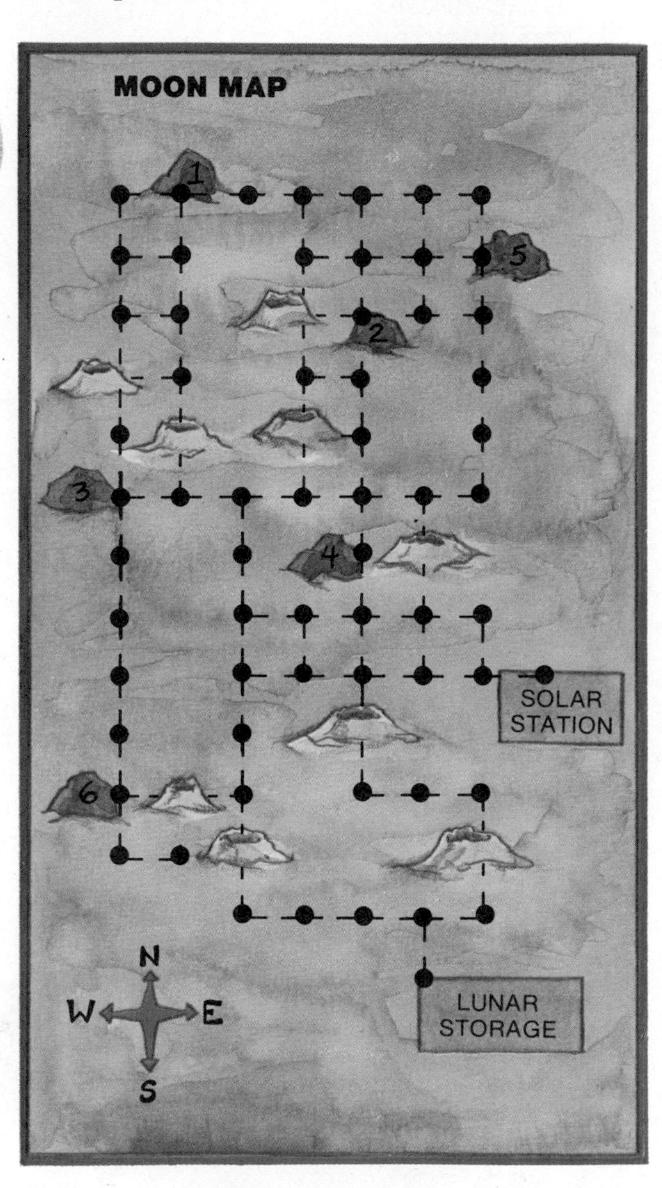

1. Use your finger to follow the path on the map.

PROCEDURE ENERGIZE
BEGIN LUNAR STORAGE
N1
E1
JUMP N
W2
JUMP N
E3
RECHARGE
GO BACK

a. What is the name of the procedure?
b. Where did you begin?
c. Where were you when the command RECHARGE was given?
d. Where did you end?

The procedure ENERGIZE is used in each program. Complete each program.

2. PROGRAM ROCK 3
   BEGIN LUNAR STORAGE
   N1
   W3
   ____a____
   N5
   __b__
   TAKE ROCK
   GO BACK
   LEAVE ROCK
   ENERGIZE
   END LUNAR STORAGE

3. PROGRAM ROCK 5
   BEGIN LUNAR STORAGE
   N1
   E1
   JUMP N
   W2
   JUMP N
   E1
   N1
   JUMP N
   E1
   __a__
   ____b____
   GO BACK
   LEAVE ROCK
   ____c____
   END ____d____

There are two mistakes in each program. Rewrite correctly.

4. PROGRAM ROCK 1
   BEGIN LUNAR STORAGE
   N1
   W3
   JUMP W
   N5
   W1
   JUMP N
   N2
   TAKE ROCK
   GO BACK
   LEAVE ROCK
   ENERGIZE
   END LUNAR STORAGE

5. PROGRAM ROCK 4
   BEGIN LUNAR STORAGE
   N1
   W3
   JUMP N
   N4
   E3
   N2
   TAKE ROCK
   GO BACK
   LEAVE ROCK
   ENERGIZE
   END LUNAR STORAGE

Write your own programs.

6. Write the shortest program that will take Rock 6 and leave it at Lunar Storage. Don't forget to Energize!

★7. Write two different programs for getting Rock 2 and returning with it to Lunar Storage. Your programs should jump only 1 crater.

SKILL TEST

	Skill					
1	Basic addition facts, sums through 18	3 +4	5 +6	8 +5	9 +8	6 +9
2	Basic subtraction facts, sums through 18	8 −3	12 −4	9 −7	16 −7	14 −5
3	Addition, no regrouping	52 +21	38 +40	226 +143	190 +406	
4	Addition, one regrouping	47 +29	38 +53	282 +154	352 +165	
5	Addition, two regroupings	78 +46	95 +68	426 +285	763 +159	
6	Addition, more than two regroupings	3821 +1689	7468 +1563	9354 +2897	6827 +3578	
7	Subtraction, no regrouping	58 −22	79 −14	526 −103	958 −325	
8	Subtraction, one regrouping	80 −26	93 −25	538 −153	746 −492	

	Skill					
9	Subtraction, two regroupings	625 −456	431 −238	758 −269	906 −358	
10	Subtraction, more than two regroupings	5263 −1884	8174 −3698	9203 −4735	5374 −2896	
11	Basic multiplication facts, products through 9 × 9	8 ×6	7 ×9	6 ×8	9 ×6	8 ×7
12	Multiplication, no regrouping	23 ×3	12 ×4	42 ×2	20 ×4	
13	Multiplication, regrouping tens to hundreds	83 ×2	40 ×5	32 ×4	74 ×2	
14	Multiplication, two regroupings	59 ×4	78 ×6	65 ×7	94 ×3	
15	Division facts, quotients through 81 ÷ 9	6)42	8)40	7)56	9)72	
16	Division with remainder	7)40	9)35	5)48	6)50	
17	Division, no regrouping	2)46	3)96	4)48	3)60	

18	Division, one regrouping	$6\overline{)84}$	$5\overline{)95}$	$3\overline{)81}$	$4\overline{)56}$
19	Division with remainder	$5\overline{)74}$	$4\overline{)81}$	$6\overline{)93}$	$3\overline{)59}$
20	Fraction of a number	$\frac{1}{2}$ of 18 = ?	$\frac{2}{3}$ of 12 = ?	$\frac{3}{4}$ of 20 = ?	
21	Fractions in lowest terms	$\frac{6}{8} = ?$	$\frac{5}{10} = ?$	$\frac{6}{9} = ?$	$\frac{3}{12} = ?$

22	Adding fractions	$\frac{3}{5} + \frac{1}{5}$	$\frac{5}{12} + \frac{1}{3}$	$\frac{2}{5} + \frac{3}{10}$	$\frac{1}{5} + \frac{2}{3}$	$\frac{3}{8} + \frac{1}{2}$
23	Subtracting fractions	$\frac{5}{9} - \frac{1}{9}$	$\frac{3}{4} - \frac{1}{8}$	$\frac{5}{8} - \frac{1}{2}$	$\frac{5}{12} - \frac{1}{3}$	$\frac{4}{5} - \frac{1}{3}$

24	Changing mixed numbers to fractions	$2\frac{1}{2}$	$1\frac{2}{3}$	$3\frac{1}{4}$	$2\frac{3}{4}$
25	Changing fractions to mixed numbers	$\frac{7}{3}$	$\frac{9}{2}$	$\frac{7}{4}$	$\frac{11}{3}$
26	Multiplication, more than two regroupings	356×4	282×8	7465×6	5698×4

	Skill				
27	Multiplying by 10	$\begin{array}{r}43\\ \times 10\\ \hline\end{array}$	$\begin{array}{r}26\\ \times 10\\ \hline\end{array}$	$\begin{array}{r}226\\ \times 10\\ \hline\end{array}$	$\begin{array}{r}438\\ \times 10\\ \hline\end{array}$
28	Multiplying by tens	$\begin{array}{r}36\\ \times 20\\ \hline\end{array}$	$\begin{array}{r}28\\ \times 40\\ \hline\end{array}$	$\begin{array}{r}152\\ \times 60\\ \hline\end{array}$	$\begin{array}{r}248\\ \times 80\\ \hline\end{array}$
29	Multiplying by a 2-digit number	$\begin{array}{r}58\\ \times 24\\ \hline\end{array}$	$\begin{array}{r}63\\ \times 37\\ \hline\end{array}$	$\begin{array}{r}268\\ \times 82\\ \hline\end{array}$	$\begin{array}{r}472\\ \times 56\\ \hline\end{array}$
30	Dividing a 3-digit number	$4\overline{)362}$	$6\overline{)753}$	$5\overline{)394}$	$8\overline{)742}$
31	Dividing a 4-digit number	$5\overline{)3658}$	$6\overline{)2174}$	$8\overline{)5974}$	$4\overline{)6081}$
32	Dividing by a 2-digit number	$13\overline{)583}$	$22\overline{)792}$	$18\overline{)964}$	$33\overline{)829}$
33	Adding decimals	$\begin{array}{r}7.2\\ +1.4\\ \hline\end{array}$	$\begin{array}{r}3.8\\ +2.9\\ \hline\end{array}$	$\begin{array}{r}0.65\\ +0.34\\ \hline\end{array}$	$\begin{array}{r}0.83\\ +0.57\\ \hline\end{array}$
34	Subtracting decimals	$\begin{array}{r}6.8\\ -2.4\\ \hline\end{array}$	$\begin{array}{r}0.86\\ -0.29\\ \hline\end{array}$	$\begin{array}{r}6.0\\ -2.8\\ \hline\end{array}$	$\begin{array}{r}0.52\\ -0.48\\ \hline\end{array}$

EXTRA PRACTICE

Set 1 **Add.**

1. $3 + 8$ 2. $6 + 8$ 3. $5 + 8$ 4. $9 + 7$

5. $6 + 6$ 6. $4 + 6$ 7. $9 + 8$ 8. $7 + 6$

9. $8 + 8$ 10. $5 + 9$ 11. $9 + 4$ 12. $8 + 3$

13. $3 + 9$ 14. $6 + 5$ 15. $7 + 3$ 16. $8 + 9$

17. $7 + 7$ 18. $9 + 9$ 19. $6 + 9$ 20. $8 + 7$

21. $7 + 9$ 22. $8 + 6$ 23. $7 + 8$ 24. $4 + 9$

Set 2 **Give each missing addend.**

1. $8 + \underline{?} = 10$ 2. $6 + \underline{?} = 9$ 3. $9 + \underline{?} = 13$ 4. $7 + \underline{?} = 15$

5. $9 + \underline{?} = 9$ 6. $9 + \underline{?} = 15$ 7. $2 + \underline{?} = 10$ 8. $5 + \underline{?} = 14$

9. $5 + \underline{?} = 13$ 10. $2 + \underline{?} = 9$ 11. $9 + \underline{?} = 18$ 12. $6 + \underline{?} = 13$

13. $5 + \underline{?} = 10$ 14. $8 + \underline{?} = 17$ 15. $3 + \underline{?} = 10$ 16. $7 + \underline{?} = 16$

17. $3 + \underline{?} = 11$ 18. $7 + \underline{?} = 14$ 19. $9 + \underline{?} = 16$ 20. $4 + \underline{?} = 12$

21. $9 + \underline{?} = 17$ 22. $4 + \underline{?} = 13$ 23. $7 + \underline{?} = 11$ 24. $8 + \underline{?} = 16$

Set 3 **Subtract.**

1. $\begin{array}{r} 8 \\ -3 \\ \hline \end{array}$ 2. $\begin{array}{r} 13 \\ -8 \\ \hline \end{array}$ 3. $\begin{array}{r} 9 \\ -5 \\ \hline \end{array}$ 4. $\begin{array}{r} 10 \\ -4 \\ \hline \end{array}$ 5. $\begin{array}{r} 8 \\ -4 \\ \hline \end{array}$ 6. $\begin{array}{r} 18 \\ -9 \\ \hline \end{array}$

7. $\begin{array}{r} 4 \\ -4 \\ \hline \end{array}$ 8. $\begin{array}{r} 10 \\ -8 \\ \hline \end{array}$ 9. $\begin{array}{r} 14 \\ -7 \\ \hline \end{array}$ 10. $\begin{array}{r} 17 \\ -9 \\ \hline \end{array}$ 11. $\begin{array}{r} 11 \\ -3 \\ \hline \end{array}$ 12. $\begin{array}{r} 10 \\ -5 \\ \hline \end{array}$

13. $\begin{array}{r} 15 \\ -8 \\ \hline \end{array}$ 14. $\begin{array}{r} 12 \\ -7 \\ \hline \end{array}$ 15. $\begin{array}{r} 15 \\ -7 \\ \hline \end{array}$ 16. $\begin{array}{r} 17 \\ -8 \\ \hline \end{array}$ 17. $\begin{array}{r} 14 \\ -9 \\ \hline \end{array}$ 18. $\begin{array}{r} 15 \\ -9 \\ \hline \end{array}$

19. $\begin{array}{r} 16 \\ -7 \\ \hline \end{array}$ 20. $\begin{array}{r} 12 \\ -4 \\ \hline \end{array}$ 21. $\begin{array}{r} 13 \\ -5 \\ \hline \end{array}$ 22. $\begin{array}{r} 15 \\ -6 \\ \hline \end{array}$ 23. $\begin{array}{r} 16 \\ -8 \\ \hline \end{array}$ 24. $\begin{array}{r} 14 \\ -6 \\ \hline \end{array}$

Set 4 How many?

1.

2.

3. 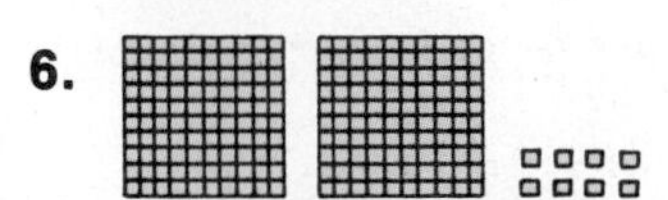

4.

5.

6.

7.

8.

9.

10.

11.

12.

Set 5 Round to the nearest ten.

1. 74	2. 29	3. 15	4. 38	5. 93
6. 67	7. 42	8. 85	9. 56	10. 99
11. 539	12. 646	13. 475	14. 693	15. 868
16. 701	17. 382	18. 807	19. 918	20. 775
21. 444	22. 725	23. 551	24. 605	25. 932
26. 733	27. 395	28. 807	29. 513	30. 697

Set 6 Round to the nearest hundred.

1. 592	2. 159	3. 746	4. 691	5. 205
6. 827	7. 658	8. 917	9. 856	10. 493
11. 646	12. 333	13. 708	14. 175	15. 349
16. 334	17. 568	18. 447	19. 773	20. 499
21. 709	22. 880	23. 645	24. 550	25. 536
26. 260	27. 942	28. 617	29. 867	30. 650

Set 7 < or >?

1. 538 (<) 542
2. 674 ● 647
3. 863 ● 859
4. 5916 ● 5961
5. 3827 ● 2927
6. 7436 ● 7099
7. 3218 ● 3219
8. 8000 ● 7999
9. 9301 ● 9308
10. 6582 ● 6483
11. 3291 ● 3219
12. 5050 ● 5049
13. 9137 ● 9173
14. 7427 ● 7328
15. 6109 ● 6110
16. 3950 ● 4000
17. 8000 ● 7999
18. 4999 ● 5000

Set 8 **Round to the nearest thousand.**

1. 3462 *3000*
2. 5578
3. 7843
4. 6803
5. 2718
6. 1605
7. 8455
8. 4500
9. 36,741
10. 52,833
11. 64,271
12. 75,672
13. 25,389
14. 76,811
15. 48,500
16. 73,999
17. 62,450
18. 47,500
19. 81,490
20. 53,261
21. 94,537
22. 39,621
23. 27,389
24. 79,500

Set 9 **Add.**

1. 42 +25 = *67*
2. 23 +42
3. 53 +30
4. 14 +65
5. 34 +51
6. 321 +208
7. 235 +302
8. 371 +406
9. 418 +240
10. 602 +271
11. 502 +341
12. 633 +132
13. 300 +574
14. 738 +220
15. 521 +400
16. 3046 +1251
17. 2113 +6250
18. 5264 +3521
19. 4360 +2538
20. 8332 +1603

Set 10 Add.

1. 64 +28
2. 36 +45
3. 79 +18
4. 56 +92
5. 83 +60
6. 561 +283
7. 728 +149
8. 364 +182
9. 593 +140
10. 726 +258
11. 426 +490
12. 635 +283
13. 529 +216
14. 704 +249
15. 652 +281
16. 358 +271
17. 496 +283
18. 374 +374
19. 207 +685
20. 856 +119

Set 11 Add.

1. 57 +79
2. 76 +84
3. 93 +39
4. 86 +17
5. 58 +98
6. 325 +195
7. 655 +278
8. 436 +394
9. 653 +179
10. 519 +296
11. 677 +165
12. 335 +186
13. 476 +265
14. 285 +285
15. 729 +167
16. 196 +196
17. 538 +284
18. 398 +389
19. 527 +277
20. 658 +152

Set 12 Add.

1. 5928 +2813
2. 7426 +1593
3. 5426 +1387
4. 2964 +1997
5. 38,219 +26,543
6. 73,691 +28,453
7. 91,782 +26,538
8. 43,027 +15,639
9. 63,849 +42,183
10. 75,368 +75,368
11. 26,854 +97,388
12. 35,485 +84,776
13. 371,829 +564,283
14. 751,208 +349,163
15. 271,846 +593,742
16. 438,296 +341,759

Set 13 Add.

1. 86 29 +57 172	2. 74 48 +25	3. 78 36 +53	4. 95 18 +68	5. 67 67 +67
6. 536 298 +174	7. 607 455 +829	8. 745 386 +537	9. 921 175 +347	10. 568 695 +868
11. 421 175 738 +296	12. 381 291 675 +256	13. 518 629 350 +476	14. 912 384 245 +195	15. 456 288 397 +965

Set 14 Subtract.

1. 89 −23 66	2. 75 −13	3. 96 −42	4. 68 −30	5. 57 −25
6. 839 −230	7. 755 −142	8. 693 −221	9. 884 −354	10. 956 −213
11. 7829 −2304	12. 5384 −1052	13. 6977 −3406	14. 8594 −2041	15. 6385 −3142
16. 58,283 −23,141	17. 46,728 −20,000	18. 73,942 −43,730	19. 65,437 −42,105	20. 83,736 −31,223

Set 15 Subtract.

1. 62 −25 37	2. 81 −39	3. 73 −48	4. 65 −19	5. 93 −58
6. 483 −126	7. 562 −248	8. 748 −319	9. 391 −158	10. 825 −550
11. 926 −472	12. 746 −328	13. 839 −440	14. 728 −376	15. 950 −248
16. 836 −294	17. 582 −356	18. 408 −264	19. 647 −455	20. 972 −348

Set 16 Subtract.

1. 524 −158	2. 627 −259	3. 834 −258	4. 928 −379	5. 703 −265
6. 824 −346	7. 921 −586	8. 602 −375	9. 763 −498	10. 901 −625
11. 453 −289	12. 726 −478	13. 865 −689	14. 942 −758	15. 835 −599
16. 631 −496	17. 857 −259	18. 703 −419	19. 640 −583	20. 810 −417

Set 17 Subtract.

1. 7128 −3482	2. 4603 −2169	3. 9528 −3785	4. 8734 −5286
5. 69,305 −43,829	6. 78,653 −26,781	7. 83,425 −42,003	8. 74,678 −58,399
9. 391,674 −176,859	10. 538,921 −252,346	11. 635,543 −382,974	12. 782,916 −634,897
13. 638,203 −479,158	14. 730,482 −396,758	15. 593,700 −265,387	16. 921,000 −246,218

Set 18 Give the time.

1. 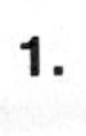 

2. 

3. 

4. 

5. 

6. 

7. 

8. 

Set 19 **Give the total value in dollars.**

1.    $1.60

2.   

3.  

4. 

5. 

6. 

Set 20 **Multiply.**

1. 3×9 27	2. 5×3	3. 4×4	4. 5×6	5. 3×4	6. 5×4
7. 2×9	8. 4×7	9. 5×5	10. 3×8	11. 5×7	12. 4×6
13. 3×3	14. 4×3	15. 0×6	16. 4×5	17. 2×8	18. 3×6

Set 21 **Multiply.**

1. 7×9 63	2. 5×7	3. 9×3	4. 8×1	5. 2×8	6. 3×6
7. 8×8	8. 4×9	9. 8×6	10. 6×9	11. 3×7	12. 7×8
13. 9×9	14. 4×8	15. 9×8	16. 9×5	17. 6×6	18. 9×0

Set 22 **Give each missing factor.**

1. $3 \times \underline{8} = 24$	2. $6 \times \underline{?} = 42$	3. $6 \times \underline{?} = 54$	4. $4 \times \underline{?} = 36$
5. $4 \times \underline{?} = 16$	6. $7 \times \underline{?} = 63$	7. $4 \times \underline{?} = 32$	8. $9 \times \underline{?} = 81$
9. $7 \times \underline{?} = 56$	10. $4 \times \underline{?} = 20$	11. $8 \times \underline{?} = 48$	12. $3 \times \underline{?} = 27$
13. $6 \times \underline{?} = 18$	14. $5 \times \underline{?} = 40$	15. $7 \times \underline{?} = 21$	16. $9 \times \underline{?} = 45$
17. $7 \times \underline{?} = 49$	18. $7 \times \underline{?} = 35$	19. $8 \times \underline{?} = 72$	20. $6 \times \underline{?} = 24$
21. $9 \times \underline{?} = 72$	22. $8 \times \underline{?} = 64$	23. $9 \times \underline{?} = 54$	24. $4 \times \underline{?} = 28$

Set 23 **Multiply.**

1. 23×2 [46]	2. 10×6	3. 20×3	4. 40×2	5. 32×2	6. 12×3
7. 30×2	8. 30×3	9. 11×5	10. 44×2	11. 22×3	12. 12×4
13. 12×2	14. 31×3	15. 11×8	16. 20×4	17. 11×6	18. 21×4
19. 10×8	20. 33×2	21. 43×2	22. 11×7	23. 21×3	24. 42×2

Set 24 **Multiply.**

1. 52×4 [208]	2. 81×5	3. 51×6	4. 52×3	5. 72×4	6. 62×3
7. 42×4	8. 60×7	9. 62×2	10. 61×6	11. 83×3	12. 74×2
13. 43×3	14. 82×4	15. 91×8	16. 73×2	17. 50×6	18. 42×3
19. 63×3	20. 71×5	21. 80×7	22. 92×4	23. 51×9	24. 30×6

Set 25 Multiply.

1. $\overset{2}{35} \times 4 = 140$
2. 83×5
3. 97×3
4. 59×2
5. 73×6
6. 53×8
7. 62×5
8. 95×4
9. 36×8
10. 47×6
11. 44×5
12. 82×7
13. 84×5
14. 48×6
15. 75×4
16. 57×3
17. 29×8
18. 64×7
19. 68×6
20. 85×5
21. 96×8
22. 49×9
23. 65×7
24. 78×6

Set 26 Divide.

1. $7\overline{)56}$ = 8
2. $2\overline{)18}$
3. $6\overline{)18}$
4. $5\overline{)40}$
5. $9\overline{)72}$
6. $2\overline{)2}$
7. $3\overline{)24}$
8. $4\overline{)16}$
9. $9\overline{)45}$
10. $8\overline{)64}$
11. $6\overline{)48}$
12. $7\overline{)63}$
13. $9\overline{)81}$
14. $2\overline{)14}$
15. $3\overline{)27}$
16. $4\overline{)36}$
17. $8\overline{)32}$
18. $8\overline{)56}$
19. $6\overline{)54}$
20. $5\overline{)15}$
21. $7\overline{)28}$
22. $4\overline{)24}$
23. $9\overline{)27}$
24. $6\overline{)36}$
25. $3\overline{)18}$
26. $4\overline{)32}$
27. $8\overline{)48}$
28. $6\overline{)42}$
29. $9\overline{)54}$
30. $7\overline{)49}$

Set 27 Give each quotient and remainder.

1. $4\overline{)23}$ = 5 R3 ($23 - 20 = 3$)
2. $5\overline{)39}$
3. $6\overline{)37}$
4. $8\overline{)38}$
5. $5\overline{)41}$
6. $5\overline{)48}$
7. $7\overline{)40}$
8. $3\overline{)25}$
9. $4\overline{)23}$
10. $8\overline{)60}$
11. $7\overline{)50}$
12. $4\overline{)18}$
13. $9\overline{)75}$
14. $6\overline{)41}$
15. $3\overline{)20}$
16. $9\overline{)43}$
17. $8\overline{)70}$
18. $6\overline{)51}$
19. $7\overline{)33}$
20. $9\overline{)56}$
21. $6\overline{)35}$
22. $5\overline{)23}$
23. $8\overline{)50}$
24. $4\overline{)35}$
25. $8\overline{)53}$
26. $7\overline{)59}$
27. $9\overline{)70}$
28. $6\overline{)43}$
29. $8\overline{)45}$

Set 28 Divide.

1. $3\overline{)39}$ (13) 2. $2\overline{)24}$ 3. $3\overline{)30}$ 4. $4\overline{)40}$ 5. $3\overline{)63}$

6. $3\overline{)93}$ 7. $3\overline{)60}$ 8. $2\overline{)42}$ 9. $4\overline{)84}$ 10. $2\overline{)88}$

11. $6\overline{)66}$ 12. $2\overline{)64}$ 13. $5\overline{)55}$ 14. $2\overline{)68}$ 15. $2\overline{)28}$

16. $3\overline{)36}$ 17. $7\overline{)70}$ 18. $4\overline{)48}$ 19. $3\overline{)90}$ 20. $8\overline{)80}$

21. $2\overline{)62}$ 22. $6\overline{)60}$ 23. $3\overline{)96}$ 24. $7\overline{)77}$ 25. $5\overline{)50}$

26. $4\overline{)88}$ 27. $3\overline{)69}$ 28. $2\overline{)60}$ 29. $3\overline{)99}$ 30. $2\overline{)84}$

Set 29 Divide.

1. $3\overline{)42}$ (14; −3, 12, −12, 0) 2. $7\overline{)84}$ 3. $3\overline{)48}$ 4. $6\overline{)72}$ 5. $3\overline{)54}$

6. $4\overline{)56}$ 7. $4\overline{)52}$ 8. $5\overline{)75}$ 9. $6\overline{)78}$

10. $4\overline{)92}$ 11. $3\overline{)72}$ 12. $6\overline{)90}$ 13. $5\overline{)95}$ 14. $3\overline{)54}$

15. $6\overline{)84}$ 16. $5\overline{)65}$ 17. $3\overline{)81}$ 18. $4\overline{)64}$ 19. $8\overline{)96}$

20. $4\overline{)96}$ 21. $3\overline{)84}$ 22. $5\overline{)80}$ 23. $4\overline{)76}$ 24. $7\overline{)91}$

25. $3\overline{)78}$ 26. $2\overline{)74}$ 27. $6\overline{)96}$ 28. $5\overline{)70}$ 29. $2\overline{)56}$

Set 30 Divide.

1. $5\overline{)63}$ (12 R3; −5, 13, −10, 3) 2. $3\overline{)59}$ 3. $4\overline{)67}$ 4. $8\overline{)90}$ 5. $6\overline{)77}$

6. $8\overline{)95}$ 7. $7\overline{)94}$ 8. $3\overline{)38}$ 9. $6\overline{)82}$

10. $5\overline{)73}$ 11. $4\overline{)57}$ 12. $7\overline{)85}$ 13. $8\overline{)99}$ 14. $8\overline{)94}$

15. $7\overline{)90}$ 16. $2\overline{)39}$ 17. $2\overline{)47}$ 18. $5\overline{)68}$ 19. $4\overline{)95}$

20. $6\overline{)79}$ 21. $3\overline{)83}$ 22. $4\overline{)75}$ 23. $3\overline{)74}$ 24. $6\overline{)95}$

25. $4\overline{)97}$ 26. $6\overline{)83}$ 27. $5\overline{)57}$ 28. $3\overline{)68}$ 29. $7\overline{)99}$

Set 31 **What fraction is colored?**

1. 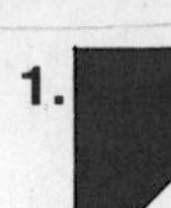 $\frac{1}{2}$ 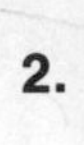
2. 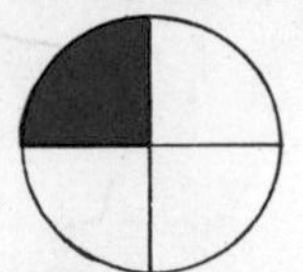
3. 
4. 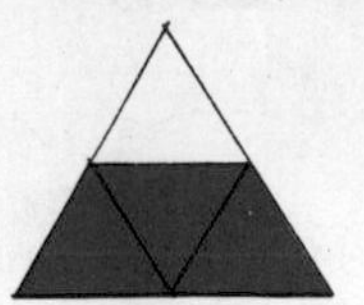
5. 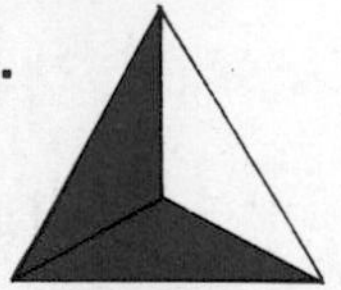
6. 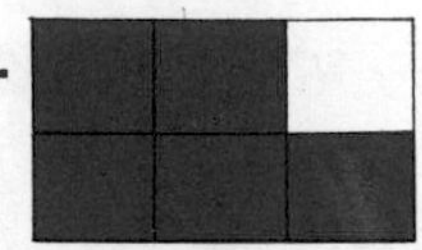
7. 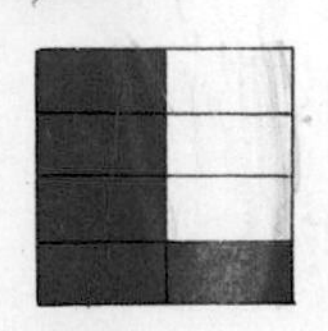
8. 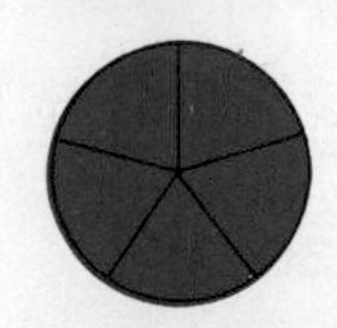
9. 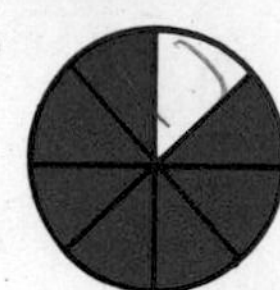
10. 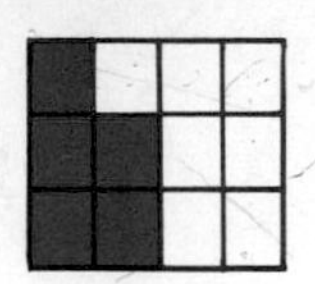
11. 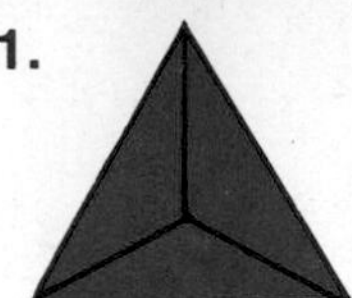
12. 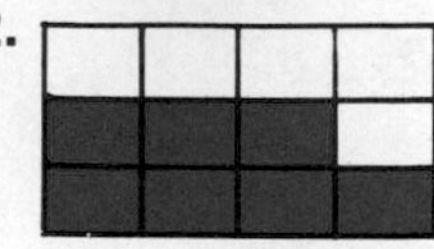

Set 32 **Complete.**

1. $\frac{1}{2}$ of 8 = 4	2. $\frac{1}{3}$ of 3 = ?	3. $\frac{1}{4}$ of 12 = ?	4. $\frac{1}{2}$ of 18 = ?
5. $\frac{1}{6}$ of 24 = ?	6. $\frac{1}{2}$ of 6 = ?	7. $\frac{1}{3}$ of 6 = ?	8. $\frac{1}{4}$ of 20 = ?
9. $\frac{1}{3}$ of 21 = ?	10. $\frac{1}{6}$ of 30 = ?	11. $\frac{1}{2}$ of 10 = ?	12. $\frac{1}{5}$ of 30 = ?
13. $\frac{1}{8}$ of 16 = ?	14. $\frac{1}{4}$ of 16 = ?	15. $\frac{1}{8}$ of 24 = ?	16. $\frac{1}{3}$ of 12 = ?
17. $\frac{1}{4}$ of 28 = ?	18. $\frac{1}{8}$ of 48 = ?	19. $\frac{1}{5}$ of 35 = ?	20. $\frac{1}{6}$ of 42 = ?

Set 33 **Complete.**

1. $\frac{2}{3}$ of 9 = 6	2. $\frac{2}{5}$ of 10 = ?	3. $\frac{3}{4}$ of 8 = ?	4. $\frac{3}{8}$ of 16 = ?
5. $\frac{4}{9}$ of 9 = ?	6. $\frac{4}{5}$ of 20 = ?	7. $\frac{2}{3}$ of 6 = ?	8. $\frac{5}{8}$ of 24 = ?
9. $\frac{2}{3}$ of 18 = ?	10. $\frac{2}{9}$ of 18 = ?	11. $\frac{5}{8}$ of 32 = ?	12. $\frac{3}{4}$ of 12 = ?
13. $\frac{3}{5}$ of 25 = ?	14. $\frac{2}{3}$ of 15 = ?	15. $\frac{2}{5}$ of 30 = ?	16. $\frac{5}{9}$ of 36 = ?
17. $\frac{3}{4}$ of 20 = ?	18. $\frac{3}{8}$ of 40 = ?	19. $\frac{5}{6}$ of 42 = ?	20. $\frac{2}{3}$ of 24 = ?
21. $\frac{3}{5}$ of 35 = ?	22. $\frac{5}{6}$ of 30 = ?	23. $\frac{3}{8}$ of 8 = ?	24. $\frac{5}{7}$ of 28 = ?

Set 34 **Complete.**

1. $\frac{1}{2} = \frac{2}{4}$ 2. $\frac{1}{3} = \frac{?}{6}$ 3. $\frac{2}{3} = \frac{?}{6}$ 4. $\frac{1}{2} = \frac{?}{8}$

5. $\frac{2}{3} = \frac{?}{12}$ 6. $\frac{3}{4} = \frac{?}{8}$ 7. $\frac{1}{2} = \frac{?}{6}$ 8. $\frac{5}{6} = \frac{?}{12}$

9. $\frac{1}{4} = \frac{?}{12}$ 10. $\frac{3}{8} = \frac{?}{16}$ 11. $\frac{2}{3} = \frac{?}{9}$ 12. $\frac{1}{3} = \frac{?}{12}$

13. $\frac{5}{8} = \frac{?}{16}$ 14. $\frac{1}{3} = \frac{?}{9}$ 15. $\frac{3}{4} = \frac{?}{12}$ 16. $\frac{1}{4} = \frac{?}{12}$

17. $\frac{2}{5} = \frac{?}{15}$ 18. $\frac{1}{2} = \frac{?}{10}$ 19. $\frac{3}{8} = \frac{?}{24}$ 20. $\frac{3}{4} = \frac{?}{16}$

21. $\frac{1}{3} = \frac{?}{15}$ 22. $\frac{5}{6} = \frac{?}{18}$ 23. $\frac{3}{5} = \frac{?}{20}$ 24. $\frac{1}{2} = \frac{?}{12}$

Set 35 **Give each fraction in lowest terms.**

1. $\frac{2}{4}$ $\frac{1}{2}$ 2. $\frac{2}{6}$ 3. $\frac{6}{8}$ 4. $\frac{2}{8}$ 5. $\frac{4}{8}$ 6. $\frac{4}{16}$ 7. $\frac{3}{9}$

8. $\frac{4}{12}$ 9. $\frac{5}{20}$ 10. $\frac{3}{6}$ 11. $\frac{8}{12}$ 12. $\frac{4}{6}$ 13. $\frac{6}{16}$ 14. $\frac{15}{20}$

15. $\frac{10}{16}$ 16. $\frac{6}{12}$ 17. $\frac{6}{9}$ 18. $\frac{3}{12}$ 19. $\frac{5}{10}$ 20. $\frac{9}{12}$ 21. $\frac{5}{15}$

22. $\frac{8}{24}$ 23. $\frac{9}{24}$ 24. $\frac{3}{18}$ 25. $\frac{8}{10}$ 26. $\frac{12}{18}$ 27. $\frac{22}{24}$ 28. $\frac{12}{16}$

29. $\frac{20}{24}$ 30. $\frac{8}{20}$ 31. $\frac{6}{20}$ 32. $\frac{15}{18}$ 33. $\frac{14}{16}$ 34. $\frac{16}{24}$ 35. $\frac{12}{20}$

Set 36 <, >, or =?

1. $\frac{1}{3} \bigcirc\!\!\!\!< \ \frac{2}{3}$ 2. $\frac{3}{4} \bigcirc \frac{1}{4}$ 3. $\frac{2}{5} \bigcirc \frac{3}{5}$ 4. $\frac{4}{5} \bigcirc \frac{2}{5}$

5. $\frac{3}{4} \bigcirc \frac{4}{4}$ 6. $\frac{3}{5} \bigcirc \frac{1}{5}$ 7. $\frac{3}{8} \bigcirc \frac{5}{8}$ 8. $\frac{7}{8} \bigcirc \frac{5}{8}$

9. $\frac{1}{5} \bigcirc \frac{1}{6}$ 10. $\frac{1}{3} \bigcirc \frac{1}{4}$ 11. $\frac{1}{5} \bigcirc \frac{1}{4}$ 12. $\frac{1}{6} \bigcirc \frac{1}{8}$

13. $\frac{1}{2} \bigcirc \frac{2}{4}$ 14. $\frac{1}{3} \bigcirc \frac{2}{6}$ 15. $\frac{1}{4} \bigcirc \frac{3}{12}$ 16. $\frac{1}{5} \bigcirc \frac{4}{20}$

17. $\frac{2}{3} \bigcirc \frac{5}{6}$ 18. $\frac{1}{4} \bigcirc \frac{3}{8}$ 19. $\frac{3}{4} \bigcirc \frac{5}{8}$ 20. $\frac{1}{2} \bigcirc \frac{3}{4}$

21. $\frac{1}{2} \bigcirc \frac{5}{6}$ 22. $\frac{2}{5} \bigcirc \frac{4}{10}$ 23. $\frac{3}{5} \bigcirc \frac{7}{10}$ 24. $\frac{4}{5} \bigcirc \frac{7}{10}$

Set 37 Give each sum in lowest terms.

1. $\frac{1}{3} + \frac{1}{3}$ $\quad \frac{2}{3}$
2. $\frac{1}{4} + \frac{1}{4}$
3. $\frac{5}{9} + \frac{1}{9}$
4. $\frac{3}{7} + \frac{2}{7}$
5. $\frac{3}{8} + \frac{2}{8}$
6. $\frac{5}{8} + \frac{1}{8}$
7. $\frac{5}{12} + \frac{2}{12}$
8. $\frac{3}{9} + \frac{2}{9}$
9. $\frac{2}{6} + \frac{1}{6}$
10. $\frac{1}{8} + \frac{2}{8}$
11. $\frac{3}{10} + \frac{4}{10}$
12. $\frac{3}{6} + \frac{2}{6}$
13. $\frac{4}{7} + \frac{1}{7}$
14. $\frac{4}{6} + \frac{1}{6}$
15. $\frac{3}{10} + \frac{2}{10}$
16. $\frac{5}{16} + \frac{3}{16}$
17. $\frac{5}{12} + \frac{5}{12}$
18. $\frac{9}{16} + \frac{3}{16}$
19. $\frac{1}{12} + \frac{5}{12}$
20. $\frac{9}{16} + \frac{5}{16}$
21. $\frac{2}{10} + \frac{3}{10}$

Set 38 Give each sum in lowest terms.

1. $\frac{2}{3} + \frac{1}{6} = \frac{4}{6} + \frac{1}{6} = \frac{5}{6}$
2. $\frac{1}{5} + \frac{7}{10}$ $\quad 10$
3. $\frac{5}{12} + \frac{1}{4}$
4. $\frac{1}{4} + \frac{1}{8}$
5. $\frac{1}{3} + \frac{2}{9}$
6. $\frac{1}{2} + \frac{3}{8}$
7. $\frac{2}{9} + \frac{2}{3}$
8. $\frac{2}{5} + \frac{1}{3}$
9. $\frac{3}{4} + \frac{1}{12}$
10. $\frac{1}{6} + \frac{4}{9}$
11. $\frac{3}{10} + \frac{1}{2}$
12. $\frac{1}{12} + \frac{2}{3}$
13. $\frac{2}{5} + \frac{1}{2}$
14. $\frac{3}{8} + \frac{1}{4}$
15. $\frac{1}{4} + \frac{1}{6}$
16. $\frac{1}{3} + \frac{1}{2}$
17. $\frac{3}{8} + \frac{1}{6}$
18. $\frac{2}{3} + \frac{1}{4}$
19. $\frac{1}{5} + \frac{2}{3}$
20. $\frac{5}{12} + \frac{1}{3}$
21. $\frac{9}{16} + \frac{1}{4}$
22. $\frac{3}{4} + \frac{1}{12}$
23. $\frac{7}{16} + \frac{3}{8}$

Set 39 Give each difference in lowest terms.

1. $\frac{4}{9} - \frac{1}{9}$ $\quad \frac{3}{9} = \frac{1}{3}$
2. $\frac{2}{8} - \frac{1}{8}$
3. $\frac{5}{8} - \frac{1}{8}$
4. $\frac{4}{5} - \frac{1}{5}$
5. $\frac{3}{8} - \frac{2}{8}$
6. $\frac{4}{6} - \frac{1}{6}$
7. $\frac{2}{6} - \frac{1}{6}$
8. $\frac{3}{6} - \frac{2}{6}$
9. $\frac{5}{6} - \frac{2}{6}$
10. $\frac{5}{10} - \frac{1}{10}$
11. $\frac{3}{7} - \frac{2}{7}$
12. $\frac{4}{10} - \frac{3}{10}$
13. $\frac{6}{10} - \frac{4}{10}$
14. $\frac{3}{4} - \frac{2}{4}$
15. $\frac{9}{16} - \frac{1}{16}$
16. $\frac{11}{15} - \frac{2}{15}$
17. $\frac{11}{12} - \frac{5}{12}$
18. $\frac{7}{10} - \frac{1}{10}$
19. $\frac{13}{16} - \frac{3}{16}$
20. $\frac{7}{12} - \frac{5}{12}$
21. $\frac{11}{16} - \frac{5}{16}$

Set 40 Give each difference in lowest terms.

1. $\frac{4}{5} - \frac{1}{10} = \frac{8}{10} - \frac{1}{10} = \frac{7}{10}$
2. $\frac{5}{8} - \frac{1}{2}$
3. $\frac{3}{4} - \frac{1}{3}$
4. $\frac{5}{6} - \frac{1}{4}$
5. $\frac{1}{2} - \frac{1}{4}$
6. $\frac{5}{8} - \frac{1}{4}$
7. $\frac{7}{12} - \frac{1}{2}$
8. $\frac{4}{5} - \frac{1}{2}$
9. $\frac{2}{3} - \frac{1}{6}$
10. $\frac{8}{9} - \frac{1}{6}$
11. $\frac{3}{5} - \frac{1}{10}$
12. $\frac{7}{8} - \frac{3}{4}$
13. $\frac{5}{12} - \frac{1}{4}$
14. $\frac{2}{3} - \frac{3}{5}$
15. $\frac{7}{10} - \frac{1}{2}$
16. $\frac{9}{10} - \frac{2}{5}$
17. $\frac{11}{12} - \frac{1}{3}$
18. $\frac{5}{6} - \frac{3}{8}$
19. $\frac{11}{12} - \frac{1}{2}$
20. $\frac{7}{12} - \frac{1}{4}$
21. $\frac{13}{16} - \frac{3}{4}$
22. $\frac{9}{10} - \frac{1}{2}$
23. $\frac{15}{16} - \frac{5}{8}$

Set 41 Write as a fraction.

1. $5\frac{1}{2}$ $\frac{11}{2}$	2. $6\frac{1}{2}$	3. $7\frac{1}{2}$	4. $1\frac{1}{3}$	5. $1\frac{2}{3}$	6. $2\frac{1}{3}$
7. $2\frac{2}{3}$	8. $2\frac{1}{4}$	9. $2\frac{3}{4}$	10. $3\frac{1}{4}$	11. $3\frac{3}{4}$	12. $5\frac{2}{3}$
13. $6\frac{1}{5}$	14. $7\frac{3}{5}$	15. $1\frac{1}{6}$	16. $1\frac{5}{6}$	17. $3\frac{3}{8}$	18. $4\frac{7}{8}$
19. $2\frac{5}{8}$	20. $1\frac{7}{8}$	21. $3\frac{2}{3}$	22. $4\frac{1}{3}$	23. $5\frac{1}{4}$	24. $4\frac{4}{5}$
25. $5\frac{2}{5}$	26. $4\frac{3}{4}$	27. $5\frac{4}{5}$	28. $2\frac{3}{8}$	29. $4\frac{2}{3}$	30. $1\frac{3}{4}$
31. $3\frac{5}{6}$	32. $2\frac{5}{6}$	33. $1\frac{9}{10}$	34. $6\frac{1}{10}$	35. $2\frac{7}{8}$	36. $6\frac{5}{8}$

Set 42 Write as a mixed number or a whole number.

1. $\frac{7}{3}$ $2\frac{1}{3}$	2. $\frac{8}{3}$	3. $\frac{9}{4}$	4. $\frac{10}{2}$	5. $\frac{9}{2}$	6. $\frac{11}{2}$
7. $\frac{5}{4}$	8. $\frac{8}{4}$	9. $\frac{9}{5}$	10. $\frac{11}{5}$	11. $\frac{15}{4}$	12. $\frac{17}{6}$
13. $\frac{18}{6}$	14. $\frac{12}{7}$	15. $\frac{14}{5}$	16. $\frac{16}{3}$	17. $\frac{17}{3}$	18. $\frac{25}{5}$
19. $\frac{8}{5}$	20. $\frac{12}{4}$	21. $\frac{19}{6}$	22. $\frac{13}{5}$	23. $\frac{20}{4}$	24. $\frac{19}{3}$
25. $\frac{11}{8}$	26. $\frac{16}{8}$	27. $\frac{15}{3}$	28. $\frac{24}{6}$	29. $\frac{13}{8}$	30. $\frac{30}{6}$
31. $\frac{18}{3}$	32. $\frac{23}{6}$	33. $\frac{19}{8}$	34. $\frac{16}{4}$	35. $\frac{19}{5}$	36. $\frac{24}{8}$

Set 43 **Multiply.**

1. $\begin{array}{r} ^{22}\\ 134 \\ \times 6 \\ \hline 804 \end{array}$
2. $\begin{array}{r} 238 \\ \times 4 \\ \hline \end{array}$
3. $\begin{array}{r} 471 \\ \times 2 \\ \hline \end{array}$
4. $\begin{array}{r} 119 \\ \times 8 \\ \hline \end{array}$
5. $\begin{array}{r} 275 \\ \times 3 \\ \hline \end{array}$
6. $\begin{array}{r} 387 \\ \times 2 \\ \hline \end{array}$
7. $\begin{array}{r} 356 \\ \times 2 \\ \hline \end{array}$
8. $\begin{array}{r} 175 \\ \times 4 \\ \hline \end{array}$
9. $\begin{array}{r} 429 \\ \times 2 \\ \hline \end{array}$
10. $\begin{array}{r} 256 \\ \times 3 \\ \hline \end{array}$
11. $\begin{array}{r} 123 \\ \times 5 \\ \hline \end{array}$
12. $\begin{array}{r} 304 \\ \times 3 \\ \hline \end{array}$
13. $\begin{array}{r} 291 \\ \times 2 \\ \hline \end{array}$
14. $\begin{array}{r} 123 \\ \times 6 \\ \hline \end{array}$
15. $\begin{array}{r} 318 \\ \times 3 \\ \hline \end{array}$
16. $\begin{array}{r} 189 \\ \times 4 \\ \hline \end{array}$
17. $\begin{array}{r} 477 \\ \times 2 \\ \hline \end{array}$
18. $\begin{array}{r} 236 \\ \times 4 \\ \hline \end{array}$
19. $\begin{array}{r} 330 \\ \times 3 \\ \hline \end{array}$
20. $\begin{array}{r} 156 \\ \times 6 \\ \hline \end{array}$
21. $\begin{array}{r} 237 \\ \times 4 \\ \hline \end{array}$
22. $\begin{array}{r} 142 \\ \times 7 \\ \hline \end{array}$
23. $\begin{array}{r} 345 \\ \times 2 \\ \hline \end{array}$
24. $\begin{array}{r} 108 \\ \times 9 \\ \hline \end{array}$

Set 44 **Multiply.**

1. $\begin{array}{r} ^{23}\\ 568 \\ \times 4 \\ \hline 2272 \end{array}$
2. $\begin{array}{r} 735 \\ \times 5 \\ \hline \end{array}$
3. $\begin{array}{r} 543 \\ \times 6 \\ \hline \end{array}$
4. $\begin{array}{r} 698 \\ \times 7 \\ \hline \end{array}$
5. $\begin{array}{r} 375 \\ \times 6 \\ \hline \end{array}$
6. $\begin{array}{r} 492 \\ \times 8 \\ \hline \end{array}$
7. $\begin{array}{r} 594 \\ \times 8 \\ \hline \end{array}$
8. $\begin{array}{r} 384 \\ \times 3 \\ \hline \end{array}$
9. $\begin{array}{r} 756 \\ \times 6 \\ \hline \end{array}$
10. $\begin{array}{r} 468 \\ \times 7 \\ \hline \end{array}$
11. $\begin{array}{r} 756 \\ \times 5 \\ \hline \end{array}$
12. $\begin{array}{r} 866 \\ \times 7 \\ \hline \end{array}$
13. $\begin{array}{r} 687 \\ \times 5 \\ \hline \end{array}$
14. $\begin{array}{r} 885 \\ \times 7 \\ \hline \end{array}$
15. $\begin{array}{r} 466 \\ \times 4 \\ \hline \end{array}$
16. $\begin{array}{r} 853 \\ \times 9 \\ \hline \end{array}$
17. $\begin{array}{r} 558 \\ \times 8 \\ \hline \end{array}$
18. $\begin{array}{r} 399 \\ \times 9 \\ \hline \end{array}$
19. $\begin{array}{r} 756 \\ \times 6 \\ \hline \end{array}$
20. $\begin{array}{r} 825 \\ \times 8 \\ \hline \end{array}$
21. $\begin{array}{r} 693 \\ \times 9 \\ \hline \end{array}$
22. $\begin{array}{r} 582 \\ \times 5 \\ \hline \end{array}$
23. $\begin{array}{r} 940 \\ \times 4 \\ \hline \end{array}$
24. $\begin{array}{r} 475 \\ \times 7 \\ \hline \end{array}$

Set 45 **Multiply.**

1. $\begin{array}{r} ^{112}\\ 2359 \\ \times 3 \\ \hline 7077 \end{array}$
2. $\begin{array}{r} 1948 \\ \times 5 \\ \hline \end{array}$
3. $\begin{array}{r} 2096 \\ \times 8 \\ \hline \end{array}$
4. $\begin{array}{r} 5736 \\ \times 5 \\ \hline \end{array}$
5. $\begin{array}{r} 7846 \\ \times 7 \\ \hline \end{array}$
6. $\begin{array}{r} 7492 \\ \times 7 \\ \hline \end{array}$
7. $\begin{array}{r} 3429 \\ \times 4 \\ \hline \end{array}$
8. $\begin{array}{r} 4524 \\ \times 7 \\ \hline \end{array}$
9. $\begin{array}{r} 4620 \\ \times 6 \\ \hline \end{array}$
10. $\begin{array}{r} 9528 \\ \times 8 \\ \hline \end{array}$
11. $\begin{array}{r} 6753 \\ \times 5 \\ \hline \end{array}$
12. $\begin{array}{r} 8786 \\ \times 5 \\ \hline \end{array}$
13. $\begin{array}{r} 3658 \\ \times 4 \\ \hline \end{array}$
14. $\begin{array}{r} 6738 \\ \times 9 \\ \hline \end{array}$
15. $\begin{array}{r} 5989 \\ \times 6 \\ \hline \end{array}$
16. $\begin{array}{r} 7236 \\ \times 8 \\ \hline \end{array}$
17. $\begin{array}{r} 5907 \\ \times 7 \\ \hline \end{array}$
18. $\begin{array}{r} 6438 \\ \times 9 \\ \hline \end{array}$
19. $\begin{array}{r} 9253 \\ \times 5 \\ \hline \end{array}$
20. $\begin{array}{r} 8671 \\ \times 6 \\ \hline \end{array}$

Set 46 Multiply.

1. 12 ×10 (120)	2. 23 ×10	3. 26 ×10	4. 34 ×10	5. 42 ×10
6. 56 ×10	7. 74 ×10	8. 65 ×10	9. 86 ×10	10. 92 ×10
11. 125 ×10	12. 136 ×10	13. 158 ×10	14. 232 ×10	15. 250 ×10
16. 294 ×10	17. 318 ×10	18. 429 ×10	19. 506 ×10	20. 740 ×10

Set 47 Multiply.

1. 24 ×20 (480)	2. 28 ×40	3. 36 ×60	4. 45 ×30	5. 56 ×70
6. 78 ×50	7. 84 ×90	8. 65 ×80	9. 92 ×40	10. 68 ×60
11. 120 ×30	12. 136 ×70	13. 245 ×20	14. 286 ×60	15. 316 ×50
16. 426 ×80	17. 538 ×60	18. 673 ×70	19. 825 ×40	20. 926 ×70

Set 48 Multiply.

1. 346 ×35 (1730, 1038, 12,110)	2. 354 ×27	3. 427 ×42	4. 529 ×57	5. 269 ×34
	6. 635 ×58	7. 562 ×64	8. 257 ×28	9. 683 ×56
10. 294 ×43	11. 362 ×75	12. 283 ×36	13. 706 ×59	14. 316 ×29
15. 445 ×48	16. 478 ×53	17. 653 ×38	18. 842 ×46	19. 937 ×17

Set 49 Divide.

1. $3\overline{)472}$ = 157 R1
   - −3
   - 17
   - −15
   - 22
   - −21
   - 1

2. $7\overline{)853}$ 3. $5\overline{)927}$ 4. $4\overline{)638}$ 5. $7\overline{)827}$

6. $3\overline{)457}$ 7. $6\overline{)946}$ 8. $4\overline{)827}$ 9. $7\overline{)904}$

10. $3\overline{)947}$ 11. $2\overline{)315}$ 12. $5\overline{)942}$ 13. $4\overline{)653}$

14. $2\overline{)358}$ 15. $6\overline{)928}$ 16. $6\overline{)708}$ 17. $3\overline{)721}$ 18. $5\overline{)850}$

19. $8\overline{)965}$ 20. $8\overline{)942}$ 21. $5\overline{)836}$ 22. $4\overline{)950}$ 23. $7\overline{)863}$

24. $3\overline{)527}$ 25. $6\overline{)820}$ 26. $4\overline{)675}$ 27. $6\overline{)917}$ 28. $5\overline{)886}$

Set 50 Divide.

1. $7\overline{)152}$ = 21 R5
   - −14
   - 12
   - −7
   - 5

2. $8\overline{)414}$ 3. $6\overline{)193}$ 4. $7\overline{)584}$ 5. $4\overline{)304}$

6. $5\overline{)271}$ 7. $7\overline{)426}$ 8. $6\overline{)246}$ 9. $8\overline{)325}$

10. $7\overline{)342}$ 11. $8\overline{)608}$ 12. $6\overline{)208}$ 13. $6\overline{)528}$

14. $5\overline{)476}$ 15. $9\overline{)627}$ 16. $9\overline{)438}$ 17. $7\overline{)623}$ 18. $4\overline{)362}$

19. $8\overline{)219}$ 20. $6\overline{)595}$ 21. $8\overline{)397}$ 22. $9\overline{)650}$ 23. $4\overline{)295}$

24. $5\overline{)395}$ 25. $6\overline{)543}$ 26. $9\overline{)608}$ 27. $7\overline{)419}$ 28. $8\overline{)658}$

Set 51 Divide.

1. $9\overline{)3628}$ = 403 R1
   - −36
   - 28
   - −27
   - 1

2. $8\overline{)4216}$ 3. $5\overline{)5374}$ 4. $6\overline{)3728}$

5. $2\overline{)5437}$ 6. $5\overline{)7928}$ 7. $4\overline{)6819}$

8. $2\overline{)4609}$ 9. $9\overline{)7034}$ 10. $3\overline{)2174}$

11. $7\overline{)2738}$ 12. $5\overline{)8555}$ 13. $9\overline{)3914}$ 14. $8\overline{)9174}$

15. $4\overline{)8326}$ 16. $6\overline{)9266}$ 17. $7\overline{)6035}$ 18. $5\overline{)2794}$

19. $8\overline{)5203}$ 20. $9\overline{)7726}$ 21. $6\overline{)5374}$ 22. $8\overline{)4429}$

Set 52 Use the given multiplication facts to help you divide.

30	30	30	30	30	30	30	30	30	30
×0	×1	×2	×3	×4	×5	×6	×7	×8	×9
0	30	60	90	120	150	180	210	240	270

1. $30\overline{)759}$
2. $30\overline{)683}$
3. $30\overline{)726}$
4. $30\overline{)594}$
5. $30\overline{)963}$
6. $30\overline{)890}$
7. $30\overline{)711}$
8. $30\overline{)806}$
9. $30\overline{)915}$
10. $30\overline{)775}$
11. $30\overline{)860}$
12. $30\overline{)943}$
13. $30\overline{)319}$
14. $30\overline{)487}$
15. $30\overline{)607}$
16. $30\overline{)374}$
17. $30\overline{)791}$
18. $30\overline{)524}$

Set 53 Divide.

1. $12\overline{)794}$
2. $33\overline{)960}$
3. $24\overline{)595}$
4. $14\overline{)673}$
5. $28\overline{)753}$
6. $35\overline{)573}$
7. $33\overline{)709}$
8. $32\overline{)673}$
9. $26\overline{)800}$
10. $16\overline{)960}$
11. $19\overline{)870}$
12. $13\overline{)863}$
13. $21\overline{)792}$
14. $31\overline{)642}$
15. $32\overline{)783}$
16. $25\overline{)658}$
17. $42\overline{)928}$
18. $36\overline{)859}$
19. $15\overline{)829}$
20. $43\overline{)952}$
21. $19\overline{)872}$
22. $18\overline{)908}$
23. $32\overline{)742}$
24. $27\overline{)708}$
25. $29\overline{)395}$
26. $42\overline{)966}$
27. $17\overline{)729}$
28. $28\overline{)846}$

Set 54 <, or >?

1. 0.8 (>) 0.7
2. 0.6 ○ 0.9
3. 0.3 ○ 0.2
4. 0.61 ○ 0.6
5. 0.5 ○ 0.54
6. 0.78 ○ 0.87
7. 0.9 ○ 1
8. 2.1 ○ 2.01
9. 0.3 ○ 3
10. 5.16 ○ 5.23
11. 8.63 ○ 8.36
12. 4.08 ○ 4.8
13. 15.6 ○ 1.56
14. 5.93 ○ 59.3
15. 68.3 ○ 6.83
16. 7.02 ○ 7.2
17. 9.6 ○ 9.06
18. 9.99 ○ 10

Set 55 Add.

1. 3.6 +2.2 (5.8)	2. 5.3 +2.4	3. 4.3 +3.5	4. 6.2 +3.4	5. 5.4 +2.3
6. 3.6 +2.8	7. 4.9 +3.6	8. 7.5 +1.6	9. 0.52 +0.38	10. 0.74 +0.18
11. 2.8 +2.8	12. 7.6 +3.9	13. 0.46 +0.38	14. 8.3 +5.7	15. 4.7 +9.0
16. 9.74 +2.65	17. 8.47 +8.47	18. 23.6 +18.7	19. 56.5 +37.9	20. 4.39 +9.75

Set 56 Subtract.

1. 7.6 −2.3 (5.3)	2. 5.9 −1.4	3. 7.4 −3.4	4. 6.4 −2.1	5. 8.1 −3.0
6. 6.3 −2.8	7. 0.85 −0.46	8. 0.72 −0.59	9. 7.3 −2.9	10. 7.0 −4.8
11. 0.95 −0.76	12. 0.82 −0.75	13. 6.3 −3.8	14. 9.8 −4.9	15. 0.65 −0.26
16. 5.36 −2.74	17. 7.11 −3.87	18. 6.05 −4.38	19. 53.4 −29.6	20. 50.1 −38.7

Set 57 Add or subtract.

1. 13.6 + 9.56
   13.6
   + 9.56
   23.16
2. 13.6 − 9.56
3. 9.3 + 4.62
4. 9.3 − 4.62
5. 12.3 + 2.74
6. 12.3 − 2.74
7. 8 + 5.2
8. 8 − 5.2
9. 6.7 + 0.09
10. 6.7 − 0.09
11. 18 + 6.9
12. 18 − 6.9
13. 23.5 + 2.94
14. 23.5 − 2.94
15. 20.6 + 5.74
16. 20.6 − 5.74
17. 53 + 6.3
18. 53 − 6.3
19. 56.32 + 41.8
20. 56.32 − 41.8
21. 8 + 7.65
22. 8 − 7.65

EXTRA PROBLEM SOLVING

Set 1 Solve.

1. Bill walked 6 blocks to George's house and then 5 blocks to the movies. How many blocks did he walk?

2. The movie theater is 8 blocks from Bill's house. The park is 5 blocks from Bill's house. How many blocks farther is the theater?

3. Bill took 5 minutes to walk to Kay's house and 7 minutes to walk back home. How many minutes did he walk?

4. Bill can run to the park in 3 minutes and he can walk to the park in 9 minutes. How many minutes less does it take him to run?

5. Bill was gone for 8 minutes. He took 5 minutes to walk to the store and back, and was in the store the rest of the time. How many minutes was he in the store?

Set 2 Solve.

1. One picture at the theater was 96 minutes long. The other picture was 107 minutes long. How many minutes long were both pictures?

2. A short subject was 12 minutes long and a cartoon was 7 minutes long. How many minutes longer was the short subject?

3. Kay took 12 minutes to walk to the theater, was 217 minutes at the theater, and took 15 minutes to walk home. How many minutes was she gone?

4. Beth left the theater early, and missed the 12-minute short subject and the 5-minute coming attractions. How many minutes early did she leave?

5. Beth walked 13 blocks to go home from the theater. She passed the park when she was 5 blocks from the theater. How many blocks was the park from Beth's house?

6. Beth's house is 13 blocks from the theater. Kay's house is 7 blocks nearer to the theater than Beth's house. How many blocks is the theater from Kay's house?

Set 3 **Solve.**

1. The Andrews family took a trip. During the first 2 hours they traveled 46 miles each hour. How many miles did they drive?

2. At their first stop they spent $15.85 for gasoline and $.95 for oil. How much change did they get from $20?

3. The whole trip was to be 425 miles. At their first stop they had driven 141 miles. How many miles did they have left to drive?

4. At their second stop they spent $11.45 for gasoline, $.85 for a magazine, and $1.80 for postcards. How much did they spend?

5. Mandy and Laura each bought a 20-picture roll of film. Mandy took 16 pictures and Laura took 17 pictures. How many pictures did they have left to take?

6. Black Cave was 425 miles from their home. By lunch time they had traveled 208 miles. How many miles were they from Black Cave then?

Set 4 **Solve.**

1. For lunch Mandy and Laura each had a hamburger for $1.95, a piece of pie for $.65, and some milk for $.40. What was the total cost of their lunches?

2. At Black Cave they could take a 50-minute tour or a 35-minute tour. How many minutes longer was the longer tour?

3. Each adult's ticket was $1.25. Each child's ticket was $.75. How much were 2 adult's and 2 children's tickets?

4. There were 2 tour guides. One guide took 19 people and the other took 23 people. How many people were taken by the guides?

5. Mandy bought 3 packages of Black Cave pictures. Each package had 12 pictures. How many pictures did she buy?

6. Mrs. Andrews bought 12 postcards. She gave 3 to Mandy and 4 to Laura. How many postcards did she have left?

Set 5 **Solve.**

1. What is the cost of 6 green beads?
2. Which costs more: 6 blue beads or 6 red beads?
3. How much more do 8 green beads cost than 8 blue beads?
4. Andrew bought 4 beads of each color. How much did he spend?
5. Benjamin bought 7 green beads and a piece of red string. How much did he spend?
6. Pamela bought 2 beads of the same color and 1 bead of another color. She spent 16¢. Which beads did she buy?

Set 6 **Solve.**

1. What is the cost of 7 pieces of blue string?
2. What is the cost of 7 blue beads and 8 red beads?
3. How much more do 5 pieces of red string cost than 5 pieces of green string?
4. Paul had 75¢. He bought 7 green beads. How much money did he have then?
5. Rosa had 90¢. She wanted to buy 8 pieces of red string. How much more money did she need?
6. Sylvia bought 1 piece of each kind of string. How much did she spend?

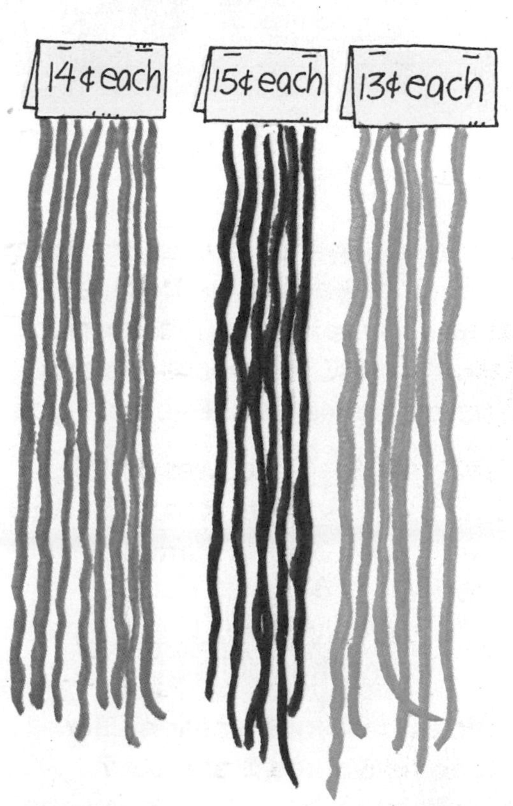

Set 7 Solve.

1. There were 467 adult tickets and 365 children's tickets sold for Fun Night. How many tickets were sold?
2. An adult ticket cost $.65. How much did 4 adult tickets cost?
3. A bingo card cost 6¢. Ellen bought some cards for 48¢. How many cards did she buy?
4. 126 people threw baseballs at bottles. Only 39 won a prize. How many did not win a prize?
5. Bob served the punch. He served 21 cups from 3 cans. How many cups did he average for each can?
6. Bob used 8 packages of punch cups. There were 24 cups in each package. How many cups did he use?

Set 8 Solve.

1. There were 42 pies. Each pie was cut into 6 pieces. How many pieces of pie was that?
2. Grace put 3 small marshmallows into each cup of hot chocolate. How many marshmallows did she need for 48 cups?
3. There were 84 raisin cookies. Arthur put 6 cookies in each bag. How many bags did he use?
4. The students had set up 16 tables with 8 chairs at each table. How many chairs was that?
5. To decorate for Fun Night, 36 students worked 3 hours each. How many hours of work was that?
6. On Fun Night the school took in $678.54. It spent $115.25 for prizes. How much money was left?

Set 9 Solve.

1. A coin-collecting club had 24 members. $\frac{5}{6}$ of them went to a pizza party. How many went to the party?

2. Each of the 24 members brought \$3 for the party. How much money did they bring?

3. There were 9 tables with 6 chairs each and 12 tables with 4 chairs each. How many chairs were there?

4. 12 of the 20 club members sat at 2 tables of 6. How many tables of 4 did they need to seat the rest of the club?

5. The Pizza Palace had been open for only 2 days. It had sold 98 pizzas in all. What was the average number sold each day?

6. The large sausage pizza was on sale for $\frac{3}{4}$ of its regular price. The regular price was \$3.60. What was the sale price?

Set 10 Solve.

1. The club ordered 8 large pizzas for \$3.60 each and 7 small pizzas for \$2.40 each. What was the total price?

2. They ordered 9 large drinks for \$.55 each and 8 small drinks for \$.35 each. How much did they spend for the drinks?

3. Each large pizza was cut into 16 pieces. Janet ate 4 pieces. What fraction of a large pizza did she eat? Give your answer in lowest terms.

4. Mike ate $\frac{1}{2}$ of a small pizza. Helen ate $\frac{1}{3}$ of the same pizza. What fraction of the pizza did they eat?

5. Maria ate $\frac{1}{2}$ of one small pizza and $\frac{1}{4}$ of another small pizza. What fraction of a small pizza did she eat?

6. Virginia ate $\frac{1}{2}$ of a large pizza and Andrew ate $\frac{1}{3}$ of a large pizza. Who ate more? How much more?

Set 11 Solve.

1. Christine listened to her records 45 minutes each day for 5 days. How many minutes was that?

2. Dotty listened to her tapes for $\frac{1}{2}$ hour on Monday and $\frac{3}{4}$ hour on Tuesday. How many hours was that? Give your answer as a mixed number.

3. Linda had $5. She bought a long-playing record. How much money did she have left?

4. Tom bought 2 cassette tapes. He gave the clerk $10. How much change did he get back?

5. What is the total cost of 3 long-playing records and an eight-track tape?

6. What is the total cost of 6 singles and 4 long-playing records?

Set 12 Solve.

1. Which costs more: 15 singles or 3 long-playing records? How much more?

2. Donna had $10. She wanted to buy 3 cassette tapes. How much more money did she need?

3. James had $8.35. He bought 1 eight-track tape. Did he have enough money left to buy 2 single records?

4. How much more money are 2 long-playing records than a cassette tape?

5. Ralph returned a bad eight-track tape and got his money back. He then bought 6 singles. How much money did he have left?

6. Judith bought a long-playing record for $\frac{2}{3}$ of its regular price. The regular price was $6. How much did she pay for it?

Set 13 Solve.

1. 4 buses went to the picnic. There were 28 children on each bus. How many children went to the picnic?

2. There were 19 adults and 112 children at the picnic. How many more children than adults were there?

3. 12 teams played tug-of-war. There were 9 children on each team. How many children played tug-of-war?

4. 92 children played in the sack race. There were 2 children for each sack. How many sacks were there?

5. 94 children wanted to play baseball. There were 9 children on each team. How many couldn't play baseball?

6. There were 9 packages with 24 paper plates in each package. 142 of the plates were used. How many plates were not used?

Set 14 Solve.

1. Each of 12 packages had 24 napkins. Each of 4 packages had 36 napkins. How many napkins was that?

2. There were 18 packages with 12 paper cups in each package. 131 of the cups were used. How many cups were not used?

3. Each of 12 boxes had 16 sandwiches. 158 of the sandwiches were eaten. How many sandwiches were left?

4. There were 7 cakes. 112 pieces of cake were needed. How many pieces did they get from each cake?

5. There were 93 cups of strawberry punch and 138 cups of grape punch. They drank 177 cups of punch. How many cups of punch were left?

6. There were 112 children at the picnic. 29 children went home in each of the first 3 buses. How many children went home in the last bus?

Set 15 **Solve.**

1. Sarah baked 16 dozen peanut-butter cookies. How many cookies was that?

2. There were 27 boxes of oatmeal cookies. Each box had 24 cookies. How many oatmeal cookies was that?

3. Each carrot cake cost $3.25. What was the cost of 2 carrot cakes?

4. Alan sold a large apple pie for $3.85. The buyer gave Alan $5. How much change did he give back?

5. Dennis had 56¢. He spent $\frac{3}{8}$ of it for cookies. How much did he spend?

6. Kay had 75¢. Oatmeal cookies cost 6¢ each. How many could she buy? How much money would she have left?

Set 16 **Solve.**

1. Carlos baked 168 peanut-butter cookies. He wanted to put 12 in a bag. How many bags did he need?

2. 18 students each baked 2 dozen rolls. How many rolls was that?

3. Terry bought 36 peanut-butter cookies for 5¢ each and 15 strawberry tarts for 18¢ each. How much did he spend?

4. Jill had $3.42. She bought 4 dozen oatmeal cookies for 6¢ a cookie. How much money did she have left?

5. Mary bought 16 bran muffins for $3.04. How much did each muffin cost?

6. Mark had $5.65. He bought a carrot cake for $3.25. He then spent the rest for 15 rolls. How much did each roll cost?

GLOSSARY

abacus A counting machine made of beads and wires.

addend A number used in an addition problem.

$$\begin{array}{r l} 9 & \leftarrow \text{addend} \\ +4 & \leftarrow \text{addend} \\ \hline 13 & \leftarrow \text{sum} \end{array}$$

adding 0 property The sum of a number and 0 is the same number.

$$\begin{array}{r} 4 \\ +0 \\ \hline 4 \end{array}$$

A.M. A symbol used for times after 12:00 midnight and before 12:00 noon.

angle A figure formed by two rays with the same endpoint.

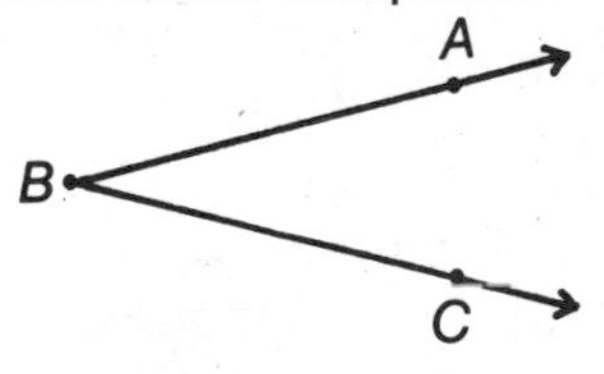

This is angle *B*.

area The number of unit squares that cover a figure.

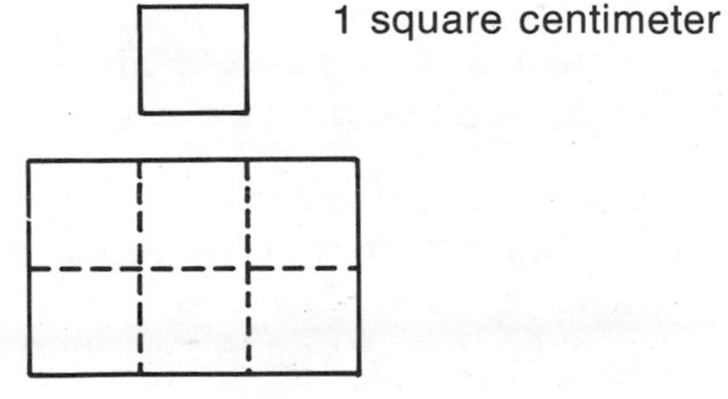

1 square centimeter

The area of this figure is 6 square centimeters.

average The average of 4, 5, 5, 7, and 9 is 6. To find the average, add the numbers and divide by the number of numbers.

Celsius temperature scale (°C) The metric temperature scale, in which 0°C is the freezing point of water and 100°C is the boiling point of water.

centimeter (cm) A unit of length in the metric system. 100 centimeters is 1 meter.

circle A curved plane figure shaped like this:

command One of the directions used in a computer program.

common factor 2 is a common factor of 4 and 10 because it is a factor of 4 and a factor of 10.

common multiple 30 is a common multiple of 5 and 6 because it is a multiple of 5 and a multiple of 6.

computer A machine that can count, keep track of information, and give out the results.

cone A solid figure shaped like this:

congruent figures Figures that are the same size and shape.

cube A rectangular solid ("box") with all edges the same length.

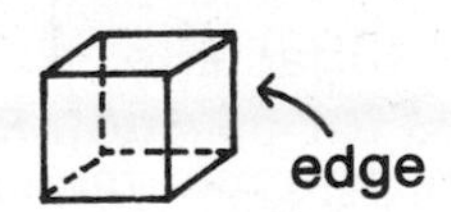

cylinder A solid figure shaped like this:

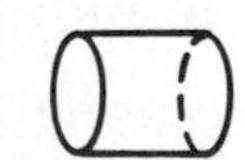

data Information used by a computer.

debug To fix mistakes in a computer program.

decimal A number such as 0.4 (four tenths) and 3.2 (three and two tenths). A decimal is sometimes called a decimal fraction.

decimal point A dot written in a decimal between the ones place and the tenths place.

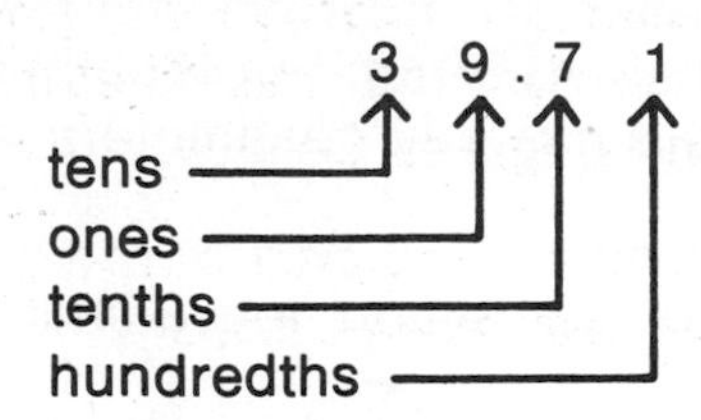

denominator In $\frac{2}{3}$, the denominator is 3.

diameter The distance across a circle through its center.

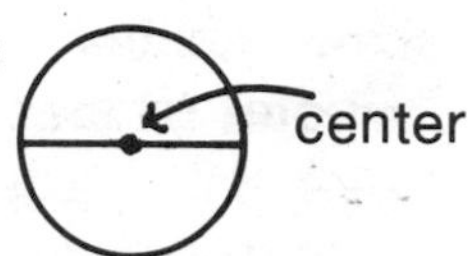

difference The answer to a subtraction problem.

$$7 - 3 = 4 \leftarrow \text{difference}$$

digit Any one of the symbols 0, 1, 2, 3, 4, 5, 6, 7, 8, and 9.

disk A round, flat object with a magnetic surface used to store computer programs.

divisor The number that is divided by.

$$\text{divisor} \rightarrow 4\overline{)36}$$

edge (of a solid figure) A segment that is a side of a surface.

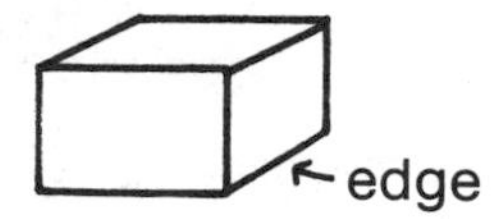

edit To change information in a computer program.

enter To send information directly to a computer.

equation A sentence with an equal sign, such as $3 \times 9 = 27$.

equivalent fractions Fractions for the same number. $\frac{1}{2}$, $\frac{2}{4}$, and $\frac{3}{6}$ are equivalent fractions.

estimate To use rounded numbers to check whether an answer is correct. For example, to estimate $47 + 32$, add $50 + 30$. The sum should be about 80.

even numbers 0 and the numbers that are multiples of 2. The numbers 0, 2, 4, 6, 8, 10, and 12 are even.

factors Numbers used in a multiplication problem.

$$\begin{array}{rl} 8 & \leftarrow \text{factor} \\ \times 6 & \leftarrow \text{factor} \\ \hline 48 & \leftarrow \text{product} \end{array}$$

Fahrenheit temperature scale (°F) The temperature scale in which 32°F is the freezing point of water and 212°F is the boiling point of water.

fraction A number such as $\frac{1}{2}$, $\frac{3}{4}$, and $\frac{4}{6}$.

gram (g) A unit of mass (weight) in the metric system. 1000 grams is 1 kilogram.

graph A picture used to show some information. A graph can be a bar graph, a picture graph, a circle graph, or a line graph.

greater than A comparison of two numbers that are not the same. The symbol is $>$. For example, $7 > 2$. (Another comparison is *less than*.)

greatest common factor The largest number that is a common factor of two numbers. 1, 3, and 9 are all common factors of 9 and 18, but 9 is the greatest common factor of 9 and 18.

grouping property of addition Changing the grouping of the addends does not change the sum.

$$(4 + 1) + 5 = 4 + (1 + 5)$$

grouping property of multiplication Changing the grouping of the factors does not change the product.

$$(4 \times 2) \times 3 = 4 \times (2 \times 3)$$

intersecting lines Lines in a plane that cross at a point.

kilogram (kg) A unit of mass (weight) in the metric system. 1 kilogram is 1000 grams.

kilometer (km) A unit of length in the metric system. 1 kilometer is 1000 meters.

least common multiple The smallest number (not 0) that is a common multiple of two numbers. 6, 12, and 18 are all common multiples of 2 and 3, but 6 is the least common multiple of 2 and 3.

less than A comparison of two numbers that are not the same. The symbol is <. For example, $3 < 8$. (Another comparison is *greater than*.)

line A plane figure that goes on and on in both directions.

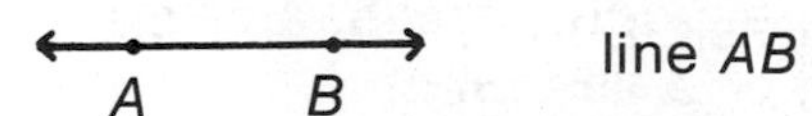

line *AB*

line of symmetry If a figure can be folded along a line so the two parts of the figure match, the fold line is a line of symmetry.

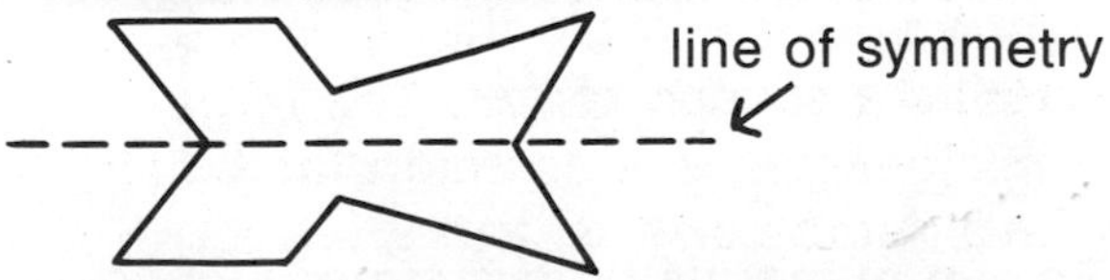

liter (L) A unit of liquid measure in the metric system. 1 liter is 1000 milliliters.

lowest terms A fraction is in lowest terms if the only common factor of its numerator and denominator is 1.

meter (m) A unit of length in the metric system. 1 meter is 100 centimeters.

metric system An international system of measurement that uses meter, liter, gram, and degrees Celsius.

microprocessor The part of the computer that makes it work.

milliliter (mL) A unit of liquid measure in the metric system. 1000 milliliters is 1 liter.

mixed number A number that has a whole-number part and a fraction part. $2\frac{3}{4}$ is a mixed number.

multiple A product. 4, 8, 12, 16, 20, and so on, are multiples of 4.

multiplying by 1 property The product of a number and 1 is the number.

$$4 \times 1 = 4$$

multiplying by 0 property The product of a number and 0 is 0.

$$4 \times 0 = 0$$

numeral A way of writing a number. These are some numerals for the number 4:

four   4   IV $6 - 2$

numerator In $\frac{2}{3}$, the numerator is 2.

odd numbers Numbers that are not multiples of 2. The numbers 1, 3, 5, 7, 9, and 11 are odd.

order property of addition Changing the order of the addends does not change the sum.

$$9 + 8 = 17 \quad 8 + 9 = 17$$

order property of multiplication Changing the order of the factors does not change the product.

$$\begin{array}{r} 2 \\ \times 5 \\ \hline 10 \end{array} \qquad \begin{array}{r} 5 \\ \times 2 \\ \hline 10 \end{array}$$

outcome A possible result of an experiment.

parallel lines Lines in a plane that do not cross.

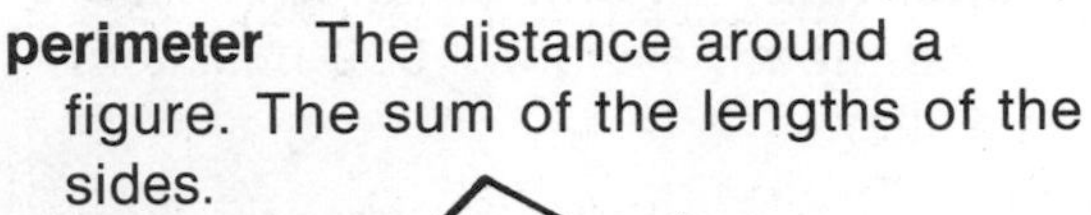

perimeter The distance around a figure. The sum of the lengths of the sides.

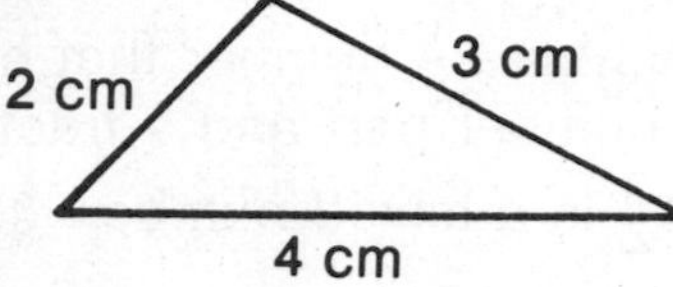

The perimeter is 9 cm.

place value The value given to the place, or position, of a digit in a numeral.

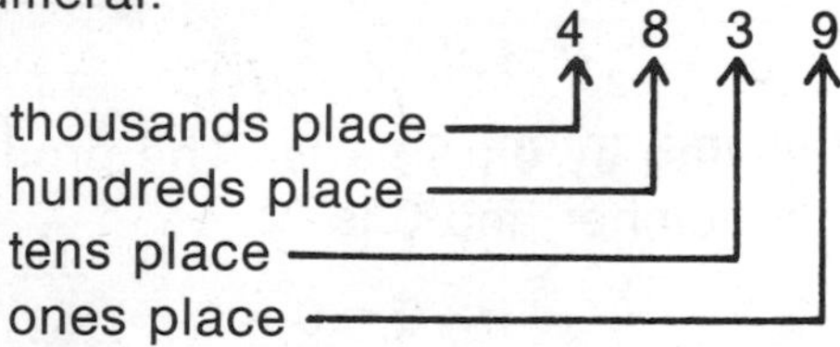

plane figures Figures that lie in a flat or level surface. Rectangles, squares, circles, and triangles are all plane figures.

P.M. A symbol used for times after 12:00 noon and before 12:00 midnight.

prime number 2, 3, 5, 7, 11, 13, and so on, are prime numbers. They have only two factors, 1 and the number itself.

procedure A list of computer commands that is used often.

product The answer to a multiplication problem.

$$\begin{array}{r} 7 \\ \times 8 \\ \hline 56 \end{array} \leftarrow \text{product}$$

pyramid A solid figure shaped like this:

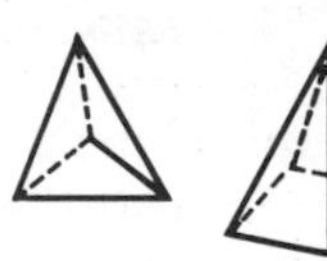

quotient The answer to a division problem.

$$\begin{array}{r} 7 \\ 8\overline{)56} \end{array} \leftarrow \text{quotient}$$

radius The distance from the center of a circle to the circle.

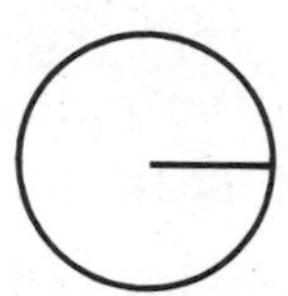

ray A part of a line that has one endpoint. This is ray *AB*.

rectangle A plane figure with four sides and four right angles.

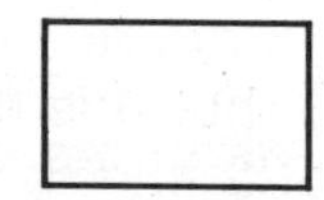

rectangular solid A box-shaped solid figure whose flat surfaces are all rectangles.

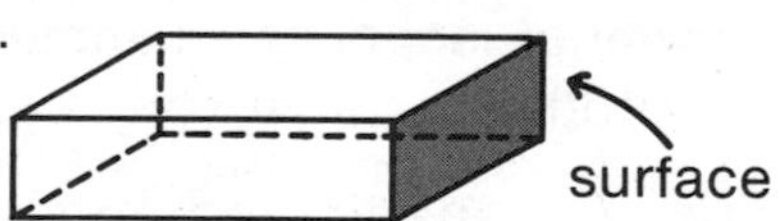

remainder The number "left over" after division.

$$\begin{array}{r} 5\text{ R}1 \\ 3\overline{)16} \\ -15 \\ \hline 1 \end{array} \leftarrow \text{remainder}$$

right angle An angle that forms a square corner.

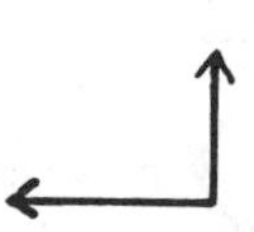

Roman numerals A system of writing numbers using letters.

I = 1 V = 5 X = 10 XXIV = 24

round To replace an exact number by another one that is easier to use.

52 rounded to the nearest ten is 50.
278 rounded to the nearest hundred is 300.

When a number is halfway between two numbers, round up.

150 rounded to the nearest hundred is 200.

segment Part of a line that has two endpoints. This is segment *AB*, segment *BA*, $\overline{AB}$, or $\overline{BA}$.

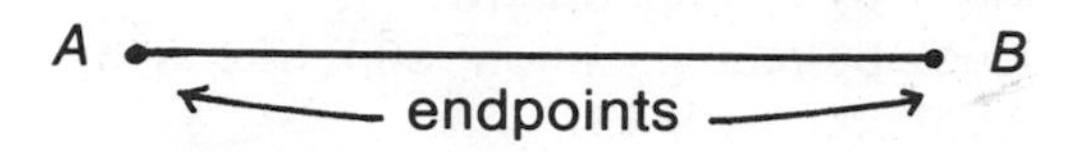

side of an angle One of the rays that make up the angle.

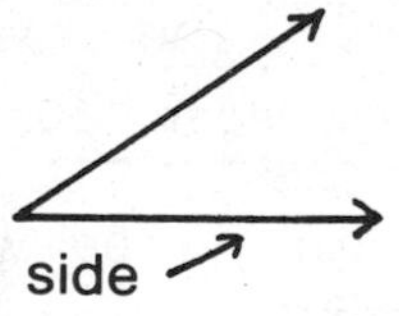

side of a plane figure One of the segments that make up a figure.

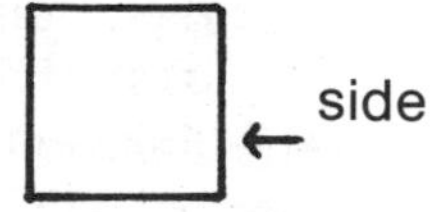

sphere A solid figure that is the shape of a ball.

square A rectangle with four sides that are all the same length.

sum The answer to an addition problem.

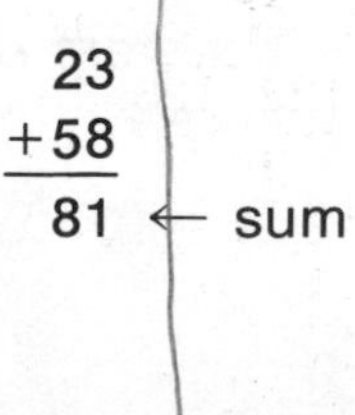

surface area The sum of the areas of all the surfaces of a solid figure.

terms (of a fraction) The numerator and denominator of a fraction.

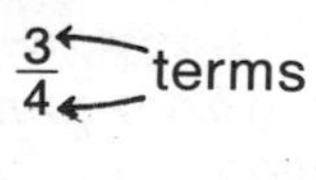

total Another name for *sum*.

triangle A plane figure with three sides.

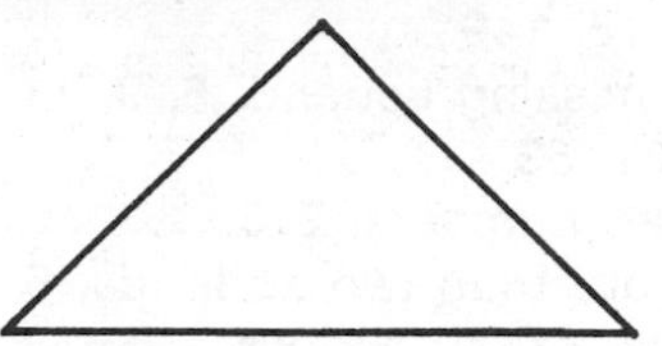

vertex The point at the "corner" of an angle, plane figure, or solid figure.

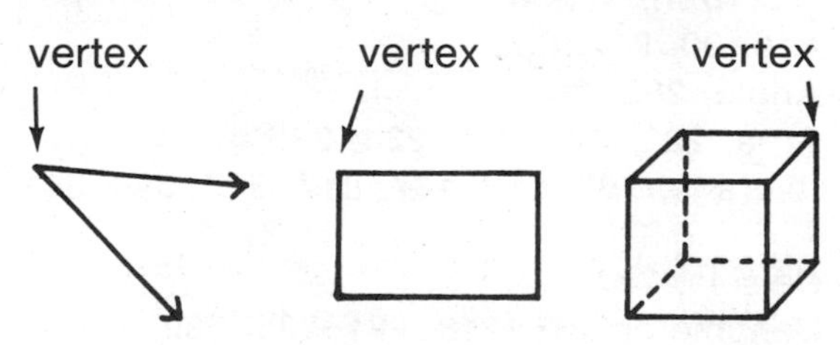

volume The number of unit cubes that fit inside an object.

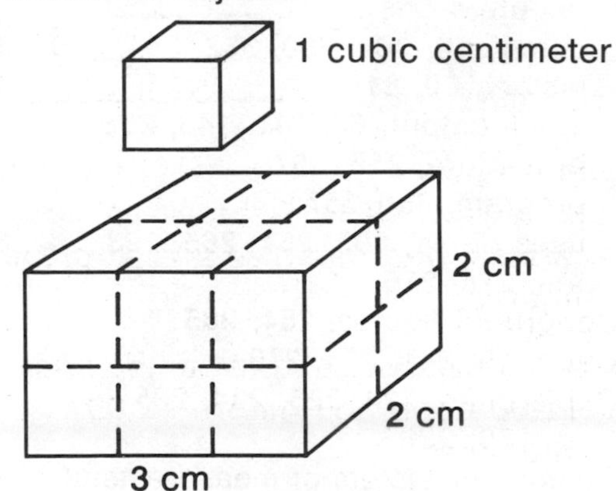

The volume is 12 cubic centimeters.

whole number Any of the numbers 0, 1, 2, 3, 4, and so on.

Index

1 2 3 4 5 6 7 8 9 0